POLYIMIDES

Synthesis, Characterization, and Applications

Volume 2

Edited by

K. L. Mittal

IBM Corporation
Hopewell Junction, New York

Plenum Press • New York and London

Library of Congress Cataloging in Publication Data

Technical Conference on Polyimides (1st: 1982: Ellenville, N.Y.)
  Polyimides: synthesis, characterization, and applications.

  Proceedings of the First Technical Conference on Polyimides, held under the
auspices of the Mid-Hudson Section of the Society of Plastics Engineers,
November 10–12, 1982, in Ellenville, New York

  Includes bibliographical references and indexes.
  1. Polyimides—Congresses. I. Mittal, K. L., 1945–   . II. Title.
TP1180.P66T43 1982   668.9   84-3316
ISBN 0-306-41670-0 (v. 1)
ISBN 0-306-41673-5 (v. 2)

Proceedings of the First Technical Conference on Polyimides, held under the
auspices of the Mid-Hudson·Section of the Society of Plastics Engineers,
November 10–12, 1982, in Ellenville, New York

©1984 Plenum Press, New York
A Division of Plenum Publishing Corporation
233 Spring Street, New York, N.Y. 10013

Printed in the United States of America

PREFACE

This and its companion Volume 1 chronicle the proceedings of the First Technical Conference on Polyimides: Synthesis, Characterization and Applications held under the auspices of the Mid-Hudson Section of the Society of Plastics Engineers at Ellenville, New York, November 10-12, 1982.

In the last decade or so there has been an accelerated interest in the use of polyimides for a variety of applications in a number of widely differing technologies. The applications of polyimides range from aerospace to microelectronics to medical field, and this is attributed to the fact that polyimides offer certain desirable traits, inter alia, high temperature stability. Polyimides are used as organic insulators, as adhesives, as coatings, in composites, just to name a few of their uses. Even a casual search of the literature will underscore the importance of this class of materials and the high tempo of R&D activity taking place in the area of polyimides.

So it was deemed that a conference on polyimides was both timely and needed. This conference was designed to provide a forum for discussion of various ramifications of polyimides, to bring together scientists and technologists interested in all aspects of polyimides and thus to provide an opportunity for cross-pollination of ideas, and to highlight areas which needed further and intensified R&D efforts. If the comments from the attendees are a barometer of the success of a conference, then this event was highly successful and fulfilled amply its stated objectives.

The technical program consisted of 69 papers (45 oral presentations and 24 posters) by about 130 authors from many corners of the globe. The purpose of a conference or symposium is to present the state of knowledge of the topic under consideration and it can best be accomplished by a blend of invited overviews and original unpublished research contributions. This is exactly what was done in this conference. The program contained a number of invited overviews on certain subtopics, and these were augmented by contributed original research papers. The invited speakers were selected so as to represent differing disciplines and interests and they hailed from academic, governmental and industrial research laboratories.

As for the present proceedings volumes, the papers have been rearranged (from the order they were presented) so as to fit them in a more logical fashion. Incidentally, these volumes also contain nine papers which were not included in the formal printed program. Also it should be recorded here that, for a variety of reasons, a few papers which were presented are not included in these volumes. It must be emphasized here that these proceedings volumes are not simply a collection of papers, but the peer review was an integral part of the total editing process. All papers were critically reviewed by qualified reviewers as the comments of peers are a desideratum to maintain high standard of publications. Also it should be recorded that although no formal discussion is included in these proceedings, but there were many enlightening, not exothermic, and lively discussions both formally in the auditorium and in more relaxed places. Particular mention should be made here regarding the poster presentations. The poster session was highly successful as can be gauged from the comments of those who participated in this format.

Coming back to the proceedings volumes, these contain a total of 71 papers (1182 pages) by 164 authors from nine countries. The text is divided into five sections as follows: Synthesis and Properties; Properties and Characterization; Mechanical Properties; Microelectronic Applications; and Aerospace and other Applications. Sections I and II constitute Volume 1, and Sections III-V grace the pages of Volume 2. The topics covered include: synthesis, properties and characterization of a variety of polyimides; metal-containing polyimides; cure kinetics of polyimides; structure-property relationships; polyimide adhesion; mechanical properties of polyimides both in bulk state as well as in thin film form; applications of polyimides in microelectronics; photosensitive polyimides; polyimides as adhesives; aerospace applications of polyimides; polyimide blends; electrophoretic deposition of polyimides; and applications of polyimides in electrochemical and medical fields.

Even a cursory glance at the Table of Contents will convince that there is a great deal of R&D activity in the area of this wonderful class of materials, and all signals indicate that this tempo is going to continue. It is hoped that these proceedings volumes which represent the repository of latest knowledge about polyimides will be useful to both the seasoned researcher (as a source of latest information) and to the neophyte as a fountain of new ideas. As we learn more about this unique class of materials, more pleasant applications will emerge.

Apropos, a comment should be made here regarding the nomenclature being used in the field of polyimides. Dr. Paul Frayer (Rogers Corp.) pointed out that there was a great deal of inconsistency in the nomenclature and different authors were using quite different terms for chemically identical species. He further suggested that a consistent method of nomenclature was needed and urged the attendees to come up with a consensus on this issue and to take the requisite action in this regard in a future conference on polyimides.

Acknowledgements. First of all I am thankful to the Mid-Hudson Section of the Society of Plastics Engineers for sponsoring this event, to Dr. H.R. Anderson, Jr. (IBM Corporation) for permitting me to organize the technical program and to Steve Milkovich (IBM Corporation) for his understanding and cooperation during the tenure of editing. My special thanks are due to the Organizing Committee members (C. Araps, P. Buchwalter, G. Czornyj, M. Gupta, J. Schiller and M. Turetzky) for their invaluable help, and in particular to the General Chairman (Julius Schiller) for his continued interest in and overall responsibility and organization of this conference. All members of the Organizing Committee worked hard in many capacities and unflinchingly devoted their time to make a conference of this magnitude a grand success. Special appreciation is extended to P. Hood, M. Htoo, R. Martinez, R. Nufer and B. Washo for their help and cooperation in more ways than one.

On a more personal note, I am thankful to my wife, Usha, for tolerating, without much complaint, the frequent privations of an editor's wife and for helping me in many ways. I am appreciative of my darling children (Anita, Rajesh, Nisha and Seema) for not only letting me use those hours which rightfully belonged to them but also for rendering home environment conducive to work. Special thanks are due to Phil Alvarez (Plenum Publishing Corporation) for his continued interest in this project, and to Barbara Mutino (Office Communications) for meeting promptly various typing deadlines. The time and effort of the unheralded heroes (reviewers) is gratefully acknowledged. Last, but not least, the enthusiasm, cooperation and contributions of the authors are certainly appreciated without which we could not have these volumes.

K. L. Mittal
IBM Corporation
Hopewell Junction, New York 12533

CONTENTS

PART V: AEROSPACE AND OTHER APPLICATIONS

CONTENTS OF VOLUME 1

PART I: SYNTHESIS AND PROPERTIES

PART III. MECHANICAL PROPERTIES

MECHANICAL AND SPECTROSCOPIC PROPERTIES OF METAL CONTAINING POLYIMIDES

Larry T. Taylor and Anne K. St. Clair[a]

Department of Chemistry
Virginia Polytechnic Institute and State
   University
Blacksburg, Virginia 24061

The incorporation of specific metal ions into polyimides is described. Detailed studies have included various compounds of copper, lithium and palladium as dopants. Addition of the metal during polymerization or after formation of the polyamic acid precedes the thermal imidization step. With many dianhydride-diamine-dopant combinations, high quality variously colored films are produced. Many metal-doped films exhibit (1) improved high temperature adhesive properties, (2) increased electrical conductivity, (3) excellent thermal stability, (4) improved acid/base resistance, (5) increased modulus in flexible films and (6) excellent high temperature tensile strength. X-ray photoelectron spectroscopic study of these films suggests that many of the additives undergo chemical modification during thermal imidization. Palladium dopants appear to be partially reduced to the metallic state; while, lithium and copper dopants are probably converted to their oxides. Ion etching experiments with Auger electron spectroscopy monitoring are discussed.

[a]Materials Division, NASA Langley Research Center, Hampton, Virginia 23665.

INTRODUCTION

The continuing demand for materials with a unique combination of properties has brought forth a sizeable research effort[1] concerning polymer modification via incorporation of a variety of non-metallic and metallic dopants into the polymer. The modification of polyimide properties is no exception in this regard. This work had its beginnings approximately 25 years ago when Angelo[2] briefly reported in a patent the addition of metal ions to several types of polyimides. The object of the invention was a process for forming particle-containing ($<1\mu$) transparent polyimide shaped structures. All of the metals were added in the form of coordination complexes rather than as simple anhydrous or hydrated salts. The properties of only one film (e.g. cast from a N,N-dimethylformamide solution of 4,4´-diaminodiphenyl methane, pyromellitic dianhydride and bis-(acetylacetonato)copper(II)) were given. These properties included: %Cu = 3.0%, dielectric constant = 3.6 and volume resistivity = 8×10^{12} ohms-cm. Comparison of metal-doped films with films cast from the neat polymer solution was not discussed. No data were given regarding thermal behavior, adhesive properties or the mode of interaction between metal and polyimide. Unfortunately further patents and published work in this regard are not available.[3]

Ten years later a report[4] appeared concerning silver incorporation into the polyamic acid derived from pyromellitic dianhydride (PMDA) and 4,4´-oxydianiline. Both Ag and $Ag(C_2H_3O_2)$ were utilized. Films containing 0.25-1.00 gram-atom of Ag per repeat unit were obtained. Electroconductivity was studied as a function of temperature and Ag content. Thermal and electrical conductivities were increased 3-7 times for the polyimide film containing dispersed Ag metal; but no change in properties was noted for the film containing $Ag(C_2H_3O_2)$. Prior to this study a patent[5] was filed covering very similar work (e.g. PMDA, 4,4´-diaminodiphenylmethane and $Ag(C_2H_3O_2)$). In this case, the Ag containing polyamic acid complex was converted to the polyimide and Ag metal by heating at 300°C in vacuo for 30 minutes. The film was stated to be tough, flexible, opaque and metallic. At this stage the film was not conducting. Further heating at 275°C in air for 5-7 hours, however, rendered it conducting although no resistivity data were reported.

More recently a very brief report[6] concerning lithium incorporation into polyimides has appeared. Superior antistatic properties have been found for a soluble polyimide (DAPI-Polyimide) film loaded with either $LiNO_3$ or LiCl. Physical properties and film smoothness remained unchanged except that electrical resistance was sharply lowered. Conductivity was increased about 20-fold or 2000% over the standard unfilled polyimide. Additional tests showed that the films were very slightly hygroscopic in the presence of lithium ions. This phenomenon was suggested to account for the lowered resistivity. The NASA Brief publication[6] did not state whether the electrical conductivity enhancement was in surface or volume conductivity.

Additional work in the area of metal-doped polyimides has centered in our laboratory during the past six years. Our efforts have focused on the extent of polyimide modification by a variety of different metal dopants. Mechanical, thermal and electrical properties have been of interest. Major fundamental questions regarding the distribution and chemical state of metal within the film have been addressed employing various spectroscopic tools. An overview of these investigations provides the framework for this discussion.

OVERVIEW

Synthesis

Polyimides derived from PMDA, IA, or 3,3´,4,4´-benzophenone tetracarboxylic acid dianhydride (BTDA), IB, and either 3,3´-diaminobenzophenone (m,m´-DABP), IIA, 4,4´-diaminobenzophenone (p,p´-DABP), 3,3´-diaminodiphenyl-carbinol (DADPC), IIB, or 4,4-oxydianiline (ODA), IIC, (and to which have been added numerous metal compounds) have been prepared. The synthetic procedure employed involved (1) formation of the polyamic acid, III, (20% solids) in either N,N-dimethylformamide, N,N-dimethylacetamide (DMAC) or diethylene glycol dimethyl ether (Diglyme), (2) addition of the metal complex to the polyamic acid, III, in a 1:4 ratio, (3) fabrication of a film of the polyamic acid-metal compound mixture and (4) thermal conversion (300°C) to the metal containing polyimide, IV. Approximately twenty metals (Al, Au, Ca, Co, Cr, Cu, Fe, Li, Mg, Mn, Na, Ni, Pd, Pt, Sn, Ti and Zn) in a variety of forms have been added to the polyamic acid solutions.[7] Several experimental problems were encountered in following this procedure which limited the number of good quality films obtained. These problems include: (1) metal complex not dissolving in an appropriate solvent, (2) gel formation or cross-linking of the polymer occurring upon interaction with the metal, (3) polyamic acid precipitating when the metal complex is added and (4) metal promoting extensive thermal oxidative degradation of the polymer film upon curing.

In general metal-doped polyimides derived from BTDA-p,p´-DABP yielded (regardless of the metal) rather poor quality, very

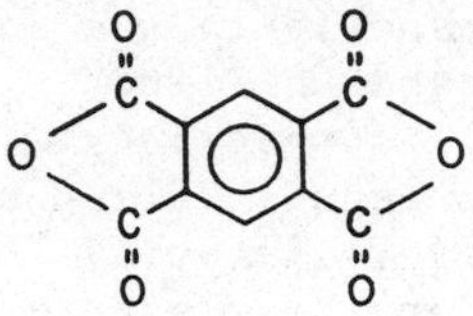

PMDA, IA

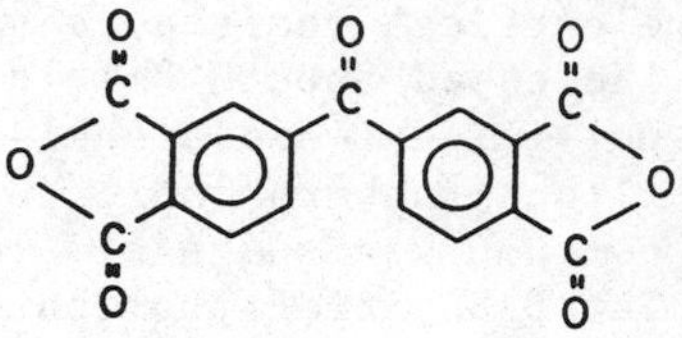

BTDA, IB

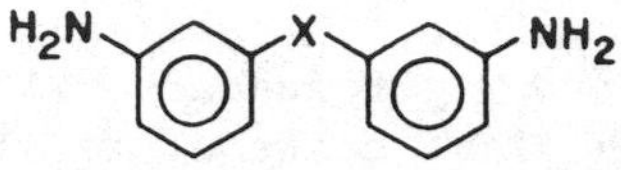

m,m'-DABP (X= C=O), IIA

m,m'-DADPC (X= CH-OH), IIB

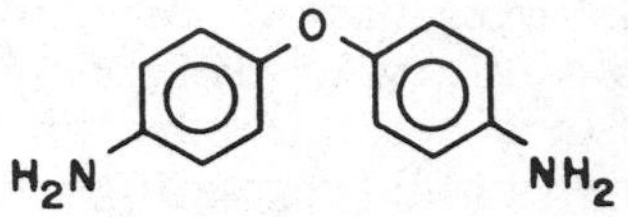

p,p'-ODA, IIC

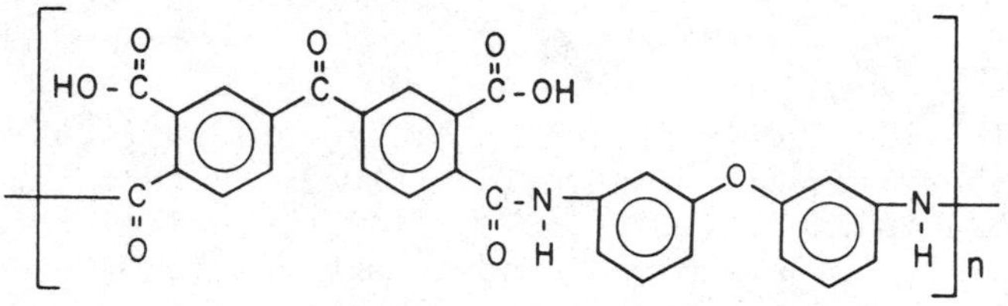

POLYAMIC ACID, III
(BTDA + m,m' - DABP)

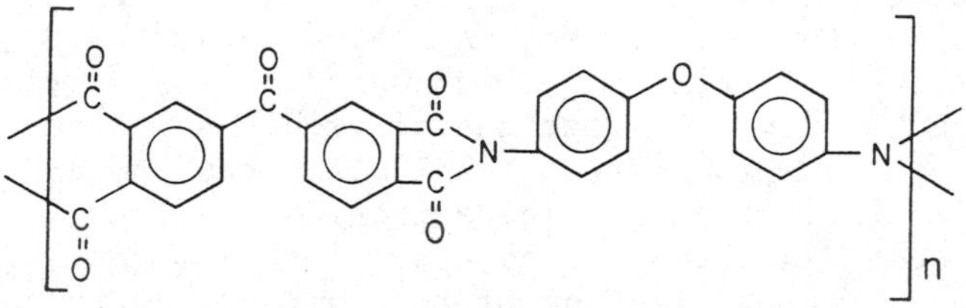

POLYIMIDE, IV
(BTDA + p,p' - ODA)

brittle films which were due in part to the low viscosity of the resulting polyamic acid solution. Other dianhydride-diamine combinations with specific metal additives yielded high quality flexible films, with ODA as a constituent being most notable. The physical appearance of the thin films (~2 mil) varied considerably with the particular metal dopant employed. Colors range from transparent pale yellow (Li, Sn) to nearly opaque red-brown (Cu, Pd).

With several additives (Cu, Pd) a readily observed deposit appeared after curing on the film side exposed to the curing atmosphere. This difference was very noticeable for the two $Pd(S(CH_3)_2)_2Cl_2$ containing films BTDA-ODA and BTDA-DABP. While the glass side had a dark red-brown appearance, the air-side possessed a definite silvery, metallic appearance. Re-heating of these films after removal from the glass plate caused the glass-side to also metallize. The metallic looking surfaces could not be produced with DADPC and Li_2PdCl_4 containing films. The presence of atmospheric oxygen during the imidization process appears crucial, since BTDA-ODA and BTDA-DABP doped with $Pd(S(CH_3)_2)_2Cl_2$ do not give rise to metallic surfaces when cured in either a dry N_2, Ar, N_2/H_2 or moist Ar atmosphere.

Determination of an optimum reaction time for preparation of the polyamic acid/metal additive solutions has been necessary in order to (1) minimize any metal catalyzed degradation of the polymer in solution and (2) maximize the solubility of the metal additives into the polymer-DMAC matrix. The optimum reaction time for the polyamic acid/metal additive solutions was ascertained by measuring the intrinsic viscosity of aliquots withdrawn from a continuously stirred reaction mixture at 7, 24 and 48 hour intervals. Aliquots of solution were withdrawn and cast as films simultaneous with the viscosity determinations. It was determined that optimum reaction time was within 7 hours from start of the synthesis. Intrinsic viscosity and thermal properties generally decreased with reaction time after approximately 7 hours.

Thermal Properties

Thermal properties of metal-ion-containing polyimide films have been studied from three viewpoints: (1) apparent glass transition temperature (T_g), (2) polymer decomposition temperature and (3) weight loss under isothermal conditions. The effect of five different dopants on T_g (Thermal Mechanical Analysis) is indicated in Table I.

Table I. Thermal Properties of Metal/Polyimide (BTDA–ODA) Films[12].

Polyimide/Metal Film	T_g(°C)	% Metal Found	Calc´d
Polyimide Alone	283	–	–
Polyimide + Al(acac)$_3$	312	Al=0.9	0.9
Polyimide + AgNO$_3$	320	Ag=4.0	4.0
Polyimide + Li$_2$PdCl$_4$	341	Li=0.1 Pd=7.7	0.5 3.8
Polyimide + AuI$_3$	320	Au=0.2	6.4
Polyimide + SnCl$_2$	283	Sn=2.7	4.3

T_g increases approximately 30–40°C compared to the polymer alone with the exception of the SnCl$_2$-containing film. The increased T_g´s do not correlate with added metal content since the mole percent (~9%) of each metal dopant was constant. T_g also does not follow the experimentally determined metal weight percentages. For example T_g increases 29°C for Al(acac)$_3$ (%Al=0.9) addition; whereas, the T_g does not change for SnCl$_2$ (%Sn=2.7) doping. Film densities have been measured at 23°C; yet, correlation with T_g is again not obvious. T_g data on copper containing films[8] are quite different from these other metal additives. Figures 1 and 2 compare different dopants with the same curing atmosphere and different curing atmospheres with the same dopant respectively. In the case of different copper dopants, copper(II) films appear to contract prior to softening; while, the Cu(I) dopant does not. The copper(II) film decomposes near the softening temperature but the copper(I) film does not. This effect is just as pronounced upon considering air and N$_2$ cured copper(II) dopants. We believe that such polymer contraction near its softening point is unusual and may be caused by a combination of polymer cross-linking and polymer decomposition promoted by copper(II)/oxygen but not by copper(I)/oxygen and copper(II)/nitrogen.[8]

While the softening temperature of these doped polyimides are in general greater than the polymer alone, significant thermal stability has been sacrificed in the $AgNO_3$ and Li_2PdCl_4 cases (Figure 3). Al, Au and Sn films possess excellent thermal stability, just as the polymer alone, as evidenced by their TGA profiles obtained in air. The lowered thermal stability of the $AgNO_3$ film may reflect the presence of the highly oxidizing nitrate ion. The film containing Li_2PdCl_4, of course, possesses two different metals as well as the highest metal weight percentage. In-depth studies with Li dopants alone[9] have shown reduced thermal stability over polymer alone; however, not to this extent. The Li_2PdCl_4 film exhibits infrared detectable moisture which may also enhance polymer decomposition. A reduction in polymer decomposition temperature occurs on copper ion addition. The contrast between curing atmosphere and copper dopant is not as dramatic insofar as thermal stability is concerned. Air and N_2 cured films yield almost identical polymer decomposition temperatures (Figure 4). In the case of copper(I) and copper(II), the differences are more pronounced. The copper(I) dopant, $[(n-Bu_3)PCuI]_4$, exhibits significant low temperature weight loss even though total decomposition occurs at a higher temperature in the copper(I) case relative to copper(II).

These observations on thermal stability are further dramatized by isothermal measurements on several of these films. Originally films were aged at 200°C for 300 hours in an inert atmosphere but none of them lost significant weight. In order to observe differences in each film, accelerated aging experiments at 300°C were conducted. The Li_2PdCl_4 film was the only film tested that did not survive this environment (Figure 5). Even though $AgNO_3$ and Li_2PdCl_4 films exhibited about the same polymer decomposition temperature by dynamic TGA, the $AgNO_3$ film is noticeably more thermally stable by isothermal TGA measurement (i.e. ~4% weight loss after 300 hours).

Reflection upon these thermal properties and the preparative thermal curing (imidization) process employed to prepare these films may reveal in part why several of these films have quite different metal percentages from their theoretical value (Table I). Au, Li and Sn films have a much lower metal content than calculated. No doubt partial vaporization of Au, Li and Sn in some chemical form (for example as the oxide, hydroxide or DMAC solvate) occurs during imidization (loss of H_2O) at 300°C in flowing air. Similar replicate analyses have ruled-out the possibility that these films are highly heterogeneous. The amount of Pd, on the other hand, is double what is predicted. The ready decomposition of the polyimide film when doped with Li_2PdCl_4 indicates that some polymer "break-down" into volatiles occurs during thermal imidization; thereby, raising the %Pd in

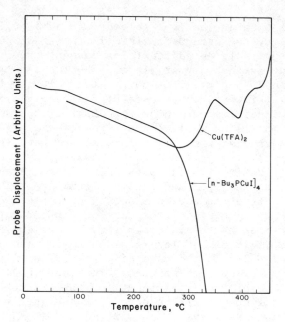

Figure 1. Plot of Thermomechanical Analysis of BTDA–ODA Film (air-cured) Doped With $[(n-Bu_3)PCuI]_4$ and $Cu(TFA)_2$.

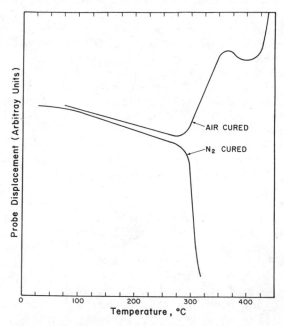

Figure 2. Plot of Thermomechanical Analysis of BTDA–ODA/$Cu(acac)_2$ Film Cured in Air and N_2.

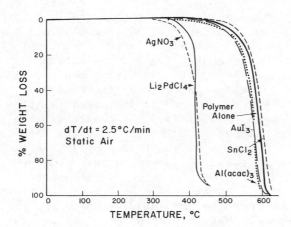

Figure 3. Dynamic TGA of BTDA-ODA/Metal Films cured at 300°C in air.

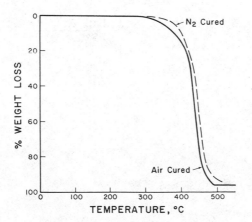

Figure 4. Plot of Thermal Gravimetric Analysis of BTDA-ODA/ Cu(acac)$_2$ Film Cured in Air and N$_2$.

the film. The curing process apparently affects adversely neither the metal additive nor polymer in the presence of $AgNO_3$ and $Al(acac)_3$.

Mechanical Properties

Certain polyimides are known to be excellent adhesives.[10] Doped and undoped polyimides derived from BTDA and m,m´-DABP have been employed in lap shear strength tests employing titanium-titanium adherends. Tests were performed at room temperature, 250°C and 275°C employing either DMAC or DMAC/Diglyme as the solvent. At room temperature, regardless of the metal ion employed, adhesive strength is decreased relative to the "polymer-alone" case (Table II). The choice of metal ion at high temperature testing is critical. The two best cases (i.e. polymer doped with $Al(acac)_3$ and $NiCl_2 \cdot 6H_2O$) were subjected to adhesive testing at elevated temperatures. Under these conditions the metal ion filled polyimides were superior. The $Al(acac)_3$ case proved to be exceptional in that it exhibited approximately four times the lap shear strength of "polymer-alone" at 275°C.[11] We feel that this enhanced adhesiveness is due in part to the increased softening temperature of the $Al(acac)_3$ filled polyimide. These results are somewhat analogous to data collected by St. Clair and Progar[10] regarding the use of aluminum metal as a filler with various polyimides. Lap shear strength was found by these workers to double at 250°C with 79% Al filled polyimide versus the unfilled polyimide.

Table II. Lap Shear Tests[a], Titanium-Titanium Adherend, BTDA-m,m´-DABP.

Metal Ion Added	Lap Shear[11] Strength[b] (psi)		
	25°C	250°C	275°C
No metal	2966 (3138)	1573	438 (496)
$Al(acac)_3$[c]	2378 (2400)	1891	1641 (1348)
$NiCl_2 \cdot 6H_2O$[c]	1800	1332	608

[a]Numbers in parentheses correspond to DMAC/Diglyme solvent mixture.
[b]Average of four tests.
[c]0.1g of metal complex (salt) per 4g of polymer (20% solids).

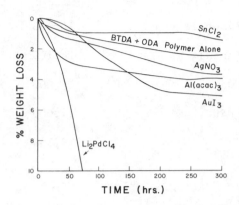

Figure 5. Isothermal TGA's of Polyimide/Metal Films at 300°C in Nitrogen.

Mechanical properties of several metal-containing polyimide films have been measured.[12] Tensile strengths for these films are given in Table III. Films were pulled in the 0 direction (as cast) at a rate of 0.2 inches/minute at both room temperature and 200 C. With the exception of the $AgNO_3$ additive, the tensile strengths are excellent. At elevated temperature all metal containing films have surprisingly increased strengths relative to the polymer alone. Incorporation of metal ions into polyimide films in order to improve their space applicability therefore offers a distinct advantage over the addition of heavy metal fillers or carbon microspheres to the polyimide. Fillers and carbon microspheres not only add unwanted weight to the film, but result in greatly lower tensile strengths making them

Table III. Mechanical Properties of Metal/Polyimide (BTDA–ODA) Films[a].

Polyimide/Metal Film	Tensile Strength[12] (psi)	
	RT	200 °C
Polyimide + Polymer Alone	16,500	6,700
Polyimide + Al(acac)$_3$	16,100	7,900
Polyimide + AgNO$_3$	12,000	8,300
Polyimide + Li$_2$PdCl$_4$	14,000	10,600
Polyimide + AuI$_3$	19,100	9,900

[a]Instron Tensile Testing

unfit for space applications. Tensile modulus and percent elongation for these films are given in Table IV. Percent elongation for each metal-containing film decreased relative to the polymer alone both at room temperature and 200 °C. Although hand-cast films are more subject to flaws and premature breaks, it is reasonable to expect tensile modulus to increase upon metal incorporation. The modulii of these films are on the order of that of DuPont's Kapton® film (400,000 psi). Further evidence of an increase in film stiffness is offered by Young's modulus data (Table V). Regardless of the test temperature or metal dopant, an increase in modulus is observed. A correlation with the weight percentage of each metal is not possible (see Table I).

Electrical Properties

In light of the excellent thermal and mechanical properties of polyimide films and the great interest currently shown in electrically conducting polymers, considerable effort has been expended in studying the electrical properties of metal-doped polyimide films. The addition of select palladium complexes to polyimides produces one of the more dramatic effects on electrical resistivity. Table VI outlines these results. For BTDA derived films four different combinations of dianhydride, diamine and palladium complex have yielded polyimides with dramatically lowered resistivities when cured in air. More

Table IV. Mechanical Properties of Metal/Polyimide (BTDA-ODA) Films[a].

Polyimide/Metal Film	Tensile Modulus[12] (psi)		Elongation[12] (%)	
	RT	200°C	RT	200°C
Polyimide Alone	284,000	178,000	10	26
Polyimide + Al(acac)$_3$	454,000	257,000	6	6
Polyimide + AgNO$_3$	486,000	215,000	3	4
Polyimide + Li$_2$PdCl$_4$	492,000	336,000	4	5
Polyimide + AuI$_3$	494,000	311,000	8	15

[a]Instron Tensile Testing

Table V. Mechanical Properties of Metal/Polyimide Films[a].

Polyimide/Metal Film	Young's Modulus,[12] $E'x10^{10}$ dynes/cm^2		
	RT	100°C	200°C
Polyimide Alone	3.52	2.81	2.02
Polyimide + Al(acac)$_3$	4.49	3.61	2.59
Polyimide + AgNO$_3$	4.89	3.97	3.10
Polyimide + Li$_2$PdCl$_4$	4.31	3.67	2.88
Polyimide + AuI$_3$	4.06	3.39	2.57
Polyimide + SnCl$_2$	4.66	3.77	2.70

[a]Autovibron

Table VI. Surface and Volume Resistivities[13] of Conductive Palladium-Containing BTDA-Derived Polyimides[a,b].

Metal Additive	ODA	DABP	DADPC	Curing Atmosphere
Li_2PdCl_4	9.5×10^5 ohm 2.0×10^6 ohm-cm	NC^c	1.3×10^7 ohm 1.0×10^7 ohm-cm	Air
$Pd(S(CH_3)_2)_2Cl_2$	$<10^5$ ohm (ATM) 1.0×10^{17} ohm (GLASS) 1.0×10^{17} ohm-cm	$<10^5$ ohm (ATM) 1.0×10^{17} ohm (GLASS) 1.0×10^{17} ohm-cm	NC	Air
Li_2PdCl_4	5.1×10^7 ohm 8.9×10^7 ohm-cm	NC	2.1×10^{10} ohm 1.4×10^{11} ohm-cm	N_2
$Pd(S(CH_3)_2)_2Cl_2$	NC	NC	NC	N_2

[a]Resistivity values are quoted for the best quality films. Values for replicate films do not differ by more than one order of magnitude.

[b]Polymer alone surface and volume resistivities are 10^{17} ohm and 10^{17} ohm-cm, respectively.

[c]NC = No change in resistivity relative to the polymer alone.

highly conductive BTDA-ODA films were produced using both palladium additives, Li_2PdCl_4 (LTP) and $Pd(S(CH_3)_2)_2Cl_2$ (PDS). Surprisingly only LTP gave lowered resistivity values with BTDA-DADPC; while, PDS with the same monomer pair exhibited values equivalent with the polymer alone. The results with BTDA-DABP, however, were totally reversed.[13]

A metallic surface on the air side characterizes the two conducting (BTDA-ODA, BTDA-DABP) PDS films. The three-point electrode method normally used in measuring the surface resistivity of these films is unsatisfactory. A four-point method yielded a surface resistivity of less than 10^5 ohms for the metallized side. The film side exposed to glass during curing gave a value similar to the polymer alone. Volume resistivity for these two films was 1.0×10^{17} ohm-cm. The PDS additive with BTDA-DADPC and all PMDA films regardless of the diamine gave no metallic surface and no resistivity lowering. This metallic surface is, therefore, apparently not a necessity for conduction, since low resistivity LTP films do not display a metallic surface.

The results on curing the films in a non-oxygenated atmosphere are equally interesting. No metallic surfaces are produced with PDS as an additive and no resistivity lowering is observed. Moist argon and forming gas (N_2/H_2) give the same unchanged results. A nitrogen curing atmosphere, however, does not change the resistivity results appreciably, from the air-cured, low resistivity LTP films. It is significant that in each case, with LTP the resistivity values are always one to three orders of magnitude higher for nitrogen cured films. We, therefore, believe that molecular oxygen is crucial for production of these lower resistivity films, however, the chemistry of LTP and PDS during the imidization process is probably quite different.

No appreciable lowering of volume or surface resistivities was observed with lithium additives except when the dopant was lithium chloride and the monomer pair was either BTDA-ODA or PMDA-ODA.[9] With the former LiCl-doped monomer pair, the typical volume resistivity range was 10^6-10^{10} ohm-cm as compared to undoped films with values greater than 10^{18} ohm-cm. These data, however, were obtained for films measured after air-drying. Upon subsequent vacuum drying at 110°C for 12 h, volume resistivities were found to increase to $10^{12}-10^{14}$ ohm-cm. Finally, when resistivity measurements were carried out in a drybox after heat vacuum drying, values in the vicinity of 10^{16} ohm-cm were observed. The surface resistivities, likewise, followed the same trend. This suggests that the primary improvement in conduction is due to the ability of the lithium ion to pick up moisture although no noticeable droplet formation was observed on the surface.

For the monomer pair PMDA-ODA, lowering of volume resistivity was only observed when dopant concentration was relatively high. For a film cured in air having 0.65% lithium, the volume resistivity was found to be 1.6×10^{10} ohm-cm. The volume resistivity, however, increased to 7.2×10^{15} ohm-cm on vacuum drying. A film cured in nitrogen containing 0.54% lithium, on the other hand, gave volume resistivities of 4.4×10^{6} and 1.9×10^{10} ohm-cm upon air-drying and vacuum drying, respectively. A possible explanation for this observation lies in the brittleness of the air-cured film vs. the more flexible nitrogen-cured film. A more homogeneous distribution of lithium might then be expected in the nitrogen-cured film, which may facilitate better electrical conduction.

The presence of water in the BTDA-ODA-LiCl film was verified by Fourier Transform infrared spectroscopy. The strong IR absorption suggests the presence of a relatively large quantity of water when contrasted to its undoped counterpart, which shows very little evidence for moisture. On extending this examination, we observed a comparable amount of water in all lithium doped films regardless of their resistivity. This implies that a further factor is necessary for reduced resistivity, perhaps the distribution of lithium throughout the film, although scanning electron microscopic study of both types of films to date reveals no major surface differences, where, typically, only a smooth surface was obtained.

Electrical resistivities in several of the copper cases are again slightly lower than for the polymer alone. In all cases the numbers reflect average of values obtained on multiple cast films. Cu(BAE),* regardless of whether the dopant was added during or after polymerization, consistently gave higher resistivity values than with Cu(TFA)$_2$.* Imperfections in the 6 cm diameter disks no doubt account for scatter in the resistivity data given similar methods of preparation. The most consistent data were obtained with Cu(TFA)$_2$ and PMDA-derived polyimide as illustrated by the two independent preparations in Table VII. It is interesting to note that the metallic-looking air side (Cu(TFA)$_2$ as dopant) always exhibits the lower surface resistivity. The resistivity lowering, however, is not at all com-

*Cu(BAE), bis(acetylacetone)ethylenediminatocopper(II)
 Cu(TFA), bis(trifluoroacetylacetonato)copper(II)

parable to the metallic-looking PDS film surfaces. In those
cases where the film is imidized in N_2 the surface resistivity
is the same regardless of the film side.

Spectroscopic Properties

X-ray photoelectron spectroscopy (XPS) has proven valuable
in studying selected metal-filled polyimides.[14] Measured XPS
binding energies (Pd $3d_{5/2,3/2}$) for the PDS film show palladium
exclusively in the zero oxidation state. This is determined by
comparing palladium binding energies in the film with analogous
data obtained on $PdCl_2$ and Pd metal. As expected, the palladium
electron count on the side exposed to air during curing (metallized), is considerably higher than for the side against the
glass (i.e. Pd(air)/Pd(glass) = 23.4). The corresponding LTP
film shows that the palladium on the film surface exists as a
mixture of zero and divalent oxidation states. The approximate
number of Pd counts is usually quite small relative to the PDS
film which is suggestive of a more homogeneous distribution of
palladium in the LTP film. This idea is supported by the fact
that the ratio of palladium electron counts on the air side to
the glass side is nearly unity. Neither of these films exhibits
evidence of differential charging at the surface as is suggested
with certain Pd-containing films.

Partial or complete palladium reduction during the
imidization process, coupled with significant film resistivity
lowering, naturally led to the idea that the presence of zero
valent metal and lower resistivities could be correlated.[14] To
test this hypothesis two experiments were designed. First, the
BTDA-DADPC-LTP film was subjected to an XPS angular dependence
study to determine the location of Pd(II)/Pd(0) on the film's
surface. Second, the surfaces of LTP and PDS conductive films
were etched with aqua regia to free the surface of palladium(0).
They were then examined by XPS and remeasured for resistivity.

The result of the angular study are shown in Figure 6. At
low incidence angle (θ), the angle between the beam of photo-
emitted electrons and the sample surface, the number of electrons
due to Pd(II) is greatly diminished, compared with those due to
Pd(0). Because the sampling depth decreases with decreasing θ,
the conclusion is reached that reduction of the metal takes place
only on the very surface of the film and that the bulk of the
polymer contains a preponderance of Pd(II). The metallic PDS
film, of course, showed no Pd(II), even at high incidence angles.
The palladium oxidation state in the bulk of this film cannot be
determined by this method.

Table VII. Resistivity Data for $Cu(TFA)_2$ Doped ODA-Derived Polyimide Films.

Dianhydride	RESISTIVITY[8] Volume (ohm-cm)	Surface (ohm)		% Cu Calc'd	Found
BTDA[a,c]	1.6×10^{15}	2.0×10^{14}	(GLASS)	2.62	2.76
		1.7×10^{12}	(ATM)		
BTDA[b,c]	8.2×10^{14}	$>10^{18}$	(GLASS)	2.62	2.82
		4.7×10^{12}	(ATM)		
BTDA[a,d]	8.6×10^{16}	4.6×10^{15}	(GLASS)	2.62	2.61
		9.7×10^{15}	(ATM)		
PMDA[a,c,e]	5.6×10^{14}	3.1×10^{12}	(GLASS)	2.94	4.08
		7.3×10^{10}	(ATM)		
PMDA[a,c,f]	4.1×10^{14}	3.0×10^{12}	(GLASS)	2.94	3.34
		9.3×10^{10}	(ATM)		
PMDA + ODA Alone[c]	$>10^{18}$	1.0×10^{16}		–	–
BTDA + ODA Alone[c]	$>10^{18}$	1.0×10^{17}		–	–

[a] $Cu(TFA)_2$ introduced after polymerization

[b] Polymerization in situ with $Cu(TFA)_2$

[c] Air cured

[d] N_2 cured

[e] Prep I

[f] Prep II

To further test this hypothesis the LTP and PDS films were subjected to surface cleaning with an aqua regia etching solution. Before treatment the LTP film had a volume resistivity of 1×10^{10} ohm-cm and a surface resistivity of 5×10^8 ohm. Acid etching of the surface of this film shows a negligible change in these values. XPS of the surface of the film before and after etching revealed a small shift in the position of the Pd $3d_{5/2,3/2}$ photopeaks to a higher binding energy which indicated the removal of most of the Pd(0). Etching in aqua regia of the PDS film completely removes the metallic surface and raises the volume and surface resistivities to those equivalent to the polymer alone (Figure 7). Binding-energy measurements show clearly that the acid-etched film contains only Pd(II) at the surface. Elemental analysis of the two films before and after acid etching indicate that a significant amount of palladium is lost from the PDS-doped film in the etching process, whereas in the film doped with LTP a negligible amount of palladium is lost. We therefore might reason that in PDS films, surface palladium accounts for the lowering of resistivity. The film behaves essentially as a metal-vapor-deposited polyimide film. The LTP-containing film consequently appears to be the only film that possesses a uniformly lowered resistivity.[15,16]

Two BTDA-ODA films, one doped with LTP and the other doped with PDS, were subjected to argon ion etching to establish the distribution of dopant components throughout the films. Figure 8 shows a plot of atom percent vs sputter time (min) for a film doped with PDS. Palladium, carbon, sulfur and chlorine were monitored continuously by Auger electron spectroscopy (AES). The estimated rate of ion etching by the Argon ion beam in this experiment is about 500A/min. The profile shows clearly that most of the palladium in the film lies in the first 1000Å of the film surface and after this point it drops off rapidly to a level similar to that of sulfur and chlorine (residual amounts from dopant complex). The concentration of carbon rapidly rises as the palladium surface is penetrated and the bulk of the polymer is reached by the ion beam. After a distance of about 5000Å all elements have reached constant concentration indicating the film is homogeneous from this point through the film. This agrees with the chemical etching data which suggests that a majority of the palladium is on the surface of the PDS film.

An LTP-doped film was ion etched over a much larger area to make possible the determination of oxidation state of components as the surface is penetrated into the bulk of the film. The film was etched by a rastered ion beam for three thirty minute sequences and XPS information obtained after each time period. The etching rate was estimated to be about 20Å per minute or a

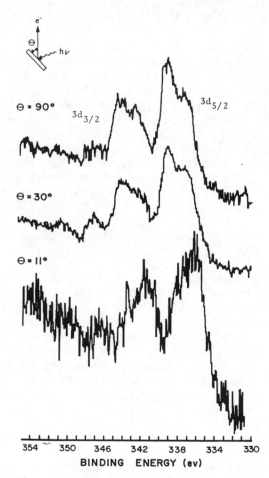

Figure 6. Angular Dependence of Palladium 3d Photoregion on a BTDA-DADPC/Li$_2$PdCl$_4$ Film.

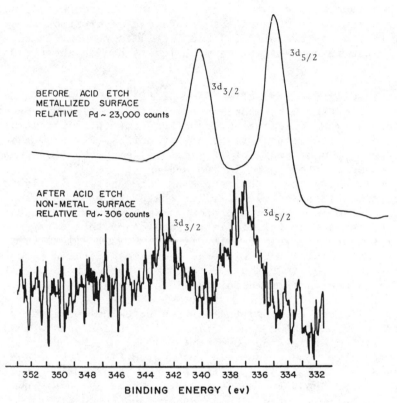

Figure 7. Palladium 3d Photoelectron Region Before and After Chemical Etching of BTDA–ODA/Pd(S(CH$_3$)$_2$)$_2$Cl$_2$ Film Surface.

total of nearly 2000Å of the surface penetrated. This study
clearly shows that after a thin layer of predominantly Pd(0) is
penetrated, the film contains a very homogeneous mixture of
Pd(0) and Pd(II) in nearly equal amounts. Also present in the
film bulk is a uniform mixture of Li^+ and Li_2O as evidenced by
the characteristic photopeaks at 55 eV and 50 eV. Chlorine is
also present in residual amounts.

A similar procedure of XPS profiling was carried out on the
PDS doped film. Over 4 hours of ion milling was necessary to
penetrate the metal layer. After this time an area about 5 mm
square had been milled away exposing clearly the dark-red brown
color of the polymer film beneath. XPS data aquisition shows
clearly that a mixture of Pd(0) and Pd(II) is present in the
film bulk but in concentrations much lower than in the LTP doped
film.

The inability of the LTP-BTDA-DABP film to conduct is per-
plexing in light of other successes with this palladium-dopant.
There are, however, indications from XPS that this film is very
inhomogeneous with respect to Pd distribution. Figure 9 shows
the $Pd3d_{5/2,3/2}$ photopeak for this film. The peaks at 350 and
346 eV are attributed to palladium and are believed to be due to
differential charging of the sample surface and are likely to be
the result of "islands" of palladium that have been isolated on
the surface, thus preventing proper contact with the spectrometer. A scan of several different samples of this film
suggests that the degree of this effect varies but is always
present. This "islanding" of metal would most certainly be
detrimental to any conduction mechanism operable in these films,
weak signals are again seen for Pd(II) and Pd(0).

These XPS and etching studies have revealed essentially
three types of palladium-doped films: (1) those doped with a
soluble lithium containing complex (LTP) and are homogeneous
and electrically conductive, (2) those with LTP which produce
inhomogeneous non conductive films and (3) those that contain a
soluble complex (PDS) but exhibit loss of a majority of the
dopant upon thermal curing, hence showing no conductivity or at
best only surface conductivity.

The conductive films (those doped with LTP) apparently have
a mixture of Pd(0) and Pd(II) together with lithium ion. It
seems likely that the coupling of Pd(0) and Pd(II) is integral to
the conduction process.

XPS has proven useful in the study of lithium speciation in

lithium-doped polymers.[9] Figure 10a shows a typical spectrum obtained for the lithium 1s photopeak. The primary lithium peak for all doped polyimides was found to be around 50.5 eV regardless of the measured resistivity. At higher dopant concentrations, a secondary peak, not always observed and much less intense, appears around 55.0 eV (Figure 10b). According to Povey and Sherwood,[17] the peak at 50.5 eV is characteristic of lithium in Li_2O, while the peak at 55.0 eV is more characteristic of lithium in LiCl. We find no evidence that spectrometer conditions can give rise to photochemical decomposition leading to lithium oxide as suggested by Povey and Sherwood.[17] A film irradiated by x-rays after 30 minutes gave the same spectrum as that of the same film irradiated after 2 hours. This would be unexpected if photodecomposition were occurring during x-ray photoelectron spectroscopic examination. It is reasonable, therefore, to conclude that lithium oxide may be generated during thermal imidization via Li^+ reaction with the released water, followed by hydrolysis and dehydration. Similar lithium XPS observations have been made on films doped with Li_2PdCl_4. The peak at 55.0 eV which appears only at high lithium concentrations arises when, no doubt, insufficient water of imidization is released to solvate all the lithium ion.

The nature of lithium on the film surface compared to the bulk is of interest. In other words is the "lithium oxide-like" material dispersed throughout the film rather than on the film surface. First of all, the atmosphere side and glass side of each film show a comparable number of electron counts for lithium. This observation is contrary to our palladium findings in that the atmosphere side of the film contains appreciably more palladium than the glass side of the film. Information regarding the lithium content and chemical state inside the lithium doped film has been obtained from angular and depth profile studies. At a 90° x-ray incidence angle, the spectrum of highly doped BTDA-ODA-LiCl is as per Figure 10b. On shifting to an incidence angle of 11°, the secondary peak is diminished significantly relative to the main peak. This suggests that "LiCl" is found only in the matrix of the polymer, whereas, "Li_2O" occupies the surface as well as the bulk. On depth profiling normally doped BTDA-ODA-LiCl for 30 minutes (argon ion sputtering), the shoulder at 55.0 eV which originally could not be discerned has grown to a distinct peak (Figure 11). The ratio of "LiCl" to "Li_2O" through the film appears to be a constant since spectra after 60 and 90 minutes sputtering are similar. One concludes, therefore, that "LiCl" is present in small quantities and that most has been converted into "Li_2O" during thermal imidization.

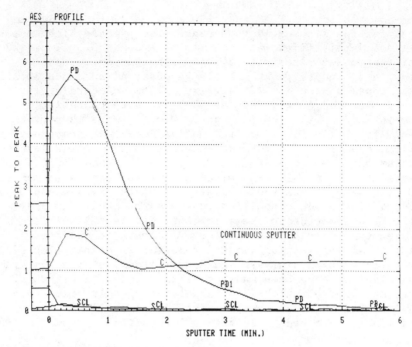

Figure 8. Auger Electron Depth Profile of Pd, C, Cl, and S for a BTDA–ODA/Pd(S(CH$_3$)$_2$)$_2$Cl$_2$ Film.

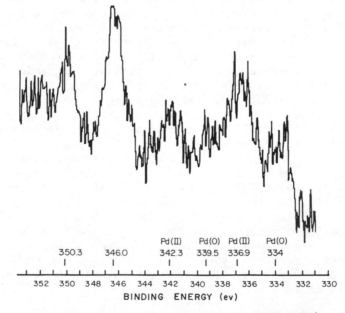

Figure 9. Differential Charging Effects on BTDA–ODA/Li$_2$PdCl$_4$ Film. Pd 3d Photoregion, Air Side of Film, Volume Resistivity= 1×10^{16} ohm–cm.

The nature of the "metallic-looking" surface of copper(II) doped films has been a major question in this investigation.[8] The surface is produced only when the film is cured in air. Longer (>1 hr @ 300°C) curing times appear to favor formation of the "mirror-like" surface regardless of the copper(II) dopant. Differences in viscosity of the polyamic acid solution and the method of preparation are not crucial, although thicker films seem to adopt the "silvery-looking" surface more readily than thin films.

XPS of pertinent photopeaks relative to copper(II) doped films have been measured. The copper 2P photopeak region in all cases affords the most information and interesting features. The $2P_{1/2,3/2}$ main peaks occur in the regions 952.5 eV and 933.0 eV respectively. These values are slightly lower than those observed in the pure copper(II) dopant, but they are higher than in copper metal. Figure 12 dramatizes these findings with $Cu(TFA)_2$ as dopant in the BTDA-ODA polyimide. The spectrum of the doped polymer is of the atmosphere (air) side of the film. Appreciable Cu(II) is observed, but surprisingly no fluorine could be detected. On the glass side of the film, the opposite conclusions were drawn (i.e. little copper but significant fluorine). Similar findings were made with $Cu(TFA)_2$ and PMDA-ODA. Since more copper was found on the atmosphere side of the film which also had the "silvery-looking" surface, one may conclude that copper in some form is contributing to this appearance. Binding energy and satellite data suggest a copper(II) rather than metallic copper state. Depth profiling (argon ion etching with Auger electron monitoring) reveals the build-up of copper on the surface relative to the bulk polymer. Figure 13 plots atomic concentration versus sputter time for graphite-coated PMDA-ODA doped with $Cu(TFA)_2$. The inverse relationship of Cu and C concentrations and the absence of F are noteworthy.

The material contributing to the "silvery surface" can also be chemically etched from the polymer film with strong non-oxidizing acid. Depth profiling of the air side of the chemically etched film reveals very little Cu whose concentration is invariant with sputter time. The "silvery" deposit therefore contains significant copper.

Frost and coworkers[18] have documented some forty-six different copper compounds and complexes and have observed three types of XPS satellite splitting patterns which appear to be dependent on the copper ion environment. Type A satellite structure exhibits two satellites in both the $2P_{1/2}$ and $2P_{3/2}$ lines. Type B has two satellites in the $2P_{1/2}$ and two over-

Figure 10. Concentration Dependence of Lithium 1s Photopeak in BTDA-ODA/LiCl Films.

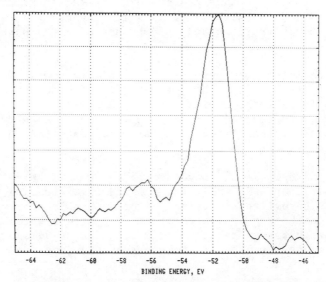

Figure 11. Lithium 1s Photoregion After Depth Profiling of BTDA-ODA/LiCl Film After 30 Minutes.

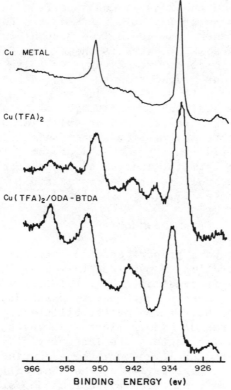

Figure 12. Copper 2P Photopeak Region of Copper Metal, Cu(TFA)$_2$ and BTDA-ODA/Cu(TFA)$_2$ Film.

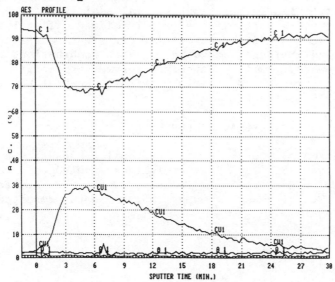

Figure 13. Auger Electron Depth Profile (C, Cu, O, F) of the Air Side of PMDA-ODA/Cu(TFA)$_2$ Film Before Chemical Etching.

lapping satellites in the $2P_{3/2}$. One satellite in each photopeak is characteristic of Type C. $Cu(TFA)_2$ and $Cu(AcAc)_2$ are of
Type A. The XPS spectra change to Type B upon curing of the
polymer film. Figure 12 illustrates these changes in satellite
structure relative to copper metal which exhibits no satellites.
Electronic transitions giving rise to these secondary photopeaks
are generally believed to be either due to charge transfer
between ligand and metal and/or excitation of metal electrons to
vacant metal orbitals (i.e. $3d \rightarrow 4s$ or $4p$). The apparent change
in satellite structure may reflect a modified environment for
copper in the polyimide relative to the dopant. Coordination of
copper(II) to the polyimide, thermal decomposition of the added
metal complex, or hydrolysis of the dopant are possible fates
for the metal in these films.

SUMMARY

In summary numerous properties of polyimides can be
modified by incorporation of metal additives at the polyamic
acid stage prior to imidization. High solubility of the additive in DMAC is required. High quality films are produced in
each case which have better thermal stabilities, higher
softening temperatures and lower electrical resistivities and
increased mechanical strength. The integrity of each additive
employed in these studies is not maintained during aerobic
thermal imidization. Segregation of the chemically-changed additive is evidenced in some cases by surface deposits on the air-
side of each film which may result from either aerobic
oxidation, thermal reduction or hydrolysis-dehydration of the
metal dopant. Clearly more stable metal additives are needed in
order to study the full effects of polyimide modification.
Additional study of different metals, different additives of the
same metal and various dianhydride-diamine combinations is
desirable toward optimizing specific polymer properties. Extension of this research to other electrically neutral polymers
would probably be fruitful.

ACKNOWLEDGEMENT

The financial support of the National Aeronautics and Space
Administration (NSG 1428) is gratefully acknowledged. Many
thanks to V. C. Carver, T. A. Furtsch, E. Khor, S. A. Ezzell, G.
Fortner, T. W. Fritz and J. Schaaf for the data presented in
this overview.

REFERENCES

1. C. E. Carraher and M. Tsuda, Editors., "Modification of Polymers", ACS Symp. Ser., No. 121, American Chemical Society, Washingtion, D.C., 1980.
2. R. J. Angelo and E. I. DuPont DeNemours and Co., "Electrically Conductive Polyimides", U.S. Patent 3,073,785 (1959).
3. R. J. Angelo, (1976), Personal Communication.
4. N. S. Lidorenko, L. G. Gindin, B. N. Yegorov, V. L. Kondratenkov, I. Y. Ravich and T. N. Toroptseva, AN SSSR, Doklady, 187, 581 (1969).
5. A. L. Endrey and E. I. DuPont DeNemours and Co., U.S. Patent, No. 3,073,784 (1963).
6. M. N. Sarboluki, NASA Tech. Brief, 3(2), 36 (1978)
7. L. T. Taylor, A. K. St. Clair, V. C. Carver and T. A. Furtsch, Ref. 1, p. 71.
8. S. A. Ezzell, T. A. Furtsch, E. Khor and L. T. Taylor, J. Polym. Sci. Polym. Chem. Ed., 21, 865 (1983).
9. E. Khor and L. T. Taylor, Macromolecules, 15, 379 (1982).
10. D. J. Progar and T. L. St. Clair, 7th National SAMPE Technical Conference, Albuquerque, NM, Oct., 1975.
11. L. T. Taylor, A. K. St. Clair and NASA Langley Research Center, "Aluminum Ion Containing Polyimide Adhesives", U.S. Patent 284,461 (1981).
12. L. T. Taylor and A. K. St. Clair, J. Appl. Polym. Sci., (1983).
13. T. L. Wohlford, J. Schaaf, L. T. Taylor, T. A. Furtsch, E. Khor and A. K. St. Clair, in "Conductive Polymers", R. B. Seymour, Editor, p. 7, Plenum Publishing Corp., New York, 1981,
14. T. A. Furtsch, L. T. Taylor, T. W. Fritz, G. Fortner and E. Khor, J. Polym. Sci. Polym. Chem. Ed., 27, 1287, (1982).
15. V. C. Carver, L. T. Taylor, T. A. Furtsch and A. K. St. Clair, J. Amer. Chem. Soc., 102, 876 (1980).
16. L. T. Taylor, A. K. St. Clair, V. C. Carver and T. A. Furtsch, NASA Langley Research Center, "Electrically Conductive Palladium Containing Polyimide Films", U.S. Patent 4,311,615 (1982).
17. A. F. Povey and P.M.A. Sherwood, J. Chem. Soc., Faraday Trans., 1240 (1974).
18. D. C. Frost, A. Ishitani and C. A. McDowell, Mol. Physics, 24, 861 (1972).

MECHANICAL PROPERTIES OF THIN POLYIMIDE FILMS

Robert H. Lacombe and Jeremy Greenblatt

IBM General Technology Division, East Fishkill

Hopewell Junction, New York   12533

A technique for investigating the thermal-mechanical properties of ultra-thin polymer films is described in some detail. The basic experiment involves a cantilevered reed of vitreous silica which is used as a substrate to support the sample film. The composite sample can be excited into vibration by an electrical drive force and the resulting frequency of vibration and damping rate can be determined. Using elementary formulae based on the mechanics of bending beams, the Young's modulus and internal friction of the sample film may be deduced as a function of temperature and frequency. Sample data on a 6 nm (6 nanometer, 60 Å) monolayer film and a 480 nm polystyrene coating are discussed. The remainder of the paper is concerned with 100 nm polyimide films which are coated on SiO_2 both with and without a silane primer interface layer. The thermal mechanical properties of silane and polyimide films are discussed. It is proposed that an internal friction peak near $-50°C$ can be assigned to interactions between the silane and polyimide layers in a laminated structure.

INTRODUCTION

Polymer resins have been a favorite insulating material of the electronics industry for a long time. One of the most popular resin insulators are the epoxies, which, combined with glass reinforcing fibers, form the cards and boards that have been the mainstay of the electronics industry in providing a packaging medium for all manner of electrical components. The good thermal and electrical properties of epoxies combined with their high resistance to chemical attack from acids and solvents have given them diverse roles as supporting substrates and hermetic sealants for computer chip modules. However, recent years have seen the rise of a newer class of materials in more demanding applications where the temperature stability of the epoxies is not good enough. These materials are the polyimides, and they may be classified as one of the emerging insulator materials of the microelectronics industry. Polyimides, in fact, are providing for the microelectronics industry what epoxies provided for the older electronics industry; namely, a high temperature insulator which can survive the processing rigors of microelectronic fabrication. Among the favorable properties of the polyimides are: thermal stability up to 400°C, low dielectric constant (~3.5), high resistance to chemical attack by solvents and acids, ability to planarize and fill small gaps in multilevel structures, and, finally, mechanical toughness of appropriately formulated molecular architectures. With such properties, polyimides have great potential for use in microelectronic structures, however, as with any other material, there are also great problems to be solved. Anyone familiar with epoxy resins is aware that the final properties achieved depend on the precise molecular formulation started with, and the details of the curing process. Polyimides are no exception. For instance, the mechanical properties of a polyimide can range from brittle to tough depending upon the initial material and the way it was cured. A further concern arises from the problem of moisture sorption. Polyimides, like epoxies, can absorb several weight percent of moisture. This can affect not only electrical and mechanical properties, but also raises the possibility of corrosion if sensitive metal structures are present. A final demon to contend with is the ever-present question of adhesion of polyimide to the various layers which must be present in any microelectronic structure.

In order to address the aforementioned problems, we must resort to analytical techniques which allow us to assess the mechanical properties or the state of cure of any particular polyimide resin. Of all the possible techniques we might use, we will be concerned with that of dynamic mechanical analysis. This technique dates back at least 40 years. An example of an early work would be the now classic study by Wert and Zener[1] on the diffusion of carbon in iron samples using the torsion pendulum technique. A thorough presentation of dynamic mechanical analysis from the point of view of the metallurgist has been given by Nowick and Berry.[2] There are two excellent texts on the subject from the polymer viewpoint by Ferry[3], and McCrum et al.[4]*

Dynamic mechanical analysis for microelectronic applications presents a special difficulty, however, which none of the aforementioned references deals with. This is the fact that very thin films on the order of 2000 nm thick are the object of study. By comparison, a typical epoxy board or card would be 1000 times thicker. Clearly, a more refined technique than those mentioned in references 2-4 must be implemented for investigating the mechanical properties of such thin films. Fortunately, we can fall back on the poineering work of Berry and Pritchet[5] on the vibrating reed experiment to provide a way out of this dilemma.

In light of the above remarks, the first part of this paper will deal with an overview of the vibrating reed technique as applied to the measurement of the thermal-mechanical properties of ultra-thin films. An example will be given of monolayer films 6 nm (60Å) thick which test the ultimate sensitivity of the technique. Following this, data will be presented on a 480 nm film of polystyrene. These results will demonstrate that for thicker films many of the results which are well known for bulk samples can be recovered. After these calibration studies, our attention turns to polyimides which are the main focus of this work. Data will be presented on the thermal-mechanical response of a 100 nm film. These results will be analyzed and compared with data from the literature on bulk samples. The final part of this work will deal with an entirely new topic: the interaction of polyimides with silane coupling agents which are used to improve the mechanical integrity of multilayer structures made with polyimide. It is in this work that the vibrating reed technique comes into its own, since the interlayer structures which are being investigated are only on the order of 8 nm thick.

OVERVIEW OF VIBRATING REED TECHNIQUE

The basic principle of dynamic mechanical analysis is quite simple. A sinusoidally varying stress is applied to the sample and the resulting response is observed.

---

*Regrettably, the latter text is no longer in print and must be obtained from a library or special book service.

In our case, the response of interest is the dynamic modulus which is a complex number of the form:

$$E^*(\omega) = E'(\omega) + iE''(\omega)$$

In the limit of small strains and anelastic behavior of the material, $E'(\omega)$ is the elastic or storage modulus commonly associated with the "Hookes Law" modulus. It represents the reversible elastic behavior of the material. $E''(\omega)$, on the other hand, is the loss modulus and is indicative of the energy dissipated in the sample per cycle.

Performing precise measurements, however, requires a good deal more sophistication than is immediately apparent. In what follows, therefore, we will review a number of the more important details of the vibrating reed experiment.

First, it should be clear that in order to test the mechanical properties of films a few nanometers thick, a supporting substrate of some kind must be employed. To this end, the vibrating reed technique employs a cantilevered beam as shown in Figure 1. As previously mentioned, the basic concept of the experiment is quite straightforward. Using the drive circuitry, an oscillatory force at a resonant frequency is induced at the tip of the reed, thus setting it into vibration. Turning off the drive force, the detection circuit is then employed to monitor the frequency and rate of damping of the now freely vibrating reed. From the frequency data and other parameters, the elastic modulus of the reed material can be determined. From the damping rate, a measure of the sample's internal friction can be obtained. These quantities then give the real and imaginary parts of the complex dynamic modulus of the material, which is the basic material parameter measured in a dynamic mechanical experiment.

Thus we have an elementary outline of the vibrating reed experiment. For more precise details, the reader should refer to references 2 and 5. To eliminate the effect of air damping, the reed must be under vacuum ($< 10^{-4}$ torr).

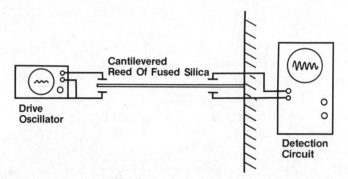

Figure 1. Schematic diagram of the vibrating reed experiment. A cantilevered reed specimen is excited into vibration by a drive oscillator. The drive signal is removed and the resulting freely decaying signal is monitored by detection circuitry.

SOME USEFUL FORMULAE FOR VIBRATING BEAMS

A reed is nothing more than a beam whose length is much greater than its width, which in turn is much greater than its thickness. We can, therefore, deduce the vibrational behavior of a reed from the classical results on beams. The dynamics of a cantilevered beam is treated in the text by Den Hartog[6]. A somewhat more refined treatment has been given by Timoshenko[7]. He includes a discussion of the higher order effects of shearing force and rotatory inertia. Some useful background material on the static deformation of beams can be found in the engineering test by Den Hartog[8]. Those who prefer a physicist's point of view will find an elightening treatment of the bending beam problem in the text by Feynman et al.[9]

Consider a cantilevered reed such as the one depicted in Fig. 2. Assuming uniformity in all dimensions, the normal mode frequencies are given by:[6,7]

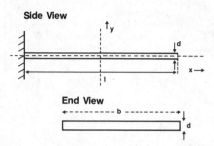

Figure 2. Geometry of cantilevered reed.

$$f_n = \frac{\alpha_n^2 d}{\sqrt{3}\pi l^2} \sqrt{\frac{E}{\rho}} \qquad\qquad n = 1, 2, \ldots \qquad (1)$$

where: the α_n are mode numbers (see ref. 2).

E = Young's modulus of reed material

ρ = Density of reed material

d = reed thickness

l = reed length

Figure 3 depicts the shapes of the first few flexural modes of vibration of a cantilevered reed. We can, in fact, think of these flexural modes in much the same way that we regard the eigenstates of a quantum mechanical system such as a hydrogen

atom. Equation (1) gives the discrete frequencies which would
correspond to energy levels in the quantum system.

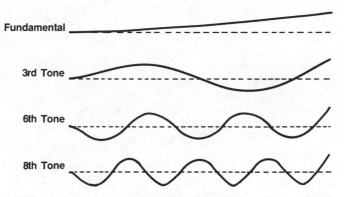

Figure 3. Shapes of the first few flexural modes of a
cantilevered reed.

We now need to know the vibrational frequencies of a coated
reed where the coating is much thinner than the underlying
substrate. The basic effect of the coating is to shift the
resonant frequency of the substrate. Thus, to first order in the
coating thickness the frequency of the composite reed is:

$$(f_n^c)^2 = (f_n^o)^2 \left(1 + \frac{2t}{d} \left(\frac{3E_f}{E_s} - \frac{\rho_f}{\rho_s}\right)\right)$$

Solving for the modulus term we get:

$$\frac{3E_f}{E_s} = \frac{d}{2t} \left(\left(\frac{f_n^c}{f_n^o}\right)^2 - 1\right) + \frac{\rho_f}{\rho_s} \qquad (2)$$

653

where:

f_n^c = frequency of composite reed at nth tone of vibration

f_n^o = frequency of bare reed at nth tone

E_f = modulus of coating; E_s = modulus of substrate

ρ_f = density of coating; ρ_s = density of substrate

d = thickness of substrate

t = coating thickness (one side)

Equation (2) is our basic formula for determining the modulus of thin coatings. We will have further comments on the use of this expression later in this paper. This equation has been presented previously by Berry and Pritchet[5].

ILLUSTRATIVE RESULTS ON ORGANIC THIN FILMS

In order to better understand the capabilities of the vibrating reed technique and to gain more confidence in its methodology, we will present results on well characterized thin films. Our first example will concern monolayer films only 6 nm thick. This study tests the ultimate sensitivity of the vibrating reed technique to detect the thermal-mechanical properties of ultra-thin films. The second study will be on a 480 nm film of polystyrene. The data from this work will confirm that the technique can reproduce results on a well-known glassy polymer.

The ability of the vibrating reed technique to detect the mechanical properties of thin films rests on the proper choice of a substrate material in fabricating composite reeds. For work on organic insulating materials, the choice of vitreous silica (amorphous SiO_2) has many advantages. The key property of silica is its very low and slowly varying internal friction behavior at acoustic frequencies above -150°C. Fig. 4 illustrates the internal friction behavior of silica which makes it such a favorable substrate. Silica also possesses a high degree of thermal and chemical stability vis-a-vis organic polymers, making it an ideal inert substrate for most applications. The suitability of silica substrates for studying the thermal-mechanical properties of ultra-thin films was tested by Lacombe and Kellner[10] who looked at 6 nm monolayer films of cadmium arachidate. The films were deposited by the Langmuir-Blodgett technique[11] and their structure is given schematically in Fig. 5. Despite their extreme thinness, an internal friction spectrum was obtained for these films which is illustrated in Fig. 6. Accompanying the mechanical damping data in Fig. 6 is dielectric loss data on stearic acid due to Careem and Jonscher.[12] The mechanical and dielectric data complement each other nicely in this example, and allow us to propose molecular relaxation mechanisms for the dispersion peaks in Fig. 6. The peak near -150°C, which dominates the mechanical spectrum but appears only weakly in the dielectric data, can be associated with a hydrocarbon backbone relaxation first studied in detail by Willbourn.[13] The peak near -20°C, however, which dominates the dielectric data but is weak in the mechanical spectrum, must clearly arise from some dipolar motion of the acid groups at the monolayer interphase. Thus it is clear from this interpretation of the data that the vibrating reed technique can detect relaxations in very thin interphases indeed!

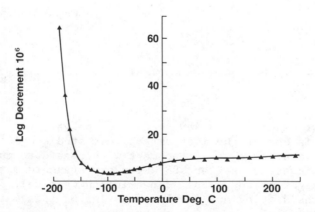

Figure 4. Internal friction vs. temperature behavior of a bare silica reed, third vibration tone at 618 Hz. Notice the low and slowly varying damping above -150°C.

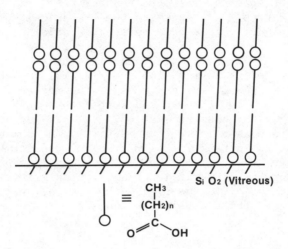

Figure 5. Schematic diagram of a three level monolayer
structure.

Our next example is a 480 nm film of polystyrene. Since
polystyrene is one of the most extensively studied glassy
polymers, it can serve as a benchmark with regard to more exotic
systems. This study used an NBS secondary standard material of
molecular weight 84,600 supplied by Aldrich Chemical.[14] Figure
7 exhibits the damping and modulus behavior of this material.
The main feature of Fig. 7 is, of course, the glass transition
peak near 120°C. We can also observe the low temperature tail of
the β relaxation in polystyrene. The main peak of this secondary
relaxation is unfortunately obscured by the glass transition
peak.

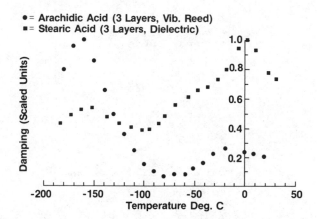

Figure 6. Mechanical and dielectric damping behavior of monolayer structures as exhibited in Fig. 5. The electrical and mechanical measurements complement each other in elucidating the nature of the damping peaks observed.

Figure 7, however, is as interesting for the features that it does not show as for those that it does. In particular, there is no relaxation peak near -100°C as is sometimes reported in the literature as a γ relaxation.[15] There is, in fact, extensive evidence that the γ relaxation in polystyrene is due to the diffusion of impurities in the sample, such as mold release agents commmonly used in sample preparation.[16] Further, we see no evidence of any relaxation peaks above the glass transition whose existence has been suggested by Boyer.[17] Considerable controversy now exists as to whether there can be some form of higher order relaxation phenomena occurring in polymer melts. The data in Fig. 7 certainly does not support such a conjecture, but because the sample is extremely thin, such a relaxation may be inhibited due to boundary effects.

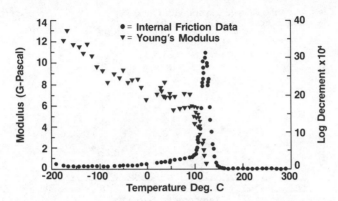

Figure 7. Damping and modulus behavior of a 480 nm film of
polystyrene, 2nd tone vibrating reed ~ 214 Hz. The modulus of
this film at room temperature is twice the value normally quoted
for bulk polystyrene samples. This may be due to a significant
amount of chain ordering in the thin film.

As a final observation on polystyrene, we note the modulus
data presented in Fig. 7 which was calculated on the basis of
Equation (2). The modulus of our sample decreases monotonically
with increasing temperature, suffering a catastrophic drop at the
glass transition temperature to values too low to measure.

At room temperature, Fig. 7 gives a Young's modulus of
~ 6.8 GPa. This is roughly twice the value normally quoted for
polystyrene.[18] We feel that this high value is quite likely
due to the ordering present in thin polystyrene films as reported
recently by Cohen and Reich.[19] A higher degree of ordering
would imply more efficient chain packing and thus higher
interchain interaction which could lead to a higher modulus for
the material. This is an apparent example of a surface effect
which would be transparent to bulk measurement techniques.

THERMAL-MECHANICAL PROPERTIES OF THIN POLYIMIDE FILMS

In this section, we will address the main topic of this paper, which is the mechanical properties of thin polyimide films. All of the preceding discussion was to clearly describe the methodology of the vibrating reed technique and to establish its capabilities via well characterized samples. In particular, it is important to verify the reliability of the technique in detecting mechanical relaxations in ultra-thin layers on the order of nanometers in thickness, because we will be attempting to look at such structures as interfaces in laminates involving polyimide, vitreous silica glass and a silane primer agent.

Mechanical Toughness vs. Brittleness

Before tackling the problem of polyimide laminate structures, we will discuss the mechanical toughness of polyimide films and whether such a property can be related to internal friction behavior.

To this end, we present in Fig. 8, damping vs. temperature data for two polyimide films, one known to be a tough material, the other brittle. The tough material, represented by circles in Fig. 8 is a pyromellitic dianhydride/oxydianiline polyimide (PMDA/ODA), whereas the brittle material is a benzophenonetetra-carboxylic acid dianhydride/methylene dianiline polyimide (BTDA/MDA). The PMDA/ODA material is a 100 nm film which will be discussed later in this paper. The BTDA/MDA material was a 2000 nm film studied by Berry and Pritchet[20] using the vibrating reed technique. The basic question is this: can we understand the substantial difference in the toughness of these materials on the basis of internal friction data which is taken at very low strains? The fracture toughness experiments of Johnson and Radon[21] on polymethylmethacrylate would answer this question in the affirmative. Furthermore, Hertzberg and Manson[22] have given an extensive review of the question covering a range of glassy polymers. The basic argument is this: the peaks one observes in an internal friction spectrum such as Figs. 7 and 8 are a manifestation of specific modes of molecular motion which are available to the polymer chains at a given temperature and frequency. In particular, above the glass transition temperature the kinetic energy of thermal Brownian motion is high enough so that nearly every possible mode of molecular reorientation is active, thus giving the material the mechanical properties of a viscous melt or a soft rubber. Thus, above the glass transition, every polymer becomes tough in the sense that it can endure large

deformations without undergoing fracture. Below the glass
transition temperature, the majority of deformation modes are
frozen out and the material thereby becomes a rigid glass. The
story does not end here, however, because a number of more
localized modes of deformation remain active below the glass
transition temperature. It is these secondary relaxations which
will control the thermal-mechanical properties of the glass. The
first major relaxation below the glass transition is
traditionally called the β transition. Referring to Fig. 7, we
see that for polystyrene a weak β peak occurs near 100°C.
Polystyrene is thus predicted to be a brittle material at
temperatures lower than 100°C, which is well known to be the
case.

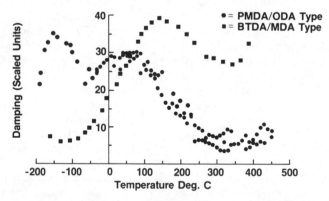

Figure 8. Internal friction behavior of two polyimide films.
The location and nature of the glassy state relaxation peaks may
account for the marked differences in the fracture toughness
behavior of these two polymers.

The BTA/MDA polyimide material, whose relaxation spectrum is given in Fig. 8, has a well defined β peak near 150°C. This material would also be predicted to be brittle near room temperature, a fact which we knew beforehand. The PMDA/ODA material, on the other hand, has a very broad secondary relaxation spanning the range from −150 → 100°C. This broad secondary relaxation has also been observed in bulk PMDA/ODA samples by other investigators.[23,24,25] A further secondary relaxation occurs near −150°C. This material is known to be tough[26] at room temperature. We presume that it is the molecular motions associated with these secondary relaxation peaks that account for the ductility of this polyimide.

Arguments based on anelastic spectra such as Figs. 7 and 8 have a basic difficulty in that such spectra are taken at very low strain levels. Toughness and fracture phenomena, however, occur at very large deformations. In spite of such an objection, we can still think of the relaxation peaks at low strains as being the "fingerprints" of latent deformation mechanisms which can come into play at larger strains. Furthermore, we can think of any large deformation as the sum total of a large number of small deformations and each small deformation can occur via a low strain anelastic mechanism. Thus, anelastic mechanisms cannot elucidate the details of large deformation behavior. However, such mechanisms may suffice to predict whether there exist sufficient internal degrees of freedom to allow for large deformations at all. For example, the existence of a relaxation peak at low temperatures may allow one to predict that a particular polymer will be tough at large strains. However, such data cannot predict any further details of the large strain behavior of the material.

Before leaving the topic of intrinsic mechanical properties of polyimide films, we present the modulus behavior of the PMDA/ODA polyimide whose damping behavior is given in Fig. 8. Because this film was only 100 nm thick, the modulus could be accurately estimated only at the highest frequency investigated which was near 2900 Hz at the 6th tone of vibration. Figure 9 presents the modulus vs. temperature data. This data actually represents a composite run where the lower points were taken going up in temperature, and the upper points were taken going down from 450°C. The data presents a small but clear hysteresis effect in the material on thermal cycling. Such behavior was also observed by Gillham et al.[27] on a wide range of PMDA based polyimides. An unusual aspect of the data in Fig. 9 is the slight decrease in the modulus on coming down from 450°C. Gillham and co-workers always noticed an increase in modulus on

cooling from 450°C. What is clear, however, is that the
mechanical properties of these polyimides are sensitive to
thermal cycling up to 400°C.

Interfacial Structures in Laminates

In this section, we come to a problem which requires the
full resolution capabilities of the vibrating reed in detecting
the mechanical properties of thin layers. A recurrent problem in
building multilayer structures from polymers or any other
materials is whether the various layers adhere to one another so
that the structure will not delaminate under typical use
conditions. A typical strategy to improve the adhesion between
layers is to employ a coupling agent at the interface. In
particular, silanes have long been employed to improve the
adhesion between glass fibers and epoxies[28]. Silanes have also
been employed to improve the adhesion between polyimide and
SiO$_2$.[29,30,31] Linde[32] has, in fact, proposed a mechanism
for the coupling of γ-amimopropyltriethoxy silane (γ-APS) to
polyimides. We have recently completed a study of the
thermal-mechanical behavior of γ-APS deposited by plasma and
solution techniques. We will report on these results as a
prelude to discussing a study of the interaction between silane
and polyimide films via internal friction measurements.

Figure 10 exhibits the internal friction behavior of two
silane films, one coated by solution techniques (dip coated .01%
aqueous silane), the other by plasma deposition.[33,34] Two features
stand out immediately, the first of which is a strong peak near
-150°C which is present in both the solution and plasma coated
films. The second feature is a peak near -50°C which shows up
strongly in the plasma deposited film, but appears only as a weak
shoulder in the solution coated film. This second peak is the
most interesting because it clearly points to a difference
between the solution and plasma coated films. The precise nature
of each of the peaks in Fig. 10 must remain obscure until further
experiments can clarify their origins. However, we can make
intelligent guesses concerning the mechanisms behind these
relaxations based on what is already known about the nature of
solution coated silane films and the differences between solution
and plasma deposition techniques. Starting with the -150°C
relaxation, we might infer that this is somehow associated with
the Si-O-Si network which we know is present in solution coated
films[28]. Because this peak is present in the solution and
plasma coated films, we can assume that the Si-O-Si network is
also present in both. The peak near -50°C is clearly most
dominant in the plasma deposited film. Plasma deposition, by its
very nature, leads to a large number of reactive species and is

662

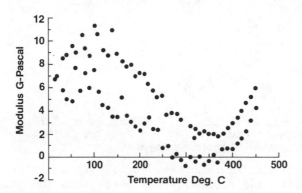

Figure 9. Modulus vs. temperature data for a 100 nm (PMDA/ODA) polyimide film, 6th tone vibrating reed ~ 3000 Hz. The lower points are increasing in temperature, the upper points are decreasing. Negative points are an artifact from using Equation (2) and the fact that the frequency shifts are at the limit of detectability for the vibrating reed technique.

thereby much more prone to inducing crosslinking reactions than the solution deposition technique.[35] It is, therefore, reasonable to speculate that the peak near -50°C is a reflection of certain crosslinked structures which apparently dominate in the plasma deposited film.

Heating silane films much above 200°C causes significant degradation and chemical reorganization. This occurs in both solution and plasma coated films.[35] The starred data points in Fig. 10 shows the effect of heating a solution coated film to 450°C. The strength of the damping is greatly reduced, indicating both a loss of material and significant structural changes in the film. The peak intensity of the solution coated film is not only greatly decreased, but also shifted to a higher temperature. Though not shown, the same effects are also observed in the plasma coated film. The plasma coated film, in

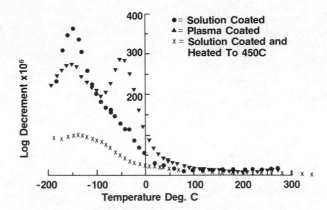

Figure 10. Internal friction behavior of 500 nm silane films coated by solution and plasma techniques. Note the strong peak near −50°C in the plasma sample which clearly distinguishes it from the solution coated film.

fact, showed a significant densification which we were able to estimate using Equation (2). We were unable to estimate the modulus of this film because we did not have accurate density data on the plasma coated material. However, because we know that the modulus must be positive, we can estimate a lower bound on the density from Equation (2) as follows:

$$\frac{d}{2t}\left[\left[\frac{f_n^c}{f_n^o}\right]^2 - 1\right] + \frac{\rho_f}{\rho_s} > 0$$

or

$$\rho_f > \frac{d\rho_s}{2t}\left[\left[\frac{f_n^c}{f_n^o}\right]^2 - 1\right]$$

All quantities are as defined in Equation (2)

Using known properties of SiO_2 and measured thickness and frequency shift data, we estimate the density of the plasma silane film after heating to 450°C is greater than 1.6 g/cc. This is a significant densification because most siloxane materials have a density near 1 g/cc. These results are also in agreement with the work of Wrobel et al.[36]. These data further point to the conclusion that silane films on heating above the limit of their temperature stability tend toward densified glassy structures more closely resembling a SiO_2 glass than a siloxane polymer structure.

Polyimide-silane Interaction

Figure 11 exhibits the internal friction behavior of two 100 nm thick, PMDA/ODA type polyimide films supported on 50-micron fused silica substrates. Both films are nearly identical in every respect except that the film represented by the circles was coated on a silica substrate that had an 8 nm (gravimetric determination) layer of γ-APS coated onto the silica substrate before deposition of the polyimide film. From Fig. 11, we see that the internal friction behavior of both films is nearly identical except for an apparent overall shift upwards in the silane containing film, and an apparent dip in the nonsilane treated film near -50°C. For our present purpose, it is the dip near -50°C in the nonsilane treated film which is the most significant, and we further enhance this feature by subtracting the spectrum of the nonsilane treated film from that which has been treated.

The result is shown in Fig. 12 which apparently gives a representation of the internal friction behavior of the polyimide-silane interfacial region in the silane treated sample. This spectrum shows a peak near -50°C reminiscent of a similar peak which appeared strongly in the plasma coated film, and much more weakly in the solution coated films shown in Fig. 10. Therefore, it is tempting to associate this peak with crosslinked structures in the silane film, or, even further, to associate it with crosslinks between the silane and the polyimide. Such conclusions, however, must be considered highly speculative, though quite plausible. Further data is clearly needed, such as independent adhesion measurements to verify whether the peak in Fig. 12 can be associated with chemical crosslinking structures between the γ-APS and polyimide coatings.

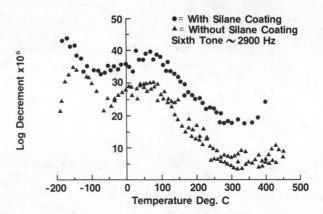

Figure 11. Internal friction behavior of identical 100 nm polyimide films of the PMDA/ODA type, one coated with and one without a silane primer. Note the dip near −50°C for the nonsilane primer film which does not appear in the silane coated sample.

SUMMARY

We reiterate the main conclusions and findings of this paper:

1. The vibrating reed experiment is a very sensitive and versatile technique for examining the thermal–mechanical properties of ultra thin organic films including polyimides.

2. Damping behavior can be detected on films as thin as 6 nm and modulus data can be determined for films down to 100 nm thick. A range of temperatures from −150°C → 450°C can be covered and a range of frequencies from 30 to 3000 Hz may be excited.

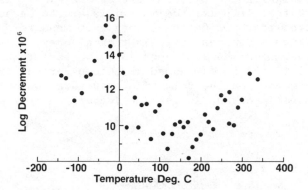

Figure 12. Difference between the two internal friction spectra in Fig. 11, 6th tone vibrating reed ~ 2900 Hz. Note that the dip near -50°C is now highlighted as a weak internal friction peak. This peak may be related to the -50°C peak for the plasma silane film in Fig. 10, in which case it should be related to chemical crosslink structures at the polyimide/silane interface.

3. Differences in the mechanical toughness of polyimide samples may be discussed in terms of secondary thermal-mechanical relaxations which occur below the glass transition.

4. Significant hysteresis in the Young's modulus of PMDA/ODA polyimide occurs on temperature cycling from 20°C to 450°C and back.

5. The presence of silane primer materials which are used to form an interphase layer between silica and polyimide may be detected by internal friction measurements. Such interphase layers give rise to an apparent internal friction peak near -50°C.

We conclude this paper where we started, with a discussion of the role of polyimides in microelectronic structures for future large scale and very large scale integrated circuit structures. The importance of polyimides to such structures has been recently highlighted by Saiki and Harada[37]. It is clear from this work that if polyimide materials are to be successful in future microelectronic devices, they must measure up to a number of stringent demands on their thermal chemical stability and overall thermal-mechanical behavior. We feel that the vibrating reed technique discussed in this paper can provide much valuable information on the thin film structures which will become even more important in fabricating microelectronic devices.

REFERENCES

1. C. Wert and C. Zener, "Interstitial Atomic Diffusion Coefficients," Phys, Rev., 76, 1169 (1949).
2. A.S. Nowick and B.S. Berry, "Anelastic Relaxation in Crystalline Solids,"Academic Press, New York 1972.
3. J.D. Ferry, "Viscoelastic Properties of Polymers," 2nd Edition, Wiley, New York, 1970.
4. N.G. McCrum, B. E. Read, and G. Williams, "Anelastic and Dielectric Effects in Polymeric Solids," Wiley, New York, 1967.
5. B.S. Berry and W. C. Pritchet, "Vibrating Reed Internal Friction Apparatus for Films and Foils," I.B.M. J. Res. Devel., 19, 334 (1975).
6. J.P. Den Hartog, "Mechanical Vibrations," 4th Ed., McGraw Hill, New York 1956.
7. S. Timoshenko, "Vibration Problems in Engineering," 3rd Ed., Van Nostrand Reinhold, New York 1955.
8. J.P. Den Hartog, "Strength of Materials," Dover, New York 1961.
9. R.A. Feynman, R.B. Leighton, and M. Sands, "The Feynman Lectures on Physics," Vol. II, Addison-Wesley, Reading, MA., 1964.
10. R.H. Lacombe and B.M.J. Kellner, "Thermal-Mechanical Properties of a Monomolecular Film Detected by a Vibrating Reed," Nature, 289, 661 (1981).
11. G.L. Gaines, Jr., "Insoluble Monolayers at Liquid-Gas Interfaces," Ch. 8, Wiley-Interscience, New York, 1966.
12. M. Careem and A.K. Jonscher, "Lattice and Carrier Contributions to the Dielectric Polarization of Stearic Acid Multi-Layer Films," Phil. Mag., 35, 1489 (1977).

13. A.H. Willbourn, "The Glass Transition in Polymers with the (CH_2) Group," Trans, Faraday Soc., $\underline{54}$, 717 (1958).

14. Aldrich Chemical Co., Milwaukee, Wisconsin, Cat. No. 182435: M_w = 321,000 (light scattering); M_n = 84,600 (GPC).

15. A discussion of the early literature on the γ relaxation has been given by Boyer and Turley, "Molecular Basis of Transitions and Relaxations," in Midland Macromolecular Monographs No. 4, D.J. Meier, Ed., Gordon and Breach, New York, 1978.

16. A. Hiltner and E. Baer, "Relaxation Processes at Cryogenic Temperatures," CRC Critical Reviews in Macromolecular Science, $\underline{1}$, 215 (1972).

17. R.F. Boyer, "Dynamics and Thermodynamics of the Liquid State $(T > T_g)$ of Amorphous Polymers," J. Macromol. Sci.-Phys., $\underline{B18(3)}$, 461 (1980).

18. D.W. Van Krevelen and P.J. Hoftyzer, "Properties of Polymers," 2nd Ed., Table 13.2, p. 264, Elsevier, Amsterdam, 1976.

19. Y. Cohen and S. Reich, "Ordering Phenomena in Thin Polystyrene Films," J. Polymer Sci.: Polymer Phys. Ed., $\underline{19}$, 599 (1981).

20. B. Berry and W. C. Pritchet, unpublished data.

21. F.A. Johnson and J.C. Radon, "Molecular Kinetics and the Fracture of PMMA," Eng. Fracture Mec., $\underline{4}$, 555 (1972).

22. R.W. Hertzberg and J.A. Manson, "Fatigue of Engineering Plastics," pp. 90-93, Academic Press, New York (1980).

23. E. Butta, S. DePetris, and M. Pasquini, "Young's Modulus and Secondary Mechanical Dispersions in Polypyromellitimide," J. Polym. Sci., $\underline{13}$, 1073 (1969).

24. T. Kim, V. Frosini, V. Zaleckas, D. Morrow, and J.A. Sauer, "Mechanical Relaxation Phenomena in Polyimide and (2,6-dimethyl-p-phenylene oxide) From 100°K to 700°K", Polym. Eng. Sci., $\underline{13}$, 51 (1973).

25. G.A. Bernier and D.E. Kline, "Dynamic Mechanical Behavior of a Polyimide," J. Appl. Polym. Sci., $\underline{12}$, 593 (1968).

26. A.S. Argon and M.I. Bessonov, "Deformation in Polyimides, with New Implications on the Theory of Plastic Deformation of Glassy Polymers," Phil. Mag., $\underline{35}$, 917 (1977).

27. J.K. Gillham, K.D. Hallock, and S.J. Stadnicki, "Thermomechanical and Thermo-gravimetric Analysis of a Systematic Series of Polyimides," J. Appl. Polym. Sci., $\underline{16}$, 2595 (1972).

28. E.P. Plueddemann, "Silane Coupling Agents," Plenum Press, New York, 1982.

29. L. Rothman, "Properties of Thin Polyimide Films", J. Electro-chem. Soc., $\underline{127}$, 2216 (1980).

30. D. Suryanarayana and K. L. Mittal, "Effect of pH of Silane Solution on the Adhesion of Polyimide to Silica Substrate", J. Appl. Polym. Sci., $\underline{29}$, 2039 (1984).

31. J. Greenblatt, C. Araps and H. R. Anderson, Jr., "Aminosilane-Polyimide Interactions and Their Implications in Adhesion", these proceedings, Vol. 1, pp. 573-588.

32. H.G. Linde, "Adhesive Interface Interactions Between Primary Aliphatic Amine Surface Conditioners and Polyamic Acid/Polyimide Resins," J. Polym. Sci.: Polym. Chem. Ed., 20, 1031 (1982).

33. N. K. Eib, K. L. Mittal and A. Friedrichs, "Wettability Studies of Plasma-Deposited Hexamethyldisilane", J. Appl. Polym. Sci., 25, 2435 (1980).

34. T. Furukawa, N. K. Eib, K. L. Mittal and H. R. Anderson, Jr., "Inelastic Electron Tunneling Spectroscopic Study of the Silane Coupling Agents: I. Aminophenyltriethoxysilane Adsorbed on Plasma Grown Aluminum Oxide and Effects of High Humidity", Surface Interface Anal., 4, 240 (1982).

35. N.K. Eib, K.L. Mittal and H.R. Anderson Jr., unpublished results.

36. A.M. Wrobel, J.E. Klemberg, M.R. Wertheimer, and H.P. Schreiber, "Polymerization of Organosilicones in Microwave Discharges. II. Heated Substrates," J. Macromol. Sci-Chem., A15(2), 197 (1981).

37. A. Saiki and S. Harada, "New Coupling Method for Polyimide Adhesion to LSI Surface," J. Electrochem. Soc. 129, 2278 (1982).

670

MECHANICAL PROPERTIES AND MOLECULAR AGGREGATION OF AROMATIC POLYIMIDES

Masakatsu Kochi, Satoru Isoda,* Rikio Yokota,**
Itaru Mita, and Hirotaro Kambe***

Institute of Interdisciplinary Research, Komaba Campus
Faculty of Engineering, University of Tokyo
Komaba, Meguro-ku, Tokyo 153, Japan

Dynamic and static mechanical properties of an insoluble polypyromellitimide(PI) and a soluble aromatic polyimide(PI 2080) were investigated. The imidization condition had a large effect on mechanical properties of PI. The relationship between the mechanical properties and initial imidization temperature is well explained in terms of the development of ordered phases. On the contrary, PI 2080 was essentially amorphous in x-ray diffraction and did not exhibit any formation of ordered phases after annealing at temperatures around its glass transition temperature(T_g). The mechanical properties were not much affected by annealing. The difference in molecular aggregation between PI and PI 2080 may be due to their different chemical structures. A more planar, regular structure of main chain of PI seems to facilitate the formation of a heterogeneous two-phase structure involving ordered phases.

---

* Central Research Laboratory, Mitsubishi Electric Corporation.
Minamishimizu, Amagasaki, Hyogo 661, Japan.
** Institute of Space and Astronautical Science.
Komaba, Meguro-ku, Tokyo 153, Japan.
*** Research Institute of Composite Materials, Faculty of Technology, Gunma University. Tenjin-cho, Kiryu, Gunma 376, Japan.

INTRODUCTION

The molecular aggregation of non-crystalline polymers is essential for an understanding of their mechanical properties. New useful properties of solid polymers can be developed by controlling their molecular aggregation. The question whether non-crystalline polymers form ordered superstructure or not is still a subject of controversy. It is certain, however, that the rigidity of the main chains favors their formation. Some rigid-chain polymers, such as poly(benzyl L-glutamate) and poly-p-benzamide are known to form liquid crystalline states.[1]

On the basis of the effect of annealing on static mechanical properties, we suggested that ordered phase was formed more easily for aromatic polymers having phenyl or/and rigid imide rings in the main chain than for vinyl polymers with flexible chains.[2] In particular, PI(Figure 1-a) exhibited a clear diffraction maximum in the small-angle x-ray scattering(SAXS) curve.[3] They were well explained by a heterogeneous two-phase model involving ordered phase.

PI 2080(Figure 1-b) was synthesized in order to eliminate processing difficulties of previous commercial polyimides. The difficulties arise from the following two reasons : (1) The soluble precursor of polyimides must go through a further imidization reaction to become the final insoluble polymer. In many polyimides, imidization reaction generates a volatile product. (2) The final form of the polyimide is infusible; therefore, once produced, it cannot be further processed by conventional method.[4]

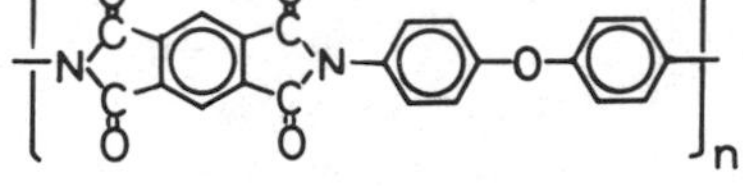

(a) Polypyromellitimide : PI

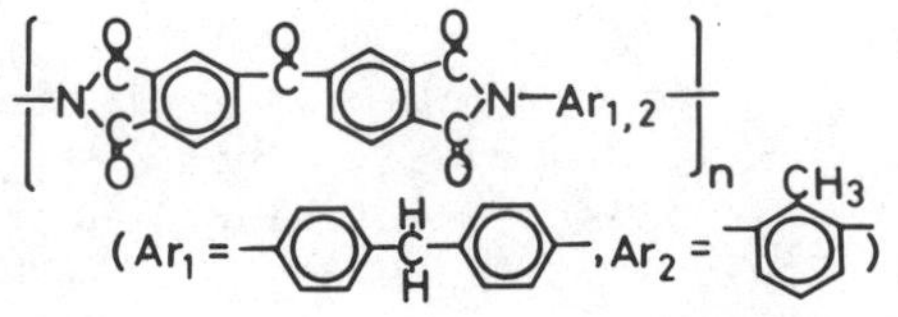

(b) PI 2080 (Ar_1 : Ar_2 = 20 : 80)

Figure 1. Chemical structures ; (a) PI, (b) PI 2080.

In the present paper, some dynamic and static mechanical properties of PI 2080 are compared with those of PI. The difference between them are explained in terms of their different molecular aggregation behavior due to their different chemical structures.

EXPERIMENTAL

Materials : Polyamic acid, the precursor for PI, was formed in dimethylacetamide(DMAc) solution by dissolving 4,4'-diaminodiphenyl-ether and pyromellitic dianhydride. The prepolymer solution was cast onto a glass plate and leveled off to a thickness of approximately 70 μm with a doctor blade. Each film was allowed to stand at room temperature for 24 h and then stripped from the glass plate. The sample films were wrapped with an aluminum foil without frame. The uncured film was then placed into a vacuum oven. Four kinds of PI films were obtained under the various imidization conditions listed in Table I.[3]

Table I.   Imidization Conditions.

PI-1 :	3.5 h at 370 K, 8 h at 430 K, 4 h at 470 K, 5.5 h at 520 K.
PI-2 :	12 h at 530 K.
PI-3 :	12 h at 630 K.
PI-4 :	2 h at 710 K.

A commercial polypyromellitimide(Kapton H film) was supplied by E. I. du Pont de Nemours Co..

PI 2080 film was supplied by Mitsubishi Chemical Industries Ltd.. Annealing of PI 2080 was performed for two hours at 573 and 603 K in a vacuum oven.

The molecular weight of PI and PI 2080 was estimated to be approximately 30,000 by measurement of inherent viscosity of solutions in DMAc and dimethylformamide, respectively.

Measurements : Dynamic tensile mechanical measurements were carried out at 3.5 Hz in air, at a heating rate of 4 K/min, using a Rheovibron DDC-II-C. Samples were thin strips of size 0.3 x 2 cm.

A Shimadzu Thermal Stress Analyzer (TSA) was used to obtain stress-strain curves and elastic recovery at high temperatures around T_g with samples of size 0.2 x 0.5 cm. An increasing load was applied at a rate of 10 g/min. Thermal distortion temperature was defined as the temperature at which a load of 100 kg/cm^2 resulted in 3 % elongation and was determined by a series of stress-strain curves measured at various temperatures.

Tensile stress-strain measurements were made at room temperature on films of size 0.6 x 5 cm with a Toyo-Baldwin Tensilon UTM-II at a constant draw rate of 1.0×10^{-1} cm/min.

T_g was determined with a Du Pont Thermoanalyzer at a heating rate of 10 K/min.

Density was measured with a xylene-carbon tetrachloride density column at 308 K.

RESULTS AND DISCUSSION

DSC Measurements : The effect of annealing temperature on DSC curves for PI 2080 is shown in Figure 2. T_g was determined by DSC as an extrapolated onset point of endothermic deflection as shown in Figure 2-a. T_g was 576 K for as-received PI 2080. An increase in T_g was observed for all the annealed PI 2080 samples. The DSC curve for PI 2080 annealed at 573 K exhibited a remarkable decrease in the half width of the endotherm due to glass trnsition. This kind of behavior is often observed for enthalpy relaxation in glassy states. Petrie[5] pointed out that because of the discontinuous

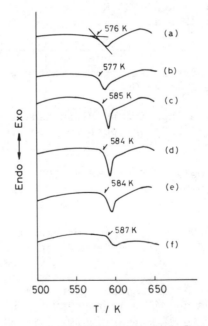

Figure 2. DSC curves of PI 2080 annealed for 2 h at various temperatures ; (a) As-received, (b) T = 563 K, (c) T = 573 K, (d) T = 583 K, (e) T = 593 K, (f) T = 603 K.

increase in the volume of the sample with onset of mobility, the molecules are able to attain rapidly the degrees of freedom associated with the fluid, and the transformation of the annealed glass to a fluid occurs over a narrower temperature than that observed for the unannealed glass. In DSC curves for PI, an endothermic deflection due to glass transition was difficult to observe.

Dynamic Mechanical Properties : The temperature dependence of dynamic tensile properties of PI samples is shown in Figure 3.[6] Two dynamic loss peaks were clearly observed for all four samples, one near 350 K and the other near 690 K. They correspond, respectively, to the β* and α dispersions. The β* dispersion is due to inter-plane slippage of aromatic and imide rings.[7] As shown in Figure 3, with an increase in the initial imidization temperature, the decrease in dynamic storage modulus E' in the α dispersion region became smaller and the corresponding peak in loss modulus E" became broader and smaller, shifting to higher temperatures. By analyzing their

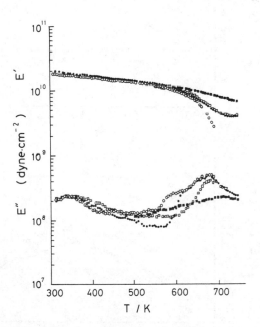

Figure 3. Temperature dependence of dynamic tensile properties of PI ; O PI-1, ● PI-2, □ PI-3, ■ PI-4.

SAXS curves, we found previously that PI-1 is amorphous but PI-2,3,4, have heterogeneous two-phase structures involving ordered phases and that these ordered phases develop with increasing initial imidization temperature.[3] Therefore, the α dispersion could be ascribed to the glass transition mode of molecular motions which are restrained by ordered phases.

Figure 4 shows the temperature dependence of dynamic tensile properties of PI 2080 as-received and annealed at 573 K. For the as-received PI 2080, two peaks appear in E" around 480 and 615 K. The broad β dispersion at 480 K may correspond to a local motion of main chains, which is observed around 400 K in polyimides based on benzophenone tetracarboxylic acid.[8] The sharp α dispersion at 615 K is due to glass transition and the corresponding decrease in E' through T_g is large like other ordinary amorphous polymers. As seen in Figure 4, the annealing of PI 2080 at 573 K leads to a slight increase in T_g.

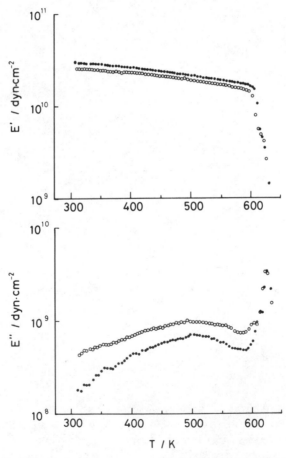

Figure 4. Temperature dependence of dynamic tensile properties of PI 2080 ; ○ As-received, ● annealed at 573 K.

Static Mechanical Properties : Figure 5 shows the TSA curve at 605 K for as-received PI 2080. Both elongation and load were recorded as a function of time. Load was removed before elongation reached the limiting value of 500 μm. The stress-strain curves were calculated from these data. After load was removed, the attained elongation of 395 μm approached asymptotically to the fixed value of 45 μm. So the elastic recovery was calculated to be about 89 %.

A series of stress-strain curves measured at various temperatures for PI 2080 as-received, annealed at 573 and 603 K are shown in Figure 6. Thermal distortion temperatures were determined from these data.

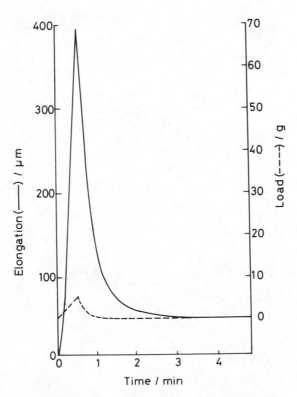

Figure 5.   TSA curve at 605 K for as-received PI 2080.

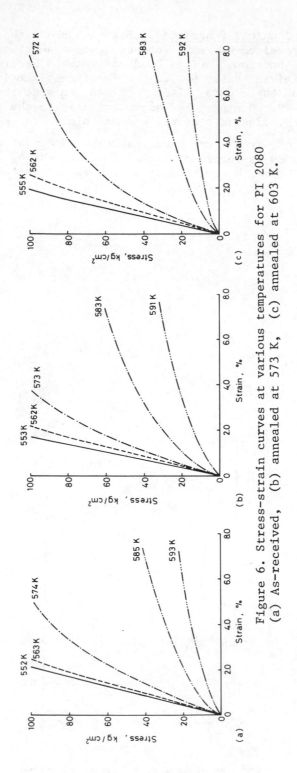

Figure 6. Stress-strain curves at various temperatures for PI 2080
(a) As-received, (b) annealed at 573 K, (c) annealed at 603 K.

In Table II are listed the static mechanical properties in addition to density and T_g by DSC. The elastic recovery was measured above T_g.

Table II. Density, T_g, and Static Mechanical Properties of PI and PI 2080.

Sample	Density (g/cm^3)	T_g (K)	T_d* (K)	Elastic recovery (%)
PI-1	1.393	------	573	42
PI-2	1.413	------	593	53
PI-3	1.420	------	623	61
PI-4	1.426	------	658	71
PI 2080 As-received	1.337	576	565	89
Annealed at 573 K	1.338	585	568	90
Annealed at 603 K	1.336	586	565	87

* T_d : thermal distortion temperature.

As shown in Table II, the density of PI increases with increasing initial imidization temperature. This is consistent with the development of ordered phases. Moreover, ordered phases in non-crystalline polymers might play the role of physical cross-linking points in some mechanical properties. A similar effect of crystallites is well known for crystalline polymers.[9] In a rubbery state, for example, an increase in elastic recovery would be expected from the development of ordered phases; and it is observed for PI (see Table II). The increase in thermal distortion temperature for PI is explained by the same mechanism, because ordered phases restrain molecular motions of main chains in the less-ordered phase.

On the other hand, Table II shows that the annealing of PI 2080 has very small effect on density, thermal distortion temperature, and elastic recovery. The change in density is negligible, which is consistent with the fact that the wide-angle x-ray diffraction for annealed PI 2080 showed a typical amorphous pattern similar to that for as-received PI 2080. The increase in thermal distortion tempera-ture for PI 2080 is small compared to that for PI and other aromatic polymers.[2] The absolute values of elastic recovery for PI 2080 are fairly large, but the changes caused by annealing is very small. Therefore, it can be concluded that the annealing of PI 2080 around T_g does not lead to the formation of ordered phases, but to a homogeneous amorphous state closer to the corresponding equilibrium glassy state. This may be attributed to two factors : One is the irregularity of copolymeric repeating units of diphenylmethane and toluenediyl, and the other is the flexible carbonyl group in benzophenone tetracarboxylic aciddiimide. According to Flory,[10] molecular order in amorphous polymers depends on the rigidity of the polymer chains and the intermolecular forces between them; and there exists a critical value of the rigidity for it.

As seen in Table II, Tg of PI 2080 increases by annealing. This kind of phenomenon is also observed for amorphous polymers with flexible main chains and is explained in terms of enthalpy relaxation in homogeneous glassy states.[11] The large elastic recovery of PI 2080 may be ascribed to the molecular entanglements which are enhanced by its flexible, bulky main chains.

Static Tensile Properties : In Table III, the static tensile properties of as-received PI 2080 are compared with those of Kapton H. With respect to the profile of SAXS curve and the temperature dependence of dynamic tensile properties, Kapton H was found to be almost identical with PI-4.[3] Accordingly it will be reasonable to interpret the mechanical properties of Kapton H by the same heterogeneous two-phase structure mechanism as applied to PI-4.

Table III. Static Tensile Properties of PI(Kapton H) and PI 2080.

Sample	Tensile modulus (dyn/cm^2)	Tensile strength (dyn/cm^2)	Elongation (%)
PI(Kapton H)	4.0×10^{10}	2.0×10^9	70
PI 2080 (as-received)	6.2×10^{10}	1.7×10^9	7

Table III emphasizes the toughness of Kapton H and the brittleness of PI 2080. On the basis of Thermomechanical Analysis(TMA), we have reported that the broad α loss peak characteristic of PI permits large tensile deformation above room temperature.[12] The α loss peak of PI 2080, as mentioned above, is so sharp that it does not seem to contribute to a large tensile deformation at room temperature. In addition, the temperature for the β loss peak of PI 2080 is higher by approximately 130 K than that of PI. This also reduces the toughness of PI 2080 film. Recently, the effect of molecular entanglements on mechanical properties of the polymer below its glass transition temperature are noted again.[13] Considering the flexible, bulky repeating units, the entanglement crosslinks in PI 2080 would be difficult to destroy during deformation. They seem to be one of the important factors which explain the brittleness and large elastic recovery of PI 2080.

REFERENCES

1. C. Robinson, J. C. Ward, and R. B. Beevers, Discussion Faraday Soc., 25, 29 (1958).
2. Z. Wu, R. Yokota, M. Kochi, and H. Kambe, J. Soc. Rheol., Japan, 9, 59 (1981).
3. S. Isoda, H. Shimada, M. Kochi, and H. Kambe, J. Polym. Sci. Polym. Phys. Ed., 19, 1293 (1981).
4. K. W. Rausch, W. J. Farrissey, and A. A. R. Sayigh, SPI-28th Annual Technical Conference (Feb. 1973), Sec.11-E.
5. S. E. B. Petrie, J. Polym. Sci., A-2, 10, 1255 (1972).
6. S. Isoda, M. Kochi, and H. Kambe, J. Polym. Sci. Polym. Phys. Ed., 20, 837 (1982).
7. R. M. Ikeda, J. Polym. Sci., B4, 353 (1966).
8. J. K. Gillham, K. D. Hallock, and S. J. Stadnicki, J. Appl. Polymer Sci., 16, 2595 (1972).
9. L. E. Nielsen, "Mechanical Properties of Polymers and Composites", Chap. 4, Marcel Dekker, New York, 1974.
10. P. J. Flory, J. Macromol. Sci. Phys., B12, 1 (1976).
11. T. Hatakeyama and H. Kanetsuna, Kobunshi Kagaku, 27, 713 (1970).
12. M. Kochi and H. Kambe, Polym. Eng. Rev., in press.
13. A. M. Donald and E. J. Kramer, J. Polym. Sci. Polym. Phys. Ed., 20, 899 (1982).

PHOSPHORUS-CONTAINING IMIDE RESINS: MODIFICATION BY ELASTOMERS

I. K. Varma*†, G. M. Fohlen and J. A. Parker
NASA-Ames Research Center, Moffett Field, CA 94035

D. S. Varma†
University of California, Davis, CA 95616

The syntheses and general features of addition-type maleimide resins based on bis(m-aminophenyl)phosphine oxide and tris(m-aminophenyl)phosphine oxide have been reported previously. These resins have been used to fabricate graphite cloth laminates having excellent flame resistance. These composites did not burn even in pure oxygen. However these resins were somewhat brittle. This paper reports the modification of these phosphorus-containing resins by an amine-terminated butadiene-acrylonitrile copolymer (ATBN) and a perfluoro-alkylene diaromatic amine elastomer (3F). An approximately two-fold increase in short beam shear strength and flexural strength was observed at 7% ATBN concentration. The tensile, flexural, and shear strengths were reduced when 18% ATBN was used. Anaerobic char yields of the resins at 800°C and the limiting oxygen indexes of the laminates decreased with increasing ATBN concentration. The perfluorodiamine (3F) was used with both imide resins at 6.4% concentration. The shear strength was doubled in the case of the bisimide with no loss of flammability characteristics. The modified trisimide laminate also had improved properties over the unmodified one. The dynamic mechanical analysis of a four-ply laminate indicated a glass transition temperature above 300°C. Scanning electron micrographs of the ATBN modified imide resins were also recorded.

* NRC–NASA–Senior Research Associate, 1979–1981.

† Permanent Address – Center of Materials Science & Technology, Indian Institute of Technology, Hauz Khas, New Delhi 110016 INDIA.

INTRODUCTION

The impact resistance of glassy polymers can be enhanced by the addition of small amounts of elastomer in the form of second phase particles. Thus the addition of an elastomer, such as carboxy-terminated butadiene-acrylonitrile copolymer (CTBN) to epoxy resins increases the fracture energy by factors of 10 to 15 over the unmodified resins. Toughening mechanisms of such systems have been extensively investigated in the past few years[1-7]. The effects of amine-terminated butadiene-acrylonitrile copolymer (ATBN) and silicone elastomers on the Tg, thermal stability, adhesive strength and fracture toughness of LARC-13 (a high temperature addition polyimide adhesive) have also been reported[8].

The particle size of the second phase is a critical factor which controls the degree of impact resistance in these elastomer modified resins. The parameters which influence the particle size are: molecular weight of elastomer, its reactivity and selectivity, mutual solubility of the elastomer and matrix resins, and curing temperature.

Attempts have been made to improve the "interlaminar fracture energy" of epoxy matrix/glass or graphite cloth composites by the addition of elastomeric toughening agents[9-10]. Scott and Phillips[11] observed a two-fold increase in the interlaminar fracture energy of CTBN modified epoxy matrix/non-woven graphite fiber composites, whereas Bascom et al[9] have reported an eight-fold increase by the addition of elastomeric modifier to an epoxy matrix/graphite fabric composite. An improvement in flexural strength and toughness of ATBN modified epoxy resin/carbon fiber composites has been reported by Gilwee and Jayarajan[10]

Although extensive work has been reported on elastomer modified epoxy resins, only scanty reports are available regarding such modifications of addition polyimides – in particular, the bismaleimides. These low molecular weight imide polymers, end-capped with reactive maleimide rings are polymerizable thermally or by nucleophilic addition of amines, thiols and phenols. Highly cross-linked heat-resistant polyimides are thus obtained without the elimination of by-products. The cured resins are somewhat brittle due to, perhaps, the high cross-link density obtained. It is of interest to investigate the modification of such resins by elastomers. In this paper we report the effects of amine-terminated-butadiene acrylonitrile copolymer (ATBN) and of an aromatic amine terminated perfluoroalkylene ether (3F) on the room temperature flexural, tensile and shear strength of phosphorus-containing imide resin/graphite fabric composites. The high chemical and thermal stability of the structures containing perfluoroalkylene ether chain segments prompted us to undertake these investigations. The morphology of these elastomer modified resins was examined by scanning electron microscopy.

A. Materials

Phosphorus-containing imide resins were prepared from maleic anhydride (J. T. Baker), benzophenonetetracarboxylic acid dianhydride (Gulf Oil Chemicals Co.), bis(m-aminophenyl)methylphosphine oxide and tris(m-aminophenyl)phosphine oxide by the procedures reported elsewhere[12,13]. The structure of the imide resins used in the present work are given below:

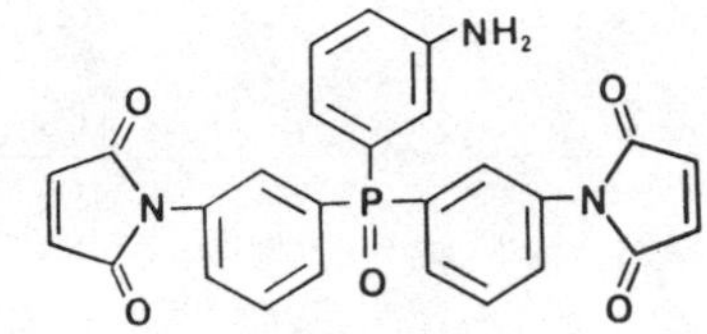

Resin I

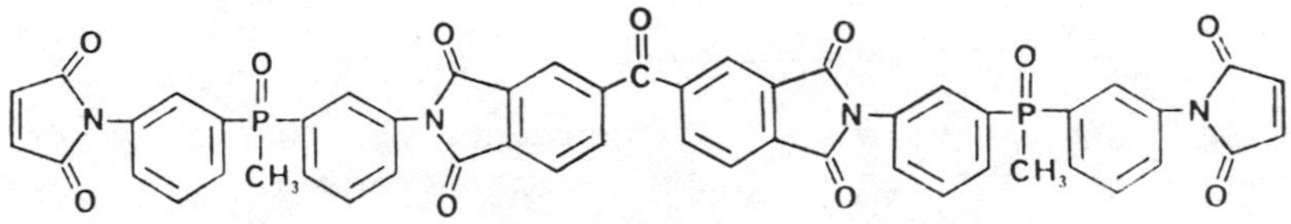

Resin II

(Throughout the text these resins are referred to by the designations shown in this sketch.)

Hycar ATBN (1300x16) is an amine terminated liquid copolymer of butadiene and acrylonitrile (B. F. Goodrich Chemical Co.). It can be represented structurally as:

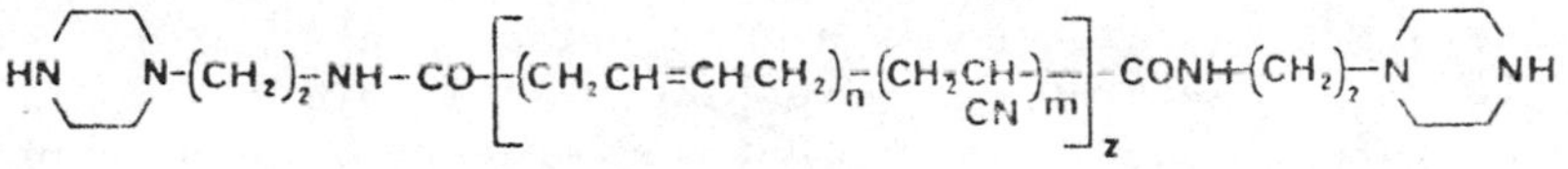

where n=5, m=1, z=10.

Hycar ATBN thus has a reactive secondary amine, an amide functional group as well as tertiary amine in each end-group structure.

The perfluoroalkylene ether diamine[15] was provided by J. Webster (Monsanto Chemical Co.). The structure of this elastomer, designated as 3F, is:

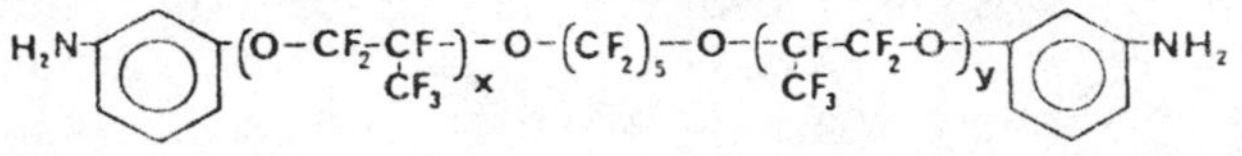

3F   x + y = 3

685

B. Procedures

Preparation of Elastomer Modified Imide Resins

Phosphorus-containing imide resins were prepared using distilled dimethylformamide (DMF) as a solvent at a concentration of
25-35% solids by weight (w/w)[12,13]. A known amount of the elastomers was added to the resin solution and mixed at room temperature
for 15-20 hrs. Part of these solutions was evaporated to dryness
in an aluminum dish. The films thus obtained were cured at
$218°C \pm 2°C$ for 16 hrs (resin I and its adduct with elastomers) or at
$280°C \pm 2°C$ for 70 mins (resin II and its modified products).

Evaluation of Properties of Elastomer Modified Imide Resins

A DuPont 990 Thermal Analyzer was used to evaluate the thermal
behavior of cured resins. All measurements were carried out in a
nitrogen atmosphere at a flow rate of 100 cc/min. A heating rate
of $10°C$/min was employed.

Scanning Electron Microscopy

A Cambridge Stereoscan S-4 instrument was used to study the
morphology of fractured surfaces of cured resins.

Composite Fabrication

Test laminates were prepared by coating graphite cloth (8-
harness satin weave cloth) designated as a style 133 fabric, with
a DMF solution of elastomer modified imide resins (I or II). After
coating, the cloth was dried for 20-30 mins at $125 \pm 5°C$. The dried
prepregs were stacked (4, 8 or 9 plies) and pressed between aluminum
plates lined with Teflon film release sheets. The cure of the
laminate was carried out in a flat platen press, using a pressure
of 140 psi. The prepregs were placed in a cold press and heated to
$180°C$ (resin I) and held at that temperature for 150 mins. In the
case of resin II curing was done at $232°C$ for 95 mins at a pressure
of 140 psi.

Post-curing of laminates was done at $218°C \pm 2°C$ (resin I) for
16 hr or $280°C$ for 70 mins (resin II). The details of composite
fabrication conditions are given in Table I.

Evaluation of Properties of Cured Graphite Cloth Laminates

Resin contents of the laminates were determined by the
hydrazine method[14]. The tests for evaluating the properties of
these laminates were carried according to the following standards:

Table I. Details of Composite Fabrication.

Sample	Resin	Elastomer	Cure Cycle
a	I	none (control)	180°C-2½ hr (140 psi) 218°C-16 hr
b	II	(control)	232°C-95 min (140 psi) 290°C-70 min
c	I	3.9% ATBN	180°C-2½ hr (140 psi) 218°C-16 hr
d	I	7.0% ATBN	180°C-2½ hr (140 psi) 218°C-16 hr
e	I	18.0% ATBN	180°C-2½ hr (140 psi) 218°C-16 hr
f	I	6.4% 3F	180°C-2½ hr (140 psi) 218°C-16 hr
g	II	6.4% 3F	232°C-95 min (140 psi) 280°C-70 min

Flexural test   — ASTM D 790-73

Tensile test   ASTM D 638-68

Short-beam shear   — ASTM D 2344-76

Limiting Oxygen Index (LOI) — ASTM D 2863-74

LOI of test laminates was determined using a Stanton-Redcroft Flammability Unit. A modified DuPont 980 Dynamic Mechanical Analyzer, which measures modulus and damping, was used to study composite systems and to evaluate glass transition temperature. The mechanical properties (flexural, tensile and shear strengths) were measured on an Instron Tester using suitable fixtures.

RESULTS

Thermal Behavior

The thermal stabilities of the neat resins were determined by dynamic thermogravimetry in a nitrogen atmosphere. Anaerobic char yields at 800°C were obtained from the primary thermograms (fig. 1). Addition of ATBN elastomer to resin I resulted in a decrease in the char yields from 61% (control) to 48% (18% ATBN). A reduction in char yield was also observed in 3F modified imide resins I and II. These results are summarized in Table II.

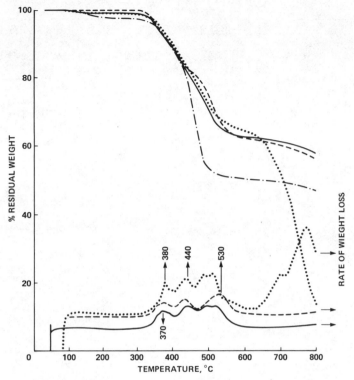

Figure 1. Thermograms of elastomer modified bisimide resin I (———)
sample c, (----) sample f, (--·--·--) sample e, in nitrogen,
(·····) sample c in air.

Table II. Thermal Characteristics of Modified Bisimide Resins.

Sample	Resin and Elastomer	Char Yield, %* Neat Resin	LOI, % O_2 Laminate
a	I (control)	61	100
b	II (control)	71	100
c	I + 3.9%ATBN	58	-
e	I + 18% ATBN	47	85
f	I + 6.4% 3F	56.5	100
g	II + 6.4% 3F	68	100

*in N_2, at $800°C$

The thermal stability of resins in air and nitrogen atmosphere was almost similar up to about 600°C. Above this temperature oxidation became important and almost complete loss in weight was observed above 800°C. Thus in the initial degradation of these elastomer modified bisimides, thermal decomposition is the predominant reaction and oxidation reactions do not contribute significantly.

The limiting oxygen indices of the laminates decreased on addition of ATBN. No difference was observed in the LOIs of the control and 3F modified resin. These tests were performed by passing oxygen upward through a chimney at a flow rate of 18 liters/min. The laminate was then ignited by a flame. On removal of the ignition flame, timer was started. The samples showed only a 15-20 sec burn. (Hence a LOI value of 100 has been reported for these 3F modified laminates). However, when this sample was re-ignited (without cutting the burnt edge), burning continued for 3 mins. In the case of the control the laminate did not burn even on the second ignition.

Dynamic mechanical analysis of a 4-ply laminate of resins (a), (b), (c), (d) and (e) was performed from room temperature to 450°C. The dynamic modulus remained constant (irrespective of the elastomer concentration) up to a temperature of ∿250°C in most of the test laminates and started decreasing above this temperature. In laminates (f) and (g) measurements were done from -100°C to +450°C. It was expected that such measurements would indicate two Tg's - one associated with elastomer and another with the cross-linked imide resin. No change in the modulus was observed indicating no effect on the Tg. The glass transition temperature of polyimide based on BTDA and perfluoroalkylene ether diamine has been reported as ∿80°C.[15]

Morphology of the Neat Resins

Morphology of neat resins was investigated by scanning electron microscopy. Distinct cell-like structures separated by the imide resin were observed in sample (d). A wide distribution of particle sizes was observed (fig. 2). Careful examination of micrograph reveals that the diameter of the matrix hole is larger than the particle size in the material. These holes may be attributed to the matrix stretching and cavitating around the particle. On increasing the ATBN concentration these cell-like structures became large and irregular (fig. 3). Sample (e) containing 18% ATBN still showed phase separation characterized by large irregular cells within the imide matrix. When the concentration of elastomer was very high (36% w/w) the fractured surface of the particles became similar to the matrix (fig. 4). In 3F modified resins (I) and (II) no phase separation was observed.

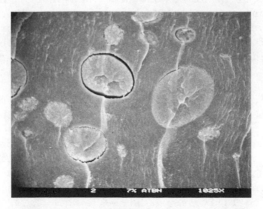

Figure 2.   Scanning electron micrograph of resin d, (I+7% ATBN).

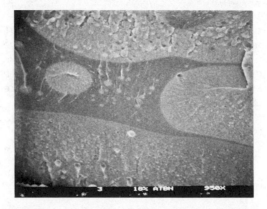

Figure 3.   Scanning electron micrograph of resin e, (I+18% ATBN).

Figure 4.   Scanning electron micrograph of resin I + 36% ATBN.

Mechanical Properties of Laminates

In Tables III and IV, the flexural strengths, moduli, energy (obtained from the integrated areas under the load versus elonga-

Table III.   Effect of ATBN & Fluoroether 3F on Properties of Phosphorylated Bisimide I.

PROPERTY	CONTROL	3.9% ATBN	Δ%	7% ATBN	Δ%	18% ATBN	Δ%	6.4% 3F	Δ%
FLEXURAL STRENGTH X 10^3, PSI	76.04	88.70	+16.6	149.62	+96.8	50.19	-33.9	-	-
FLEXURAL MODULUS X 10^6, PSI	21.42	16.30	-23.9	25.21	+17.7	13.34	-37.7	-	-
ENERGY, FT-LB	14.32	26.93	+88	40.39	+182	11.83	-17.4	-	-
TENSILE STRENGTH X 10^3, PSI	45.79	49.18	+7.4	61.91	+35.2	43.90	-4.1	48.4	+5.7
SHEAR STRENGTH X 10^3, PSI	2.94	3.58	+21.8	5.35	+98.9	0.983	-66.6	5.95	+105.4
LOI, % O_2	100	-		-		85	-15	100	0
RESIN CONTENT, %	18.3	28.9		26.6		20.3		24.2	
RESIN CHAR YIELD, % AT 800°C IN N_2	61	58	-4.9	-		48	-21	59	-3.3

Table IV.   Properties of Phosphorylated Bisimide II Modified With Elastomer 3F.

PROPERTY	BISIMIDE II CONTROL	+6.4% 3F	EPOXY*	% IMPROVEMENT OVER CONTROL	OVER EPOXY
FLEXURAL STRENGTH X 10^3, PSI	109.45	138.64	92.26	+26.7	+ 50.3
FLEXURAL MODULUS X 10^6, PSI	19.4	20.04	7.1	NO CHANGE	+ 182
ENERGY, FT-LB	28.3	41.3	-	+ 45.9	-
TENSILE STRENGTH, X 10^3, PSI	59.8	76.9	82.6	+ 28.6	-6.9
SHEAR STRENGTH X 10^3, PSI	5.38	10.2	7.74	+ 89.6	+32.5
LOI, % O_2	100	100	36	NO CHANGE	+++
RESIN CONTENT, %	18-20	22.5	25.3	-	-
RESIN CHAR YIELD, %	71	68	20	- 4.2	-

*MY 720 CURED WITH DDS

tion curves), tensile strengths and shear strengths of the elastomer-modified imide resins (I) and (II)/graphite cloth laminates are listed. An improvement in the mechanical properties of the laminates was observed up to 7% w/w of the elastomer. A further increase in the concentration of elastomer decreased the modulus and flexural strength of the laminates.

Moisture Absorption

The gain in weight of test specimens after boiling in water for 96 hrs was determined. The samples were surface dried by wiping before weighing. These results are given in Table V.

Table V. Water Absorption of Laminates After Boiling for 96 Hrs.

Sample No.	Gain in Weight, %
a	1.01
b	6.3
c	2.25
d	3.5
e	1.5
f	2.5
g	4.6

DISCUSSION

Imide resins toughened with reactive elastomers consist of a continuous rigid polyimide phase and a dispersed rubber phase. These rubbery domains impart some ductility to the system which leads to an improvement in crack resistance and toughness.

Chemistry of Dispersed Phase

The double bonds in bismaleimides are highly electron deficient due to the flanking carbonyl groups and hence are capable of undergoing nucleophilic additions with primary and secondary amines[16,17]. Such an addition has been found to follow second order kinetics and proceeds under relatively mild conditions. For example good yields are obtained when piperidine and N-arylmaleimides are reacted at room temperature in benzene solution[18].

Based on the above arguments, it seems logical to propose the following Michael-type addition reaction between maleimides, imides and elastomers containing primary or secondary amino groups:

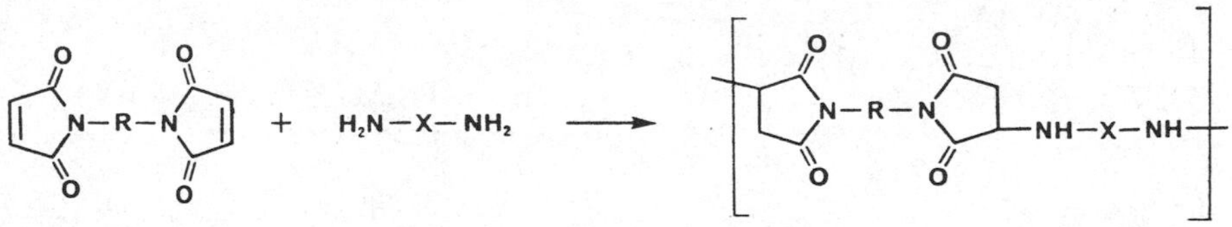

Such a linear copolymerization of imide resins and amino functional groups would increase with the concentration of elastomer. This will lead to flexibilization of imide resins, and hence a decrease in the modulus. Particle size of the second phase may increase and ultimately the system might become indistinguishable from the matrix due to the solubility of such a copolymer in the imide resin. The results of SEM support the solubility of elastomer at high concentration.

Maximum toughening is reported under the conditions of combined shear and craze deformations. It has been suggested that shear deformations are dominant in resins toughened with small rubbery particles. Crazing, which is associated with polymer whitening and microvoid development, is dominant in resins toughened with large particles[2]. The SEM micrographs of laminate (d) indicate the presence of particles of varying sizes. Graphite cloth laminates fabricated from this sample also exhibited the maximum improvement in mechanical properties. On the basis of this information, it may be concluded that the particle size of elastomers and their distribution in the imide matrix is a critical factor in toughening of the resins and laminates. Whether the morphology of elastomer modified imide resins is influenced by the presence of graphite fibers is not well understood at this stage.

In conclusion, improvements in the mechanical properties of the imide resins/graphite cloth laminates were observed by the incorporation of amine terminated elastomers in the resins. Only marginal effects on thermal properties of the laminates were observed when perfluoroalkylene ether diamines were used. Such laminates had good mechanical properties and thermal characteristics.

ACKNOWLEDGEMENTS

The authors thank Mr. Mark Rosenberg for help with the DMA work, Dr. James A. Webster of Monsanto for supplying the 3F fluoro-elastomer, and Mr. Kenneth Algers of University of California at Davis, for doing the scanning electron microscopy.

693

REFERENCES

1. F. J. McGarry, Proc. Roy. Soc. London, A, 319, 59 (1970).
2. J. N. Sultan and F. J. McGarry, Polymer Eng. Sci., 13, 29 (1973).
3. R. C. Laible and F. J. McGarry, Polym. Plast. Technol. Eng., 7, 27,(1976).
4. A. C. Meeks, Polymer, 15, 675 (1974).
5. C. B. Bucknall and T. Yoshii, Brit. Polym. J.,10, 53 (1978).
6. J. E. Sohn, ACS Org. Coatings and Plastics Chem. Preprints, 44, 38 (1981).
7. W. D. Bascom, R. L. Cottington and C. O. Timmons, J. Appl. Polym. Sci., Appl. Polym. Symp., 32, 165 (1977).
8. A. K. St. Clair and T. L. St. Clair, Proceedings of the 12th National SAMPE Technical Conference, Oct. 7-9, 1980, p.729.
9. W. D. Bascom, J. L. Bitner, R. J. Moulton and A. R. Siebert, SAMPE Symposium, 24, (2) 1952 (1979).
10. W. J. Gilwee and A. Jayarajan, Proceedings, National Symposium on "Polymers in the Service of Man", June 9-11, 1980, Washington, D. C. Published by The American Chemical Society, 1980.
11. J. M. Scott and D. C. Phillips, J. Materials Sci., 10, 551 (1975).
12. I. K. Varma, G. M. Fohlen, M.-S. Hsu and J. A. Parker, Proceedings, U. S. -Japan Joint Polymer Symposium, Nov. 22-24, 1980, Palm Springs, California, in "Contemporary Topics in Polymer Science, Volume 4", (1984) Plenum Press, New York.
13. I. K. Varma, G. M. Fohlen and J. A. Parker, J. Macromol. Sci.-Chem., A19(2), 209-224 (1983).

14. "Determining Resin/Fiber Content of Laminates", NASA Tech. Brief, LAR-12442, vol. 4,No. 2, 1979, p. 227.
15. J. A. Webster, J. M. Butler and T. J. Morrow, in "Polymerization Reactions and New Polymers", N.A.J. Platzer, Editor, Adv. in Chem. Series No. 129, 61 (1973).
16. J. V. Crivello, Polymer Preprints, 14, (1) 294 (1973).
17. J. V. Crivello, J. Polym. Sci., Polym. Chem. Ed., 11, 1185 (1973).
18. A. Mustafa, W. Asker, S. Khattab and J. Zayed, J. Org. Chem., 26, 787 (1961).

MEASUREMENT OF STRESSES GENERATED DURING CURING

AND IN CURED POLYIMIDE FILMS*

P. Geldermans, C. Goldsmith, and F. Bedetti

IBM East Fishkill General Technology Division

Hopewell Junction, New York 12533

We have measured the stresses generated in two kinds of
polyimide films. One was the material produced from the
ethyl ester of benzophenone tetracarboxylic acid and
methylene dianiline (EBTA-MDA), with a number average
molecular weight ($\bar{M}_n$) of approximately 500. The other
material was pyromellitic dianhydride-oxydianiline
(PMDA-ODA), with a weight average molecular weight ($\bar{M}_w$)
of approximately 30,000. We found that the final stress
in the films was a function of the cure cycle and that
the stress in a fully cured film is due to the difference between the expansion of the substrate and polyimide film. The behavior during the curing process was
found to be very different between the two materials.

---

*   A part of this paper was presented at the American
    Vacuum Society meeting in Baltimore, Maryland,
    November 1982.

INTRODUCTION

Polyimides find wide application in the electronics industry as a dielectric for both semiconductors and packages. They are stable up to 400°C and have a low dielectric constant.

In our work, we have evidence of tensile stresses, which in one case were high enough to create radial cracks around features in the fully cured material. To insure long-term reliability, it is necessary that the stresses in the several materials of a complex structure are less than both the cohesive strength of the materials and the adhesion strength between the layers. The ultimate tensile strength of polyimides is in the range of 130-200 MPa (18-29 kPSI). The strain at which failure occurs varies widely from material to material. Some materials fail at 1-2% strain; others permit almost 100% elongation before failure occurs.

Considering the relative coefficients of thermal expansion (CTE) between a ceramic substrate and a polyimide, one can expect the polyimide film to be strained to approximately 1% at room temperature due to the cooling from the cure temperature to 25°C. For some materials, this results in cracking around features that may give rise to stress concentration.

This paper reports on the experiments we did to gain some understanding of the mechanisms underlying the observed stresses.

When polyamic acid is applied to a surface by spinning or spraying of a solution, a "cure" process is required to obtain the desired properties of the resultant polyimide. During this curing, two main processes take place: loss of solvent and a chemical reaction between the constituents. Both these processes result in loss of film volume and can produce stresses in the film if the losses occur after the film has become (semi-)rigid. In addition, stresses in the film can arise when there is a mismatch between the coefficient of expansion of film and substrate as the film cools from its curing temperature to room temperature.

Many organic materials exhibit a "glass transition temperature." We will define this as the temperature where the material becomes plastic and relieves internal stresses. For polyimide films cured at 400°C, the glass transition temperature is approximately 400°C, as we show later in this paper. Since CTE of polyimides is approximately 50×10^{-6}/°C and the CTE of silicon is approximately 3×10^{6}/°C, one can expect that cooling from the glass transition temperature will generate stresses in the polyimide film. As a matter of fact, by plotting the stress as a function of temperature, one can determine the glass transition temperature using the definition given above.

From a standpoint of structural integrity and long-term reliability, it is important to know more about these stresses - when they start to build-up, how they behave as a function of temperature variations, etc.

EXPERIMENTAL

Stress Measurement

Stress measurements in films are customarily done by applying the film to the surface of a substrate that is mechanically flexible enough to undergo a deformation as a result of the stress.

After proper calibration, the deformation can then be used as a direct quantitative measure of the stress. In the field of paint technology, there is extensive literature on this subject (e.g., Reference 1). We would have availed ourselves of these techniques except that we wanted to measure the stresses at temperatures from room temperature to 400°C in an inert gas environment. The latter requirement was because we wanted to measure how stresses develop during the curing process.

The lattice curvature ("wafer bending") technique, available in this laboratory, was ideally suited for our purpose. It combines these desirable attributes: it is noncontact; it is easy to conduct both in an inert atmosphere and at elevated temperatures. The method has been described in References 2 and 3. In the following, only a brief description will be given.

Stresses can be measured using an x-ray double-crystal lattice curvature (substrate bending) technique. Figure 1 shows a schematic of the double-crystal technique in the transmission-transmission mode. Figure 1a shows the detector with an unstrained crystal in position two, with each crystal aligned to obtain Bragg diffraction in transmission from a set of planes. Translation of these crystals occurs without change in the double-diffracted x-ray intensity because the Bragg angle remains constant across the crystals. In Figure 1b, the crystal in position two has a curvature due to the stress generated by a thin film. Again, the crystals are aligned to obtain Bragg diffraction in transmission from a set of planes. Translation of these crystals results in a sharp decrease in the double-diffracted X-ray intensity due to the lattice curvature. Consequently, the Bragg angle θ has to be adjusted continuously during translation to maintain diffracted intensity. This required adjustment can be used as a measure of the crystal curvature where the radius of curvature $(R) = L/\theta$, where L is the translation distance.

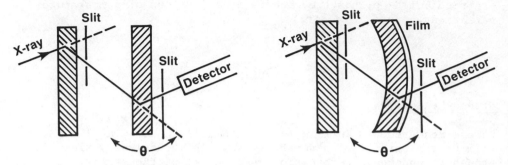

Figure 1.   Schematic of the double-crystal camera arrangement.

1a.   with unstrained crystals
1b.   with a strained crystal

The lattice curvature of the substrate is a direct result of the film stress expressed by the following equation, assuming that the film thickness is much smaller than the wafer thickness.

$$\sigma_F = \pm \left(\frac{t_s^2}{6\, t_f\, R} \right) \times \left(\frac{E_s}{1-\nu_s} \right)$$

σ_F = film stress
t_s = substrate thickness ($\sim 400\,\mu m$)
t_f = film thickness ($\sim 15\,\mu m$)
R = radius of curvature
E_s = Young's modulus of the substrate
ν_s = Poisson's ratio of the substrate
$\pm$ = sign of the film stress (+ = tension)

The value of $E/(1-\nu_s)$ for [100] silicon is 1.805×10^5 MPa (see Reference 5). Figure 2 shows the convention used to describe tensile and compressive stress. Considering the experimental error in measuring film thickness and variations in the film thickness over the whole wafer, the accuracy of our stress calculation is ·±15%.

Compression   **Tension**

Figure 2.   Convention showing tensile and compressive stresses.

The in-situ measurements were made by enclosing the crystal in an insulated chamber. It was heated by flooding the chamber with hot nitrogen. The nitrogen was heated by flowing over a resistance heater controlled by a variable transformer. Temperature was measured using a thermocouple placed near the center of the crystal.

We compared the wafer bending measured by the x-ray method with what we obtained using the optical method described by Schaible and Glang.[6]

In this method, the tip of a fiber optics bundle travels in a straight line just over the surface of a bent wafer. The bundle contains both light-sending and light-receiving fibers. The amount of reflected light (into the receiving fibers) is a func-tion of the distance between the tip of the bundle and the wafer. By plotting the intensity of the reflected light as a function of the position of the bundle, one can obtain a plot of the bend in the wafer.

Stresses measured on the same wafers, using x-ray and the above optical technique, gave results within 10% of each other. The experiment included four different kinds of polyimides, with stresses ranging from 41 to 64 MPa.

Film Preparation

The films were prepared using the following steps:

1.   Clean and dry the silicon wafers.

2.   Spin-apply a 1% aqueous solution of γ-aminopropyl-triethoxysilane as an adhesion promoter followed by rinsing in

Table I. Stress vs. Film Thickness.

Film Thickness (μm)	Stress (MPa)$^{\pm}$ 15%
3.2	75.8
9.6	75.8
23.7	69.0

deionized water. This silane has been shown to be a good adhesion promoter for polyimide (References 7-9).

3. Dry the adhesion promoter layer at 85°C for 10 minutes in air.

4. Spin-apply a solution of polyamic acid in NMP (N-methyl-2-pyrrolidone). A single coat, fully cured, yields a film thickness of approximately 5 μm at 3000 rpm.

5. Cure the polyamic acid using the following sequence:

    a.   85°C   –   10 minutes in air

    b.   200°C   –   30 minutes in air

    c.   300°C   –   30 minutes in nitrogen

    d.   400°C   –   30 minutes in argon

Multiple coats were required to produce polyimide film thicknesses greater than approximately 5 μm. These multiple coats were applied by curing each separate coat through Step 5b. After the last coat, the total film was fully cured (Step 5c & 5d).

For in-situ curing in the x-ray camera, the film was initially cured through Step 5a prior to transferring to the heating chamber for x-ray diffraction measurements. For these experiments we used films of only one coat 5 μm thick.

RESULTS

EBTA-MDA

All the initial work was done on EBTA-MDA. The first experiments were done to determine whether the stress in the film is a function of the thickness for fully cured material. The results are shown in Table I.

The values obtained were, within the experimental error, independent of the thickness.

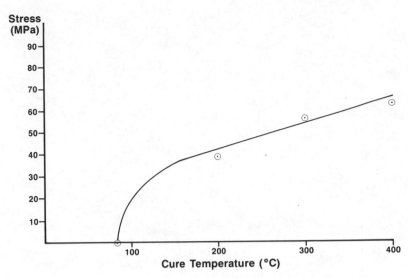

Figure 3.   Stress at room temperature vs. standard film cure
cycle of EBTA-MDA.

The next step was to determine how the stress at room temperature
depends on the curing cycle.   This is shown in Figure 3.

As shown, there is a linear increase in stress with increasing cure
temperature.   The shape of the curve below 200°C is unknown.   For
example, we know the stress is zero at 85°C, but it may also be
zero at 100°C.   In another experiment, we added a cure step of
150°C for 30 minutes.   These results are shown in Figure 4.   The
stress measured in this case and that measured in Figure 3 are
within the experimental error range of ±15%.

The stress versus cure temperature is linear.   The stress with the
extra cure step is (within the experimental error) the same as that
shown in Figure 3.   The curve shape below the 150°C cure step was
not determined.   This curve indicates that the film becomes rigid
at some point between the 85°C cure step and the 150°C cure step,
allowing a stress build-up.

The addition of the high-temperature chamber to the x-ray camera
allowed stress measurements at elevated temperatures.   The first
films measured were fully cured polyimides, 15.7 µm thick.   The
stress versus temperature curves are shown in Figure 5.   The
stresses decreased linearly with temperature approaching zero

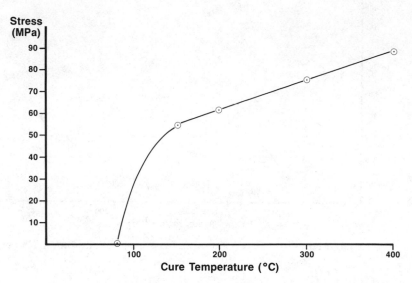

Figure 4.   Stress at room temperature vs. cure temperature of EBTA-MDA with added 150°C cure step.

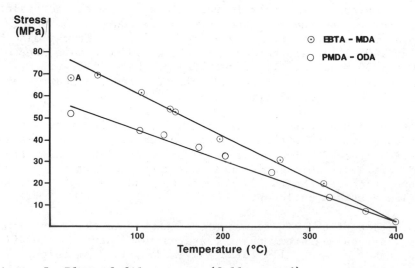

Figure 5. Plot of film stress (fully cured) vs. temperature.

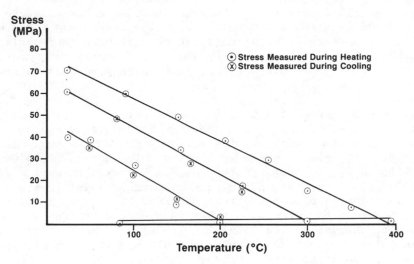

Figure 6.   In-situ measurement of stresses generated during
cure cycle of EBTA-MDA.

stress at 400°C (the final cure step).   This curve indicates that
the stress is due to CTE mismatch between film and substrate.   The
point "A" in Figure 5 is "off the curve."   This may be due to the
plasticizing effect as a result of absorbed water (the film was
stored in a plastic container in an uncontrolled room environment) after the final curing at 400°C, but before the measurements
were taken.   We have evidence in separate experiments, to be
discussed later in this paper, that water absorption reversibily
reduces stress in a fully cured film.

The final experiment was to measure in situ the stresses generated
during the curing cycle.   The in-situ stresses were corrected for
changes in film thickness during the curing by measuring the
change in film thickness on a separate film and applying this
correction to the measured stress of the film being cured in the
x-ray hot stage.   Figure 6 shows the in-situ stresses generated in
an EBTA-MDA film during curing.   We started with a film cured at
85°C (zero stress) and heated to 200°C, holding for 30 minutes
(stress is still close to zero).   The film was cooled to room

temperature with stress measurements being taken at various points during cooling. The stress begins to increase immediately upon cooling and linearly increases to approximately 42 MPa at room temperature. This stress is comparable to the stress observed in Figures 3 and 4 at the 200°C cure step. The sample was reheated to 200°C, with stresses being measured at various points during heating. The stress changes reversibly and linearly to approximately zero at 200°C, following the same curve generated during cooling.

The sample was then heated to 300°C, held for 30 minutes, and cycled 300°-25°-300°C with stress being measured at various points. Again, the stress varied linearly during the cooling and heating cycles, following the same curve. The stress at 25°C cooled from 300°C is approximately 14 MPa higher than when cooled from 200°C (100°C higher ΔT for TCE mismatch). Again, this stress is comparable to the stress observed in Figures 3 and 4 at the 300°C cure step. The sample was then heated to 400°C, held for 30 minutes, and cooled to 25°C. The stress was measured at various points during cool down from 400°C. Again, the stress increased linearly to approximately 70 MPa at 25°C.

The stress-thermal history indicates that the stress is due to TCE mismatch. The history also indicates that this polyimide has a glass transition temperature that is the same as or close to the highest cure temperature. Thus, each time the polymer reaches a higher temperature than any previous excursion, it softens and relaxes all stresses. Upon cooling, the polymer becomes rigid and stress starts to rise due to the TCE mismatch.

PMDA-ODA

We also studied how PMDA-ODA behaves during the curing cycle. Its behavior is much more complex (see Figures 7-a to 7-d). As in the previous case, we started with a film cured at 85°C. At that temperature, the film already showed a substantial stress as shown in Figure 7-a. While the wafer with film was stored in the chamber with flowing nitrogen, the stress rose to 23 MPa over a period of approximately 1 hour. This rise is probably due to further evaporation of the solvent. The temperature was then raised to 150°C without an observed increase in stress. The film was then cooled to room temperature and it incurred a substantial increase in stress. Next, the film was heated again and a measurement was taken at 150°C. The film was then heated to 200°C. As shown in Figure 7-b, the film was cooled to 25°C and reheated to 300°C. Figure 7-c shows what happened when the film was then cooled to 25°C, followed by reheating to 400°C. The last step is shown in Figure 7-d; the film traces the previous curve. Repeated cycling leaves the shape unchanged.

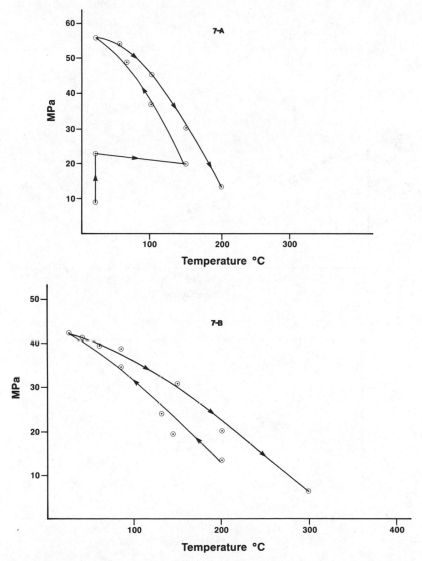

Figure 7. In situ measurement of stress generated during cure cycle of PMDA–ODA.

(continued)

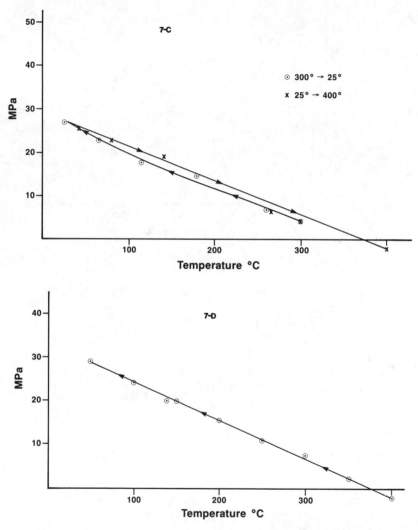

Figure 7 (continued)

Notice that the stress at the cure temperature decreases with increasing cure temperature and that the slope of the cooling curve (cure temp. to 25°C) is steeper from lower cure temperatures than from higher cure temperatures. A steeper slope means that the product of the modulus of the material and the thermal expansion over that particular range is higher.

<u>Film Exposure to Solvents</u>

Since for our processing the films had to be immersed in NMP and water, we also performed an experiment to evaluate the effect of this on the stresses in the film. Films of 15 μm thickness of either material on silicon wafers were exposed to NMP for varying times. The amount of absorbed NMP was measured by weighing, and the stress was measured using the wafer bending. The measurement accuracy is low since evaporation of the NMP took place while measuring; and, consequently, we were unable to correct for variations in the film thickness. The results are, however, accurate enough to give a general impression of what happens (see Figure 8). By heating to 200°C, the NMP evaporated and the film stresses returned to their original values.

A similar experiment was conducted with water, but less extensively. It indicated the same result: substantial reduction of stress, with complete return to original stress values after the films were dried out.

DISCUSSION

In Figure 9-a we have plotted the stress at room temperature as a function of the cure temperature for EBTA-MDA. Figure 9-b shows the stress at various cure temperatures. The same is done for PMDA-ODA in Figures 9-c and 9-d.

PMDA-ODA behaves according to an intuitive model that will be discussed using Figure 10. Here we show how the viscosity, the volume, and the stress of the film behave at three temperatures: approximately 100°C, approximately 150°C, and at 300°C or higher. On the viscosity line, we identified the "Rigidity Point" at which the film becomes sufficiently rigid to support its own stresses.

Starting at a low temperature, the PMDA-ODA film will, as a function of time, lose solvent but retain enough to prevent it from becoming rigid. Also, the rate of imidization is low enough to keep the film viscous. In this film, one would expect absence of stress at the "cure" temperature and at room temperature.

At approximately 150°C, the PMDA-ODA film would lose solvents and some curing would take place. As a result, an "intrinsic" stress

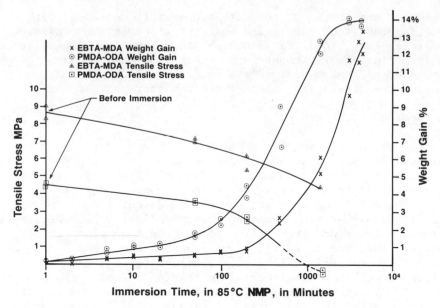

Figure 8. Weight gain and stress in EBTA–MDA & PMDA–ODA as a function of time immersed in NMP.

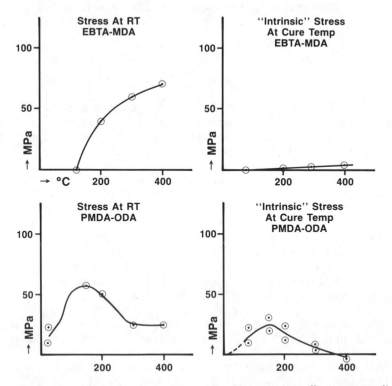

Figure 9. Stress at room temperature and "intrinsic" stress for EBTA–MDA and PMDA–ODA films.

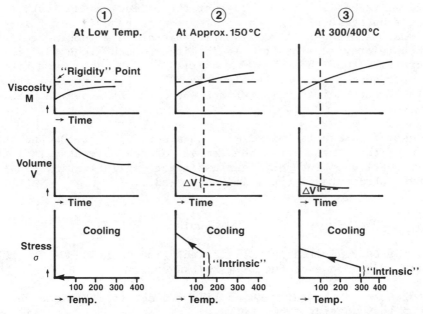

Figure 10. Viscosity, volume, and stress in a PMDA–ODA film as a function of time and temperature for a proposed model for generation of stress.

is built up. Since the film is rigid, the stress in the film will increase when cooled to room temperature due to the added stress as a result of TCE mismatch.

When the temperature is again raised , the PMDA–ODA film will soften, as polymers do; but further imidization and a (smaller) loss of solvents will again result in an intrinsic stress. This intrinsic stress is lower, however, because there is a smaller change in volume. Upon cooling to room temperature, one finds a total stress, which is the result of adding the intrinsic stress to the stress due to the mismatch in thermal expansion. Figure 10 shows the magnitude of the intrinsic stress and the slope of the cooling curves in the same manner as they occur in the actual measurements in Figure 7.

A qualitative explanation for the steeper cooling curve might be that the less completely reacted material has a higher coefficient of expansion due to the presence of unevaporated NMP and a looser molecular structure, both of which lead to a higher TCE.

709

We lack the necessary materials properties at the various temperatures for calculations to give our model a more quantitative basis. We can suggest the following qualitative explanation, however, for the difference in behavior of EBTA-MDA and PMDA-ODA. The EBTA-MDA material has a low molecular weight and, therefore, retains its flexibility to much higher temperatures than the PMDA-ODA material, which has a much higher molecular weight. Consequently, the EBTA-ODA can relieve the intrinsic stresses at temperatures where the PMDA-ODA is unable to do so.

The loss of stress in the films with absorbed solvents, and the return of the stress after the removal of the solvents, indicates that the solvents act only as plasticizers while leaving the internal structure of the films unchanged.

CONCLUSIONS

The stresses generated in fully cured films of EBTA-MDA and PMDA-ODA are independent of the film thickness.

The stresses in the fully cured film are due to TCE mismatch between the substrates and films.

For the EBTA-MDA film, the glass transition temperature is the same as the highest cure temperature the film has experienced. The PMDA-ODA film has a much more complex behavior.

In the EBTA-MDA film, the stress due to drying and/or curing is negligible. Again the PMDA-ODA film shows a more complex behavior.

The maximum stress observed in the EBTA-MDA film is of the order of 75 MPa. The PMDA-ODA film shows a lower stress of the order of 30 MPa.

Exposing either of the two films to NMP and water shows a substantial reduction of the stresses measured at room temperature. These stresses are restored by evaporation of the solvents.

ACKNOWLEDGEMENTS

The authors acknowledge the assistance of W. Druschel, P. Kok, and W. Scanlan in building the high-temperature x-ray chamber. We had helpful discussions with E. Hearn, A. Baise, and R. Lacombe.

The films were made by E. Lodato and E. Koblinski.

REFERENCES

1. D.Y. Perera and D. Van den Eynde. J. Coatings Technol. 53, No. 677, 39, June 1981.
2. E.W. Hearn, Adv. X-Ray Anal., 20, 273 (1977).
3. C.C. Goldsmith and G.A. Walker, Adv. X-Ray Anal., (1983), in press.
4. G.G. Stoney, Proc. Royal Soc. (London) A 82, 172 (1909).
5. W.A. Brantley, J. Appl. Phys. 44, 534 (1973).
6. P. M. Schaible and R. Glang, in "Thin Film Dielectrics" (F. Vratny, Ed.) pp. 577-587, Electrochem. Soc. Inc., New York (1969).
7. L.B. Rothman, J. Electrochem. Soc. 127, 2216 (1980).
8. J. Greenblatt, C. Araps, and H. R. Anderson, Jr., These proceedings, Vol. 1, pp. 573-588.
9. D. Suryanarayana and K. L. Mittal, J. Appl. Polymer Sci. 29, 2039 (1984).

PART IV. MICROELECTRONIC APPLICATIONS

USE OF POLYIMIDES IN VLSI FABRICATION

Arthur M. Wilson

Texas Instruments Inc.

P.O. Box 225621, MS 947, Dallas, Texas 75080

The functional requirements of overcoats and
multilevel insulators for very large scale integrated
circuits (VLSI) are outlined. The moisture barrier
properties of polyimide films are reviewed. Polyimide
performance vs plasma enhanced chemically vapor
deposited (CVD) silicon nitride overcoats are compared.
The topological and via forming advantages of polyimides
vs plasma enhanced CVD silicon oxide as a multilevel
insulator are cited. The temperature and voltage
field induced electronic charge transport and trapping
at oxide interfaces is cited as the most serious limitation to the use of polyimides as multilevel insulators
on VLSI chips.

INTRODUCTION

Polyimide films have an enviable history as encapsulants,
insulators and flexible substrates for printed circuits in the
electronics industry. In all these applications films have been
several mils (50 to 100 microns) thick. At these dimensions the
electrical and physical properties of polyimide films are adequate
to perform as a dielectric insulator. The chemical purity of these
thick films, although always important in electrical applications,
is not as demanding as it is in the thin films required for
integrated circuits (IC's). The advent of very large scale integrated circuits (VLSI's), because of their even smaller dimensions,
places even more stringent demands on dielectric insulators.

An appreciation for scale requirements placed on VLSI
dielectrics can be obtained by referring to Figure 1. Figure 1
shows the packing density trends of one of the semiconductors most
advanced IC technologies, the dynamic random access memory (DRAM)

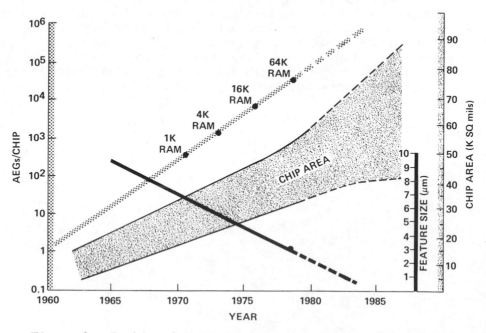

Figure 1. Packing density of Texas Instruments MOS DRAM's.

as built by Texas Instruments. Other semiconductor manufacturers'
DRAM technologies are progressing along similar density versus
time trend lines. The definition of VLSI is 100,000 active
electrical gates (AEG's) per chip. Thus, until the first 254K bit
DRAM has been built, VLSI has not been achieved. Secondly, without
improvement in process technology the area of the chip must get
larger and larger. There are practical reasons why larger DRAM
chips are not economic. Large computer systems are sold on the
basis of the speed with which they process data. As the chips gets
larger the interconnections get longer and the RC time constants
of these transmission lines increase to the point where the access
times to the individual memory cell become prohibitively long.
These slow parts then limit the speed with which larger computer
systems can process data. Large chip area is not economic for
the semiconductor manufacturer because the yields are inversely
proportional to the chip size. There are several reasons why yields
are inversely proportional to chip size. The simplest explanation
involves the concept of killing defects. The number of defective
chips on a wafer depend upon the factory cleanliness and the type
of process technology used. For a given factory and process
technology the killing defects per unit is constant. Thus larger
chips have a higher probability of failure and provide lower multiprobe and final test yields. For the present state of manufacturing
technologies 40 to 80K sq. mils chips represent the upper economic
limit.

Third concept implied by Figure 1 is the improvements
in process technology that were necessary to achieve the increase
in AEG's per chip. The gate structure for the 1K DRAM was a metal-
oxide-silicon (MOS) field effect transistor (FET). The metal was
also used to contact the source and drain of the transistor, and
to provide a single level interconnection for the circuit. Since
memory FET's do not operate at high currents it was possible to
switch to higher resistivity polysilicon as the gate and source-
drain contacts. A heavily doped phosphorous glass, which was able
to be densified and reflowed, was introduced as the insulator and
a metal level placed on top of that insulator to provide circuit
interconnection. With the advent of the 16K and 64K DRAM's a double
level polysilicon plus metal (poly/silicon oxide/poly/phosphosili-
cate glass/aluminum)interconnection system was necessary to achieve
compatibility with the small feature sizes of those designs.

The final trend line in Figure 1, feature size, implies a
dramatic improvement in lithographic (masks, photoresists and
printer) technology. The higher resolution lithography was the
result of better critical dimension control on stepped working
plate masks, the switch from negative to positive resists and
from contact printing to focused, projection type printers. The
improvements were manifold and not describable here. Lithographic
advances will continue to be critical to advances in MOS DRAM
technology.

There are many other VSLI technologies, such as micro-
processors, programmable logic arrays, analog to digital units,
etc., all of which require passivation overcoats and multilevel
insulators. The use of polyimides as insulators for multilevel
interconnections has been reviewed elsewhere.[1] Can polyimides
perform these functions for these VLSI technologies? It is the
purpose of this paper to review the requirements for overcoats
and multilevel insulators. Then to compare the properties of
commercially available polyimides to semiconductor dielectrics
which are presently used for these VLSI applications. Having
compared their strengths and weaknesses, an attempt will be made
to forecast the opportunities for polyimides to replace these
materials.

POLYIMIDES AS VLSI OVERCOATS

The overcoat for a VSLI chip is required to perform a
variety of functions. The overcoat must be formed at or below
the 450°C device anneal temperature. Since the chip has a soft
metal as its final interconnection the overcoat must provide
protection from scratches after it leaves the wafer fabrication
area and the wafer proceeds to the multiprobe area. Continued
scratch protection must be provided as the good electrical chips
proceed to and are processed in the final assembly area. Thus
the overcoat must have good mechanical strength. Since the chip
will be subjected to temperature cycling in the chip mount, wire
bond and plastic or ceramic packaging operations, the overcoat
should have a low value for Young's modulus so that chip deforma-
tional stresses are minimized and the overcoat will not crack.
In order to maintain its integrity as an overcoat and not peel,
the overcoat must have good adhesion to the metal and primary
passivation layer, silicon oxide. Finally the chip must be
protected from ionic contaminants which are introduced during the
final assembly operations or from plastic molding compounds themselves. Also, since molding compounds are not notably good moisture
barriers the overcoat must perform this function.

Polyimide-isoindroquinazoline-dione, PIQ, developed by
Harada, Saiki and coworkers was the first polyimide formulated
for semiconductor overcoat and multilevel applications.[2] Hitachi
Chemical Co. Ltd. published a table comparing the physical and
electrical properties of PIQ with some materials used in semiconductor fabrication. These data are reproduced in Table I.
The electrical data imply that PIQ is comparable to silicon dioxide
in resistivity, dielectric breakdown strength and dielectric
constant and that both of these materials are superior to silicon
nitride and aluminum oxide as multilevel insulators. More will
be said about this in the second section of this paper.

TABLE I.   Comparison Between Properties of PIQ and Some
Inorganic Materials (from PIQ Bulletin, Hitachi Co.).

ITEM / MTL	COEFF THERMAL EXPANSION $(°C)^{-1}$, $\times 10^6$	YOUNG'S MODULUS $(dynes/cm^2)$ $\times 10^{-10}$	THERM CONDUC- TIVITY Cal/cm- Sec-$°C$	MELTING POINT $(°C)$	DIELECTRIC CONSTANT	DIELECTRIC BREAKDOWN STRENGTH (MV/cm)	RESIS- TIVITY $(\Omega\text{-cm})$
PIQ	20–70	3	4	450–500	3.5–3.8	1	$>10^{16}$
Si	2.3	160	720– 3400	1420	11.7–12	0.1	
SiO$_2$	0.3–0.5	74	50	1710	3.5–4.0	1–10	$>10^{16}$
Si$_3$N$_4$	2.5–3	155	280	1900	7–10	1–10	10^{12}
Al$_2$O$_3$	9	373	780	2050	7–9	0.1	10^{14}
Al	25	70	5700	660	–		2.5×10^{-6}

In an overcoat application the insulator of choice would have the
highest thermal conductivity, the lowest Young's modulus and a
thermal coefficient of expansion which most nearly matches that
of the thickest element in a wafer, silicon. Based on thermal
conductivity the order of choosing overcoat materials would be:
Al$_2$O$_3$ > Si$_3$N$_4$ > SiO$_2$ > PIQ. Based on Young's modulus the order
would be exactly reversed.   The thermal expansion coefficient
data indicate silicon nitride films would have the least stress
induced as silicon chips are temperature cycled in the final
assembly operations. However such a simplistic analysis ignores
the fact that thin films have intrinsic stress at the temperature
of formation as well as induced stresses, and that the total
stress is the algebraic sum of the intrinsic, s_i, and thermally
induced s_{th}, stress, i.e.

$$s_t = s_i + s_{th} \tag{1}$$

Furthermore the thermally induced stress is given by:

$$s_{th} = \frac{E}{1-p} \int_{T_s}^{T} (a_{Si} - a_f) \, dT \tag{2}$$

where E is Young's modulus, p is Poisson's ratio, T_s is the temperature of film formation and a_{Si} and a_f are the thermal expansion coefficients of silicon and the thin film respectively.[3]

The semiconductor industry has used a variety of glassivations for final passivation overcoats: sputtered quartz, low temperature chemically vapor deposited (CVD) silicon oxide (both phosphorous doped and undoped) and plasma enhanced (for low temperature deposition)CVD silicon nitride (both tensive and compressively stressed films). Phosphorous doped low temperature CVD silicon oxide can not be used as a final overcoat in plastic molded packages because it is a potential source of phosphoric acid, which will promote aluminum lead corrosion. Kern and Rosler have reviewed the properties of inorganic semiconductor insulators deposited by a variety of reactors and conditions.[4] Sinha et al have measured the thermal stresses and estimated the relative cracking resistance of plasma CVD silicon nitrides on

TABLE II.   Intrinsic and Thermal Induced Stresses in Thin Insulator Films on Silicon Wafers

Material	Formation Temp,°C	Formation Stress,dynes/cm^2	Stress,dynes/cm^2 25°C	Stress,dynes/cm^2 450°C	ref
Silicon Oxide					
Steam grown	1050	-0.5×10^9	-3.5×10^9	-2.5×10^9	3
CVD	480	3	± 0	hysteresis	3
Silicon Nitride					
LPCVD	840		12 to 18		4
Plasma CVD "Tensile"	280	1.5	1.3	1.8	3
"Compressive"	280	-1.3	-1.8	-0.9	3
Polyimide	350	+	-1 to 3 +		5,20

Stress conventions: Tensile (+), Compressive (−)

720

silicon substrates.[3] The intrinsic and total stress induced
at room temperature and at a typical assembly temperature of
450°C are shown in Table II. Except for low pressure chemical vapor
deposited (LPCVD) silicon nitride, the stresses are low. Sinha
has estimated the order of crack resistance to be highest
for compressive and lowest for tensile plasma enhanced CVD
silicon nitrides. Overcoats are formed on top of patterned metal
leads with an underlying layer of steam grown thermal oxide. Modeling the stresses induced by this more complex structure is difficult
and prediction of crack resistance from fundamental film properties
is even more tenuous. However, if the overcoat has good adhesion
to the substrate, the film is completely supported and there is
no free surface across which the tensile stress can operate to
initiate a crack. Compressive films are preferred;however tensile
plasma CVD silicon nitride overcoat films have been found to be
effective moisture and mobile ion barriers.

Moisture Absorption and Stress in Polyimides

In a previous paper the moisture barrier properties of
polyimides were described.[5] The isothermal chronogravimetric
data for Dupont P12545 is reproduced in Figure 2. In this study
4 micron thick polyimide films were spun onto pre-weighed
oxidized and patterned silicon wafers and cured at 205°C to drive
off the polyamic acid resin solvents. After storage in a silica
gel containing desiccator for 24 hours the samples were subjected
to anneal temperatures from 350°C to 450°C in either nitrogen or
oxygen atmospheres and reweighed. The results are shown as dotted
lines in Figure 2. The 5% weight gain of all samples during the
24 hour storage in the silica gel containing desiccator is
indicative of the desiccant strength of polyimide films. It
was later found calcium chloride is a better desiccant and it
was used to protect cured polyimide samples in all succeeding
studies. The same samples were then exposed to a room temperature
humidifier for 113 hours and the samples again reweighed. The
results are shown as solid lines in Figure 2. Based on the weights
of the repeating unit (mer) of the P12545 polymer, it can be
shown that a weight change of 3.5% corresponds to a gain or loss
of one water molecule per mer. The 205°C cured films gain
approximately 1.5 molecules of water per mer. Assuming all the
weight loss on annealing the polymer is due to water loss, the two
waters formed during the polyamic acid to polyimide conversion are
only driven out of the film after an anneal of 450°C in nitrogen
for 30 minutes or more. Depending upon the anneal conditions the
films can reabsorb 3 to 5% water. The higher the temperature of
the anneal the lower the final amount of water reabsorbed. The
annealing thus causes some structural transformation in the polymer
which blocks availability of hydrogen binding sites.

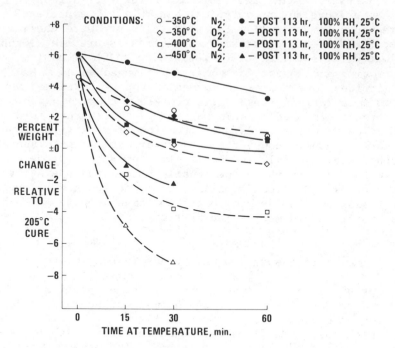

Figure 2. Isothermal Chronogravimetric Anneal & Moisture Absorption of Dupont P12545.

Since these films were not coupled to the oxide with any adhesion promoter, there was an opportunity to observe the polarity of the stress created by the absorption of the additional water in the case of the samples cured at 350°C in nitrogen. The stress was contractile, i.e., the polymide film experienced a tensile stress upon water absorption. In all cases the polyimide film at the edge of the wafer was displaced from the wafer geometry toward the center of the wafer. An example is shown in Figure 3. Thus the additional bound water must be forming an intermolecular bridge in order to cause a volume or linear shrinkage of these polyimide films. This experiment clearly shows the need to couple polyimides to silicon oxide surface in order to overcome thermal and moisture induced stresses.

Polyimides are known to adhere well to aluminum surfaces. However, in using polyimides as overcoats, the final metallization interconnection is not always aluminum. Furthermore, even when aluminum is the interconnection, a considerable area of the chip is the underlying silicon oxide insulation. Saiki and coworkers have formed thin coupling films of aluminum oxide by spinning

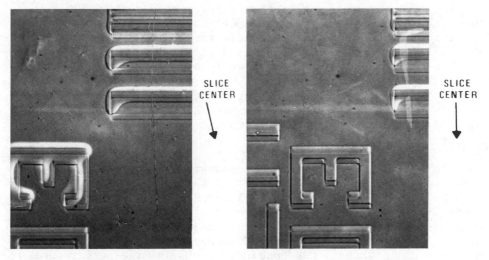

Figure 3.   Polyimide Film Displacement Due to Tensile Stress of Hydration.

organic aluminum chelate solutions on wafers and converting the aluminum chelate to the oxide in a 350°C air environment.[6]   They have shown polyimides, coupled in this fashion, to maintain excellent peel strength after pressure cooker exposure up to 300 hours.   The excellent adhesion of polyimides to aluminum surfaces must also be explained by the native oxide formed on all vacuum deposited aluminum films once exposed to air.   Polyimides can also be coupled to silicon oxide surfaces with organosilanes.[7]   An analytical technique to quantitatively measure the peel strength of polyimides on wafer surfaces has been described by Rothman.[8]

The Moisture Barrier Properties of Polyimides

In order for an overcoat film to be an effective moisture barrier it must be free of holes.   Holes due to cracks have already been addressed.   With proper cleanliness of the wafer and pretreatment with couplers, the formation of "fish-eyes" due to de-wetting have been eliminated.   The remaining hole formation is due to contamination by particulates.   The main source of particulates are: the wafers themselves, the film forming chemicals and the environment in which the film is formed.

Wafer cleanliness is a constant no matter whether a CVD reactor or
spin coater is used. The cleanliness of a CVD reactor is more
easily controlled by routine maintenance schedules than monitoring the environmental air in the vicinity of a coater. Figure 4
shows the pinhole density of a variety of commercially available
polyimides coated in class 100 clean room environment as a function
of polyimide thickness. As-received material is compared with some
material filtered with a natural wool chemical filter. We have
since switched to an inorganically bonded, fibreglass, depth type
filter. It provides high flow rates and defect densities equal
or less than "as received" material. The as-received material
curve is comparable to what is obtained for a clean photoresist
in the semiconductor industry.

In a study of various plastic passivants for beam lead
circuits, White found that by coating the space between two leads
with polyimides or silicones the leakage current, induced by a
water drop on top of the film, was reduced to the same value as
that of the dry control.[9] After a 4 hour pressure cooker exposure
the silicone performance remained unchanged, but the polyimide performance was degraded. He concluded the polyimide performance
was degraded because the pressure cooker stress had caused the

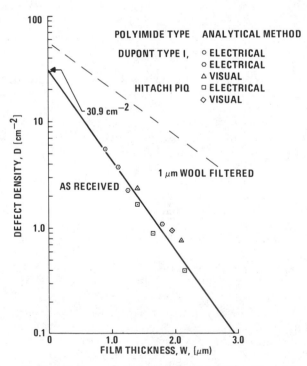

Figure 4. Dependence of Polyimide Pinhole Density on Film
Thickness.

polyimide to lose adhesion and a small liquid interfacial
leakage path had been introduced between the polyimide and the
ceramic substrate. In their study of the effects of epoxide and
silicone molding compounds on conductor corrosion in biased 85°C-
85% relative humidity environments, Sims and Lawson found that a
polyimide passivation film yielded the longest cathodic and
anodic mean time before failure when compared with silicon nitride,
phosphosilicate glass and no passivation overcoat, if all were
encapsulated in the same epoxide mold compound.[10] In a previous
study the permeability of Hitachi PIQ, Dupont PI2545 and PI2555
were measured as a function of anneal temperatures in nitrogen
and oxygen environments.[5] Under optimum cure PIQ has a permeability
of 31 pg-μm/hr-μm^2 and the Dupont films have similar permeabilities,
about 60 to 70 pg-μm/hr-μm^2. Assuming the water transport to the
surface of an aluminum lead is the rate determining step for
corrosion and that only a stoichiometric amount of 3 moles of
water are needed to corrode 2 moles of aluminum, it was shown
that 2 to 4 micron thick polyimide films could not provide a
corrosion barrier for aluminum for more than an hour. In the
same paper the pressure cooker failure rate data (shown in Figure
5) was presented comparing the performance of a polyimide versus a

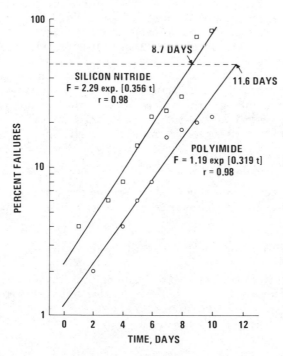

Figure 5. Pressure Cooker Failure Rates of μ747 Operational
Amplifier Circuits in Morton 526A Plastic Package.

standard tensile plasma enhanced CVD silicon nitride overcoat. Compressive plasma enhanced CVD silicon nitride was not available at the time these tests were performed. The important point to be made is that properly coupled and coated polyimides can be used as overcoats to increase the moisture environment performance of sensitive integrated circuits.

POLYIMIDES AS VLSI MULTILEVEL INSULATORS

Polyimides become the insulator of choice, as do low temperature plasma enhanced CVD silicon oxides, only after the first level aluminum interconnection is deposited on the wafer. Metal silicide interconnection systems also require a low temperature insulator. Polysilicon interconnection systems are inherently high temperature processes; interlevel oxide is grown from the first level polysilicon and a multilevel insulator is supplied by a heavily doped phosphosilicate glass deposited by CVD methods. This multilevel insulator is densified and reflowed in order to obtain smooth topography before the first aluminum layer is deposited.

Some of the requirements for a multilevel metal interconnection system can be readily understood by reference to Figure 6. The main topographic elements of a multilevel interconnection system are the regions where the second lead crosses over the first lead (cross-overs) and where the second lead makes contact through a hole (via) in the insulator. The main function of the insulator is to planarize or smooth underlying topography at cross-overs so that the second level metal is not thinned due to shadowing effects in the case of evaporated metal or irregular growth patterns in the case of sputtered metal. Via walls must also be properly sloped or incomplete metal coverage can result. A slope of 45° with respect to the wafer surface is ideal, however wall slopes as high as 60° can be covered well by sputtered metal.

In addition to these topographic requirements the insulator should have dielectric characteristics comparable to thermal silicon oxide.

Polyimide Solutions to Multilevel Topography Problems

One of the main problems with plasma CVD silicon oxide insulators is shown in Figure 7. Since the film is grown from the gas phase at high nucleation rates there is a preferred

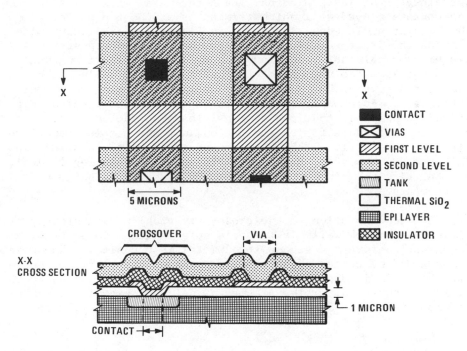

CONTACT
VIAS
FIRST LEVEL
SECOND LEVEL
TANK
THERMAL SiO$_2$
EPI LAYER
INSULATOR

5 MICRONS

CROSSOVER VIA

X-X
CROSS SECTION

1 MICRON

CONTACT

Figure 6. Components of a Two Level Interconnection System.

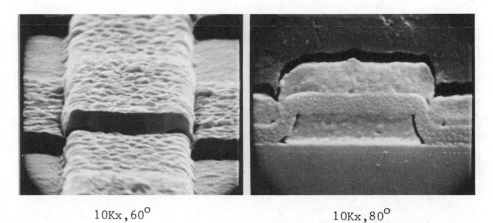

10Kx,60°   10Kx,80°

Figure 7. Scanning Electron Micrograph of Second Lead Step
          Coverage Problem.

growth rate at sharp edges or projections. Thus with conventionally patterned first level aluminum leads the metal cross-section shows the concave sloped wall which terminates at the top surface in a sharp right angle edge. As shown in Figure 7 the silicon oxide softens this feature slightly but the second lead deposition system is unable to avoid shadowing by the steeply rising wall and it exaggerates the edge into what has been aptly called a "bread loaf cross-section." One solution is to slope the first level metal wall. An example is shown in Figure 8. One disadvantage of this process is that the design rules must include extra lead width to compensate for the reduction in cross-section associated with the subtractive etch process used to slope the aluminum lead. This process has been used since 1975 on all double level metal, digital circuit, bipolar programmable only memories built by Texas Instruments.

A second solution to the cross-over problem with steeply etched first level metal walls is to use a spin-on glass to smooth the topography before the second level metal is deposited. The spin-on glass could be coated either before or after the plasma enhanced CVD silicon oxide is deposited. A schematic representation where the glass is spun on after the plasma CVD silicon oxide is deposited is shown in Figure 9. The main problem with spin-on glasses is

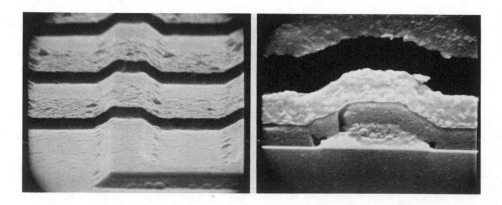

5Kx, 75°   10Kx, 80°

Figure 8.   Scanning Electron Micrograph of Second Lead Step
            Coverage Solution: Slope Wall of First Level Metal.

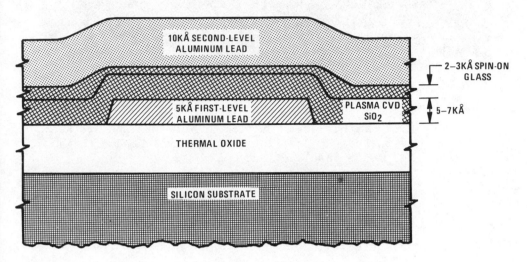

Figure 9.   Schematic of Second Lead Step Coverage Solution: Use
of Spin-On Glass.

they are notoriously variable in chemical purity and quality from
lot to lot. Secondly, they need to be densified at 900°C or more
before the dielectric properties and etch rates approach those of
plasma CVD silicon oxide. The higher etch rate of the spin-on
glass causes badly over etched vias in composite insulator
structures.

One of the polyimides inherent properties is its planarizing
capability. This property most strongly recommends polyimides as
a VLSI multilevel insulator.

Polyimide Via Patternability

Vias can be patterned in polyimides by wet etching, plasma
ashing or reactive ion etching.[11-14]   With properly designed
processes all these techniques provide positive slope to the via
wall, such that good second level metal coverage is achieved.
Using partially cured films and alcoholic tetramethyl ammonium
hyroxide it is possible to achieve the results shown in Figure
10 for 5 micron vias and Figure 11 for 3 micron vias.[15-16]

Figure 10.   Micrographic Cross Section of Two Vias and Two Second
Level Leads Contacting a First Level Lead with
Polyimide Insulator and Overcoat.

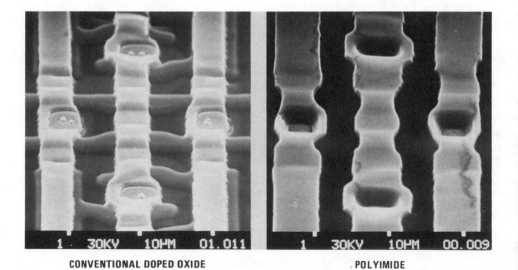

CONVENTIONAL DOPED OXIDE   POLYIMIDE

Figure 11.   Scanning Electron Micrograph of Three Second Level
Leads and Four-3 Micron Vias on MOS Memory Structure.

The cross section of Figure 10 clearly shows the smooth first level lead coverage by the polyimide insulator. The ideal 45 to 60 degree slope of the via wall in the polyimide film is also evident. In Figure 11 the underlying topography has been more completely planarized by polyimide than by the conventional reflowed phosphosilicate glass. In addition the second level lead coverage is better in the vias of the 2 micron thick polyimide than in the vias of the one micron thick doped oxide insulator. This is due to the better control of the slope of the via walls by the etch process for polyimide than for the doped oxide.

Dielectric Problems of Polyimides

The electrical properties of polyimide films including dielectric constant, dissipation factor, dc conduction and dielectric strength have been measured by Brown[17], Rothman[8] and Samuelson[18]. In his first paper Brown used a classical metal/polyimide/oxide/silicon structure in a thick oxide field effect transistor configuration to measure the thick field voltage instability to voltage and temperature stress. He demonstrated that this field threshold voltage shift is described by the relation:

$$V_{TX} = - \frac{C_o V_s}{C_p} \left(1 - \exp\left(- \frac{t}{R_p (C_o + C_p)}\right)\right) \quad (3)$$

where C_o and C_p are the oxide and polyimide capacitances, R_p is the polyimide resistance, V_s is the stress voltage and t is the time of the stress. In a later paper Brown measured the low field (less than 10^5 v/cm) dc conduction.[19] Resistivities as low as 10^{15} Ω -cm were observed at 125°C. The resistivity measured this way agrees well with values calculated from the field threshold voltage shift Equation (3) above. In an attempt to separate this high electronic conduction from mobile ion drift effects Brown used an integrated charge-ramp voltage technique. By plotting the mobile ions at 290°C vs the polyimide thickness he was able to compute a bulk ion content for Dupont PI2545 of 3 x 10^{16} ions/cm^3 from the slope. If this ionic charge is totally due to sodium this would correspond to 0.8 ppm, which is the manufacturers' specification, It is quite apparent from Brown's work that polyimide films are relatively leaky dielectrics. The value of the parallel resistor element is dependant upon temperature and voltage stress. If the conduction mechanism is due mostly to mobile ions then perhaps decreasing the ionic contamination to the ppb range will increase the resistivity to the point where the interfacial charge storage will not represent a reliability hazard.

CONCLUSIONS

1. As a final passivation overcoat polyimides provide superior moisture barrier protection for VLSI circuits. Proper use of adhesion promotors is important to achieve this performance. Plasma enhanced CVD silicon nitride is well established as a passivation overcoat. The newer compressively stressed films can be deposited "crack-free" at film thicknesses of one micron or more. Since compressive silicon nitride is a better ion barrier at elevated temperatures and more scratch resistant, it will be hard to replace it with polyimide in VLSI overcoat applications. There are some circuits which are particularly sensitive to the stresses of a thick compressive silicon nitride film. If the parametric values cannot be retrieved by some process modification then polyimide could be substituted for silicon nitride.

2. Polyimides seem to be the ideal multilevel insulator because of the low temperature cures, the planarizing property and the ease of patterning 3 micron vias. However, the less than ideal dielectric properties makes them doubtful contenders for VLSI multilevel insulators. Without significant improvements in dielectric properties polyimides cannot be recommended for MOS VLSI circuit applications. The most reasonable multilevel application would be for low voltage, bipolar VLSI circuits.

ACKNOWLEDGEMENTS

It is a pleasure to thank K. L. Mittal for inviting me to deliver this paper. I would like to thank George Brown for valuable discussions and for permission to show several figures from his work. Finally, without the last minute artwork of Jim Chambers and typing and word processor efforts of Jean Shay and Peggy Wamble, this manuscript would never have met the deadline.

REFERENCES

1. A. M. Wilson, Thin Solid Films, 83, 145 (1981)
2. M. Tomono, A. Abe, S. Harada, K. Sato, T. Takagi, Y. Oya and A. Saiki, "Discrete Semiconductor Device Having Polymer Resin as Insulator and Method for Making Same", U.S. Patent 4,017,886, Apr. 12, 1977.
3. A. K. Sinha, H. J. Levinstein and T. E. Smith. J. Appl. Phys., 49, 2423 (1978).
4. W. Kern and R. S. Rosler, J. Vac. Sci. Tech., 14, 1082 (1977).
5. A. M. Wilson, Electrochem. Soc. Extended Abst. Vol. 80-2 Electrochemical Society, Princeton, N.J., 1980, p. 1236.
6. A. Saiki and S. Harada, J. Electrochem. Soc., 129, 2278 (1982).
7a. A. Saiki, K. Sato, S. Harada, T. Tsunoda, Y. Oka, "Insulating Protective Film for Semiconductor Devices and Method for Making Same", U.S. Patent 4,001,870, Jan. 4, 1977.

7b. Y. K. Lee, paper presented at the Electrochemical Society
    Meeting in Minneapolis, MN, May 1981.

7c. D. Suryanarayana and K. L. Mittal, J. Appl. Polym. Sci., 29,
    2039 (1984).

8. L. B. Rothman, J. Electrochem. Soc., 127, 2216 (1980).

9. M. L. White, Proc. IEEE, 57, 1610 (1969).

10. S. P. Sims and R. W. Lawson, Proc. 17th Annual Reliability
    Physics. Symp. IEEE, New York, 1979, p. 103.

11. J. C. Yen, Electrochem. Soc. Extended Abst. Vol. 75-1,
    Electrochemical Society, Princeton, N.J.,1975, p. 444.

12. H. J. Fick, U. S. Patent 3,395,657, July 30, 1968.

13. A. Saiki, T. Mori and S. Harada, U. S. Patent 3,846,166,
    November 15, 1974.

14. T. Herndon and R. L. Burke, Proc. Kodak Interface Conf., Oct.
    25, 1979, p. 146.

15. P. Shah, D. Laks and A. Wilson, Proc. Int. Conf. on Elect.
    Devices, 1978, IEEE, New York, 1979, p. 465.

16. A. M. Wilson, D. W. Laks and S. M. Davis, "Process for Etching
    Sloped Vias in Polyimide Insulators", U.S. Patent 4,369,090,
    Jan. 18, 1983.

17. G. A. Brown, in "Polymer Materials for Electronic Applications",
    E. D. Feit and C. Wilkins, Jr., Editors, ACS Sym. Series No.
    184, Amer. Chem. Soc., Washington, D.C., 1982, p. 151.

18. G. Samuelson, ibid, p. 93.

19. G. A. Brown, Proc. Int. Reliability and Phys. Symp. IEEE, New
    York, 1981, p. 282.

20. G. Samuelson and K. Ginn, Personal Communication, unpublished
    data, 1980.

NOVEL METHOD FOR THE ELECTROLESS METALLIZATION OF POLYIMIDE: SURFACE OXIDATION OF ZINTL ANIONS[†]

Robert C. Haushalter[*] and Larry J. Krause[†]

Materials Science and Technology Division

Argonne National Laboratory, Argonne, Illinois 60439

A facile general scheme for the metallization of polyimides (PIm) and certain other organic polymers is presented. The driving force for the metallization is the extremely strong reducing power of the polyatomic main group anions, the so-called Zintl anions.

The polyimide can be metallized with many main group elements (e.g., germanium, tin, lead, arsenic, antimony), by simply treating the PIm with solutions obtained from the extraction of potassium-main group alloys. The metal films are formed by the topochemical oxidation of the anionic clusters by the polyimide. The reaction proceeds by a reduction-intercalation-deposition mechanism where the PIm is reduced to monoanions and dianions, the Zintl anion's alkali metal countercation intercalates the polymer to balance the charge from the reduction and the clusters are oxidized to the metallic state on the polymer surface.

[†]Work performed for the Division of Materials Science, U.S. Department of Energy at Argonne National Laboratory

[*]Author to whom correspondence should be addressed at Exxon Research and Engineering Co., Clinton, NJ 08809

[†]Current address: 3M Center, St. Paul, MN 55144

The polyimide metallization has been extended to certain transition metals by employing a method that is conceptually the reverse of the main group metallizations discussed above (i.e., using the PIm as a reducing agent toward an oxidized metal species in solution). Specifically, the treatment of PIm with methanol solutions of K_4SnTe_4 gives reduced intercalated material, $K_x \cdot PIm$, with no surface metallization. The reaction of $K_x \cdot PIm$ with solutions of transition metal cations with reduction potentials more positive than that of $K_x \cdot PIm$ results in metal deposition.

The preparation of surfaces containing two metals, by a combination of the above two methods, will be discussed.

INTRODUCTION

Much effort has recently been devoted to the use of polyimides, PIm,[1] in the fabrication of large scale integrated circuits, mainly as a support and insulating dielectric layer.[2-6] We report here a new, general method for the electroless metallization of PIm[8] based on the topochemical[9] oxidation of the highly reduced, anionic main group metal clusters known as the Zintl anions which is fundamentally different than other methods for polymer metallization.[7] As explained below, the basic features of the Zintl anion/PIm system that allows the metallization reactions to proceed are: (1) the extremely strong reducing power of the Zintl anions, which, once oxidized, deposit on the surface of the oxidant as the zero-valent element and (2) the fact that PIm can (a) form stable reduction products, and (b) readily intercalate alkali metal cations.

All of the reactions discussed below depend on the high chemical reactivity of the Zintl anions. The Zintl anions are easily obtained from extraction of a Zintl phase[10] (an alloy between an alkali or alkaline earth metal and a main group element) or by electrochemical methods.[11] The Zintl anions have been known to exist in solution for over 60 years but only recently have species such as Sn_9^{4-},[12] Bi_4^{2-},[13] Sb_7^{3-},[14] and $Tl_2Te_2^{2-}$[15] been isolated and structurally characterized by Corbett and coworkers.

RESULTS AND DISCUSSION

All of the metallizations discussed in this work use PIm either as an oxidant to deposit anionic metal species on the surface of the polymer or as a reductant to deposit metal cations from solution. The metallizations are grouped into three types.

736

Type (1): main group metals are deposited by oxidation of Zintl anions onto the polymer surface. The Zintl anion is oxidized to metal and the PIm is reduced and intercalated by alkali metal cations. Type (2): metal cations are deposited by first reducing and intercalating the PIm followed by reduction of a cation from solution which is accompanied by oxidation and deintercalation of the polymer. Type (3): bimetallic surfaces are prepared by sequential application of Type (1) and Type (2) metallizations. The three types of metallizations are illustrated schematically in Figure 1 and will be discussed separately.

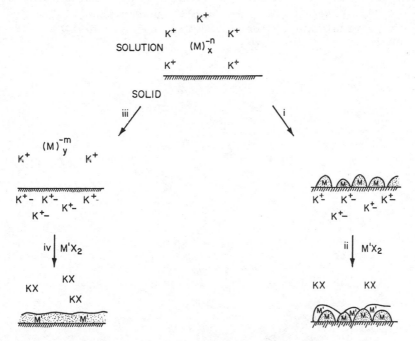

Figure 1. Generalized schematic for the formation of thin films/surface crystallites using the solid support as a redox reagent. (i) topochemical oxidation of Zintl anions onto reducible solids, (ii) treatment of the metallized/reduced surface from (i) with a reducible metal cation to give potassium halide and a bimetallic surface, (iii) reduction and intercalation without surface metallization in oxidation of the solid with metal reduction from solution. In the text, the reaction (i) corresponds to Type 1 metallization, (iii) and (iv) to Type 2, and (ii) to Type 3, respectively.

A. Type (1): Topochemical Oxidation of Zintl Anions by PIm

Zintl anion metal clusters are ideal candidates for the preparation of metals because (a) unlike transition metal clusters, each vertex of the Zintl cluster has an unshared pair of electrons precluding coordination of electron donating exopolyhedral ligands. Being unincumbered by external ligation should facilitate metal-metal bond formation during oxidation of the anionic cluster to the metal. (b) The Zintl anions are very strong reductants and thus are readily oxidized. (c) Since no chemical bonds need to be broken when forming the metal, only electrons are transferred, the reactions are very rapid. (d) Once oxidized, the zero-valent element is insoluble and must deposit from solution.

The Zintl anions deposit on the PIm by simply immersing the polymer into ethylenediamine (en) solution of Zintl anions such as K_4Pb_9, K_4Sn_9, K_3As_{11}, K_4Ge_9 or K_3Sb_7. As shown in Figure 2, which shows the rate of deposition of Sn from K_4Sn_9 solutions, the reaction is very rapid.

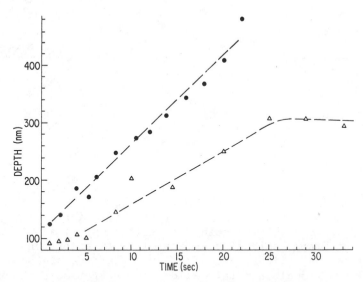

Figure 2. Plot of gravimetrically determined depth of Sn, deposited from solutions of ~ 0.2 M K_4Sn_9 onto 5 mil PIm film, versus time in the plating bath. The circles represent reactions in pure en, the triangles, reactions in 1:1 (vol:vol) en:toluene all at 20° C.

Based on the evidence presented below, the PIm is metallized
by the Zintl anions according to a mechanism where the Zintl anion
is oxidized to the metal, the polyimide is reduced and the alkali
metal cations diffuse into the polymer (intercalate) to balance
the charge from the reduction. It is thus necessary for the PIm
to form stable reduction products (i.e., stable radical anions)
and to be permeable to alkali cations.

The site of the reduction in the PIm is the easily reduced
bisimide functionality in the polymer. Figure 3 shows the repeat
unit in PIm[8] and a monomeric bisimide (1) compound which models
the bisimide moiety in PIm. The site of reduction in PIm was
determined to be the bisimide functionality by comparison of the
physical properties of bisimide (1) with those of reduced PIm.

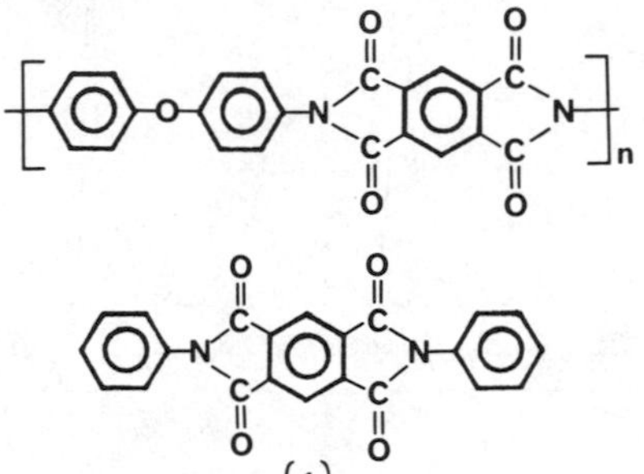

(1)

Figure 3. Comparison of the polyimide structure with that of
bisimide (1).

When PIm is metallized with Zintl anions, the metallized polymer
that has been handled anaerobically displays an intense epr
spectrum with g = 2.00 confirming the presence of S = 1/2 organic
radical anions (reduction products) in the polymer. Although
neutral PIm is yellow and insoluble in DMF, extraction of freshly
metallized PIm shows the presence of a green and a purple compound
upon examination by visible absorption spectroscopy (VAS). Furthermore, bulk electrochemical oxidation of the purple compound
converts it to the green compound whereas bulk electrochemical
reduction of the green compound generates the purple material.
Thus the purple compound is a reduction product of the green
compound. Also, the purple compound is converted to the green one
in the presence of excess neutral PIm. Treatment of model
bisimide (1) with excess reductant (K_4Sn_9) in DMF generates a
purple anion, (2), while reducing bisimide (1) (with K_4Sn_9) in the
presence of excess (1) generates a green anion, (3). In addition,
treatment of purple (2) with (1) generates (3) demonstrating
that (3) is a reduction product of (1). As shown in Figure 4,
comparison of the VAS of the green compound, in DMF extracts of

metallized PIm, with the green reduction product of (1) shows them
to be identical as is the purple extract from PIm and reduction
product (2). Thus, the bisimide functionality in PIm is the site
of the reduction.

Furthermore, green compound (3) is paramagnetic with S = 1/2
and g = 2.00 at 25°C. The epr spectrum (Figure 5) is easily interpreted in terms of interaction of the unpaired spin with two
equivalent I = 1/2 ^{1}H nuclei (a_H = 0.5 gauss) and two equivalent
^{14}N nuclei (a_N = 1.0 gauss) consistent with the monoanion (3).

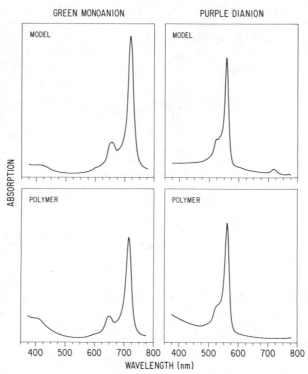

Figure 4. Comparison of the visible absorption spectra of the
purple and green anions of metallized polyimide with the
corresponding anion from bisimide (1) identifying the bisimide
ring as the reduction site in the polyimide.

The green DMF extract from metallized PIm shows an epr spectrum
with identical hyperfine coupling (Figure 5) providing additional
evidence for the site of the reduction being the bisimide portion
of PIm. The values of a_H and a_N indicate the bulk of the unpaired
electron density is on the carbonyl groups and essentially none on
the outer phenyl rings. The fact that bisimide (1) can form two
one-electron reduction products is also demonstrated by cyclic
voltammetry (Figure 6).

Other facts supporting the reduction-intercalation-deposition
scheme (Figure 1-i) include the elemental analysis of the metallized polymer. The ratio of the alkali metal:main group element
in the metallized polymer reflects the ratio of the two elements
in the Zintl anion solution used for the metallization. For
example, when PIm is metallized with solutions of K_3Sb_7, the ratio
of potassium to antimony in the metallized polymer is found to be
0.432 (calculated for K_3Sb_7, 0.429). This provides evidence for
the K^+ intercalation of the polymer, that is, one K^+ enters the
polymer per electron added. The metallization of PIm proceeds at
an appreciable rate only when the counter-cation of the Zintl
anion is an alkali metal. Larger cations, such as $(crypt-K)^+$ and
tetraalkylammonium cations, are apparently too large to inter

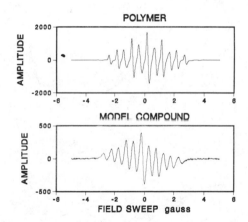

Figure 5. The ~ 9.1 GHz EPR spectra (20°C) of DMF extracts of PIm
freshly metallized with Sn_9^{4-} (upper) and the green monoanion of
bisimide (1). For both compounds the center of the signal is at
g = 2.00.

calate the polymer. The deposited metal films have been identified by their powder x-ray patterns, which show the films to be
free of crystalline impurities.

In most cases, examination of the deposited metal films by
scanning electron microscopy (SEM) shows the films to be of a

granular nature rather than completely continuous. As discussed
in more detail elsewhere[16], a mechanism has been proposed to
explain the granular morphology where the initial reaction of the
Zintl anion creates nucleation sites on the solid's surface. As
the reaction proceeds the Zintl anions preferentially decompose
onto the surface of the growing metal crystal and transfer their
electrons through the metal crystal to reduce the substrate
beneath the crystal. The cations then diffuse into the substrate
around the edges of the growing metal crystal. This is
illustrated schematically in Figure 7a.

This type of proposed deposition reaction may provide a
possible explanation for the fact that the electrical properties
of the metals films deposited by this Zintl method have been
observed to vary greatly as discussed below.

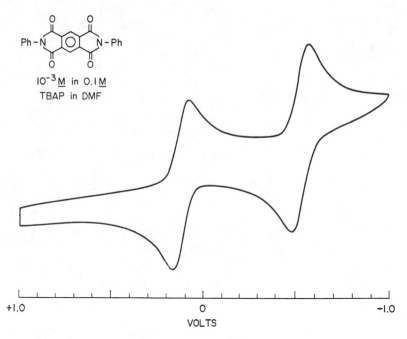

Figure 6. Cyclic voltammogram of 10^{-3} M ($\underset{\sim}{1}$) in DMF with tetra-n-
butylammonium perchlorate electrolyte at a Pt working electrode.
Volts are relative to a Pt wire quasi-reference electrode. Sweep
rate = 100 mV/S.

The electrical properties of the deposited metal films can
vary depending on the conditions used for the deposition. Under
proper conditions, the films approach the resistance of the bulk
metal when sufficiently thick, as shown in Figure 8. In the
K_4Sn_9/PIm system, the resistance of the metal film has been

observed to vary over several orders of magnitude from that of insulators to metallic conductors, depending on the deposition conditions. Scanning electron microscopy (SEM) shows (Figure 9) that the more insulating films are island films with metal particles surrounded by insulating reduced polymer, while the conducting films have overlapping metal grains. The island films could possibly be formed as shown in Figure 7a. Note, however, that both types of films are microporous which allows K^+ diffusion. A reasonable hypothesis explaining the formation of the insulating island films is the cleavage and dissolution of the polymer under the strongly reducing conditions of the metallization. This is supported by: (a) the fact that aromatic imide linkages can be cleaved by reducing agents in polar solvents has been known for nearly 60 years;[17] (b) a reducing agent (N_2H_4) in en is used to etch PIm in the manufacture of integrated circuits;[5] (c) as discussed above, radical anion-containing polymer fragments can be extracted from freshly metallized PIm with DMF, whereas PIm

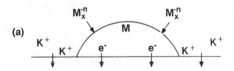

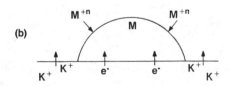

Figure 7. Schematic illustration comparing (a) anion oxidation/solid reduction/solid intercalation with (b) cation reduction/solid oxidation/solid deintercalation.

is insoluble in DMF demonstrating polymer cleavage; (d) commercial polyimide is known to contain 1-3% H_2O when received. Metal films obtained by using dried PIm (200°C, several hours) always show

lower resistance than "wet" films. Presumably the H_2O is partici-
pating in the base catalyzed amide hydrolysis causing polymer
cleavage during the formation of the insulating island films.
Thus, more highly conducting films are obtained under conditions
of minimum polymer degradation, i.e., rapid deposition from low
polarity solvents onto smooth, dry polymer surfaces. As shown in
Figure 8, the decreased resistance observed when using less polar
solvents is consistent with diminished solubilities of the charged
radical anion polymer fragments. Thus, it could be possible, in
principle, to vary the electrical resistances of the deposited
metal films by controlling the exact deposition conditions.

    To summarize this section, the reaction of Zintl anions with
PIm, to produce various metal films is shown to proceed by a
reduction-intercalation-deposition scheme, Figure 1-i.

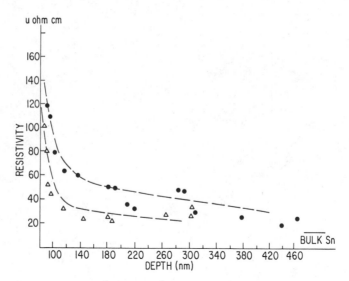

Figure 8. Resistivity (μohm·cm) of various depths of Sn on PIm.
Circles and triangles are the same as in Figure 2.

                B.   Type 2:   Reduction of Metal Cations by Reduced,
                          Intercalated Polyimide

    Rather than using PIm as an oxidant to deposit metal films
from anionic metal clusters, it has also been possible to use PIm
as a reductant to deposit metal films from metal cations in solu-

tion. This is conceptually the reverse of the Zintl anion oxidations and again depends on the ability of the PIm to form stable reduction products and on the alkali metal cation permeability of PIm.

In this scheme, illustrated schematically in Figure 1:iii and iv, the polyimide is first reduced and intercalated <u>without</u> surface metal deposition and then this material is used to reduce metal cations from solution while simultaneously oxidizing and deintercalating the solid.

Treatment of PIm films with methanol (MeOH) solutions of the heteropolyatomic Zintl phase K_4SnTe_4[18] at 25°C rapidly colors the PIm green. The green color is due to the formation of the K^+ salt of the monoanion (1-) oxidation state of the PIm. X-ray fluorescence analysis shows the presence of K but no Sn or Te. Since there is no surface metal deposition this implies the $SnTe_4^{4-}$ is

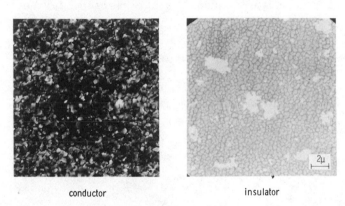

conductor   insulator

Figure 9. SEM of insulating and conducting Sn on polyimide. The conductor has overlapping grains of metal while the insulator has metal grains surrounded by polymer.

oxidized to a soluble species ($SnTe_4^{2-}$?) and remains in solution. Furthermore, unlike the more strongly reducing Zintl anions such as Sn_9^{4-}, the $SnTe_4^{4-}$ anion appears incapable of reducing the PIm to the purple dianion state under the conditions used. Presumably there are other reductants, with properly chosen reduction potentials and countercations, which will be able to perform this reduction/intercalation. If too strong a reductant were used, the purple dianion would be formed which often appears to be accompanied by cleavage of the polymer.

The green, reduced/intercalated PIm acts as a reductant toward metal cations in solution. When immersed in solutions of metal halides or nitrates, usually in DMF solution, the green color fades as metal is deposited on the PIm surface. The green $K_x \cdot PIm$ reduces Co^{2+} to Co and could therefore reduce metals with more positive reduction potentials (> 0.28 V vs SHE).

The metal films deposited by this method show varied properties depending on the element and amounts deposited. For example, reaction of $K_x \cdot PIm$ with Pt^{2+} or Pd^{2+} in DMF rapidly gives uniform, highly reflective films with resistivities approaching that of the bulk metal. In contrast, Ag^+ ions, noted for their high mobility in solids, give films with resistances several orders of magnitude higher than the Pd films containing similar amounts of metal. Apparently, the Ag^+ ions can diffuse into the solid at a rate roughly comparable to the diffusion of the K^+ and electrons to the surface of the polymer. The polymer is presumably partially metallized in the bulk solid. The bulk versus surface metallization is illustrated schematically in Figure 10. As expected, when the metal film becomes thick enough, the rate of diffusion of products and reactants through the film impedes further deposition.

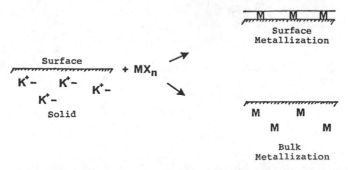

Figure 10. Comparison of reduction of cations on the surface of the polymer to reduction in the polymer bulk.

C. Type 3: Formation of Bimetallic Films with the Polymer Acting Sequentially as an Oxidant then Reductant

It has been possible to electrolessly deposit two metals on the surface of a polymer by using the polymer first as an oxidant to deposit a metal from a Zintl anion, then using this reduced/intercalated metallized polymer as a reductant toward a metal cation in solution. This is a combination of the Type 1 and Type 2 metallization discussed above and is illustrated schematically in Figure 1:i and ii. Using polyacetylene, rather than polyimide as an example, it is observed that K_4Sn_9 reacts very rapidly with polyacetylene (PAc) to give highly conductive Sn coated, reduced polyacetylene, $Sn[K_x \cdot PAc]$. The $Sn[K_x \cdot PAc]$ can then act as a reducing agent and deposit metal cations from solution giving bimetallic surfaces $M \cdot Sn[PAc]$. The reaction of $K_x \cdot PAc$ and $Pd(NO_3)_2$ in DMF gives Pd deposits on the elemental Sn,

$$Sn_xPd_y(CH)_n$$
Palladium Rich Region

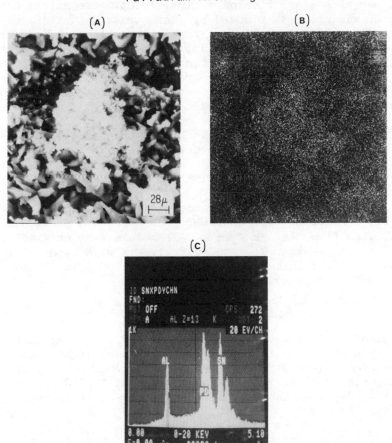

Figure 11. Deposition of an island of Pd granules on a background of Sn needles. (A) SEM, (B) Pd x-ray emission map, and (C) x-ray emission from $Pd \cdot Sn[PAC]$ (Al is from sample holder).

Pd·Sn[PAc]. This product is shown in Figure 11 where the morphology of the two deposited metals is quite different with Pd granules deposited on needles of Sn. As contrasted in Figure 7, it is suggested that the cation reduction-solid oxidation/deintercalation (7b) proceeds in a manner that is a reverse of the Zintl anion oxidation-solid reduction/intercalation scheme (7a).

SUMMARY AND CONCLUSIONS

The techniques described here are general and should allow the preparation of a large number of novel surface-modified materials. Several methods for the chemical metallization of certain reducible organic polymers, particularly polyimide, are described. These methods depend on the very strong reducing power of the Zintl anions. Main group films can be prepared easily by treating the polymer with a solution of a Zintl anion. The main group deposition was shown to proceed by a substrate reduction/intercalation-deposition mechanism. Transition metal films can be prepared by the reaction of the appropriate metal cation with unmetallized, reduced, intercalated polyimide, K_x·PIm. Finally, unique bimetallic surfaces can be prepared by a combination of these methods using the polymer first as an oxidant, then a reductant.

The general reaction types described here also occur with many other substrates besides organic polymers.[16]

EXPERIMENTAL

All manipulations were performed under high purity argon (1 ppm O_2). Solvents were thoroughly degassed and ethylenediamine was distilled from CaH_2 or K_4Sn_9. All metals were obtained from Cerac, Inc., Milwaukee, WI, and all polymers were commercial samples.[8] The polyimide film was heated at 200°C in air for several hours before use.

A. Preparation of Alloys

Most of the alloys were prepared by fusion of the main group element with potassium in quartz ampoules under argon. Less reactive elements (Ge, Si) were alloyed in evacuated, vycor-jacketed, electron-beam-welded Ta capsules at 1000°C for 24 hours followed by rapid quenching. All alloys were finely powdered before use.

B. Metallization of Polyimide with Zintl Anions

Care should be taken that the surface of the polymer is clean, dry and free of surface imperfections. The film is immersed in en or en:toluene (1:1) solutions of the potassium

748

Zintl phase (~ 0.02 $\underline{M}$) until the desired depth of coverage is obtained, followed by a toluene rinse.

C. Preparation of $K_x \cdot PIm$ (green monoanion)

The reduced, K^+ intercalated polyimide, $K_x \cdot PIm$, is prepared by immersion of 5 mil polyimide film into 0.1 - 0.5 $\underline{M}$ solutions of K_4SnTe_4 in MeOH. The immersion is varied from a few seconds up to hours depending on the extent of reaction desired. The inter-calated film is washed with MeOH after reaction.

D. Metallization of $K_x \cdot PIm$ (green monoanion)

The metal films are prepared by immersion of $K_x \cdot PIm$ into concentrated DMF solutions of the metal salts (usually the nitrate or halide). The reaction is continued until the green color of $K_x \cdot PIm$ disappears (very thin films) or until the rate of diffusion through the deposited metal film becomes negligible.

E. Synthesis of Bisimide (6)

Under argon, pyromellitic anhydride (Aldrich, Milwaukee, WI) is dissolved in dry 1,2-dimethoxyethane and two equivalents of dry aniline added. The precipitate is filtered and heated slowly to 250°C under 0.01 Torr pressure. The off-white powder is annealed for several hours at 200°C in air.

ACKNOWLEDGEMENTS

We are grateful to M. Wasielewski for the epr spectra and to B. Tani for assistance with the SEM. We are also grateful to the Division of Materials Science, Office of Basic Energy Sciences, U.S. Department of Energy for financial support under contract W-31-109-Eng-38. Some of the Figures and text in this paper were previously published (Ref. 19) and were reproduced by permission of the publisher.

REFERENCES

1. Abbreviations used: PIm = polyimide, DMF - N, N-dimethyl-formamide, en = ethylenediamine, MeOH = methanol, SHE = standard hydrogen electrode, SEM = scanning electron microscopy.
2. A. M. Wilson, Thin Solid Films, $\underline{83}$ 145 (1981).
3. G. Samuelson, in "Polymer Materials for Electronic Applications", ACS Symp. Ser., No. 184, p. 93, American Chemical Society, Washington, D.C., 1982.
4. Y. K. Lee and J. D. Craig, $\underline{ibid}$, p. 107.
5. A. Saiki, K. Mukai, S. Harada and Y. Mujadera, $\underline{ibid}$, p. 123.

6. (a) A. M. Wilson, D. Laks and S. M. Davis, ibid, p. 139;
    (b) G. A. Brown, ibid, p. 151.
7. For references see K. L. Mittal, J. Vac. Sci. Technol. 13, 19 (1976).
8. In this paper, polyimide (PIm) refers to Dupont's Kapton® polymer which is a copolymer of pyromellitic anhydride and 4,4'-diaminodiphenyl ether.
9. For a review of diverse topochemical reactions, see R R. Schöllhorn, Angew. Chem. Int. Ed. Eng., 19, 983 (1980).
10. H. Schäfer, B. Eisemann and W. Müller, Angew. Chem., 9, 694 (1973).
11. S. B. Pons, D. J. Santure, R. C. Taylor and R. W. Rudolph, Electrochim. Acta, 26, 365 (1981).
12. J. D. Corbett and P. A. Edwards, J. Am. Chem. Soc., 99, 3313 (1977).
13. A. Cisar and J. D. Corbett, Inorg. Chem., 16, 2482 (1977).
14. J. D. Corbett, D. G. Adolphson, D. J. Merryman, P. A. Edwards and F. J. Armatis, J. Am. Chem. Soc., 97, 6267 (1975).
15. R. C. Burns and J. D. Corbett, J. Am. Chem. Soc., 103, 2627 (1981).
16. R. Haushalter, Angew. Chem. Int. Ed. Eng., 22, 558 (1983); R. Haushalter, Angew. Chem. Suppl., 766 (1983).
17. H. Ing and R. Manske J. Chem. Soc., 2348 (1926).
18. R. Teller, L. Krause and R. Haushalter, Inorg. Chem., 22, 1809 (1983).
19. R. Haushalter and L. Krause, Thin Solid Films, 102, 161 (1983).

RELIABILITY OF POLYIMIDE IN SEMICONDUCTOR DEVICES

Gay Samuelson and Steve Lytle

Motorola, Inc.
2200 W. Broadway
Mesa, Arizona   85202

The levels of absorbed water and the degree of polarization are a function of polyimide chemistry. More
rigid structures are less polarizable and less hygroscopic. Three types of absorbed water are identified
from the dehydration conditions required to restore
original C-V characteristics. Finally, the conditions
to achieve optimum cure of polyimide films are discussed.

INTRODUCTION

Polyimide is a class of high temperature stable organic polymer whose properties as a dielectric/passivant are compatible with
the requirements of the semiconductor industry. It was originally
introduced in Japan as an interlevel dielectric for multilevel
metal transistors[1] or alternatively, as a final passivant[2]. More
recently, its interlevel role in silicon and aluminum MOS technology has been demonstrated[3].

The advantageous properties of polyimide include its adher-
ence and planarizing effect[4] and its dielectric performance and
patternability[5].

Some limited reliability data on polyimide films has been described in the literature. For example, the effect of incomplete
cure on device performance was discussed by Gregoritsch[6]. The via
resistance and leakage current of a multilevel bipolar LSI chip was
reported by Mukai et al[7]. Finally, the effects of ionic or polar
contamination and electrical conduction in polyimide was discussed by Brown[8].

The present work describes the correlation between polyimide
chemistry and water absorption monitored by weight gain, dielectric
properties and C-V analysis. The results of C-V analysis can be

interpreted in terms of three forms of water: so-called "free" water, "unassociated molecular" water, and H-bonded water. Dehydrated films also exhibit C-V shifts that suggest that dipoles intrinsic to the polyimide backbone itself cause polarization effects. The relaxation times of such polarization events are strongly a function of polyimide chemistry. Finally, recognizing the role of cure conditions in the determination of maximum film reliability, the conditions to achieve optimum cure are discussed.

EXPERIMENTAL

1. Kinetics of Cure

Adhesion promoter γ-amino propyltriethoxy silane was spin coated on Sb doped (111) Si wafers of resistivity .005-.02Ωcm. Next, polyimide (DuPont PSH61453, PI2562 or PI2555) was spin coated at 5K rpm and baked for various times at temperatures in the range 150-450°C using air convection ovens. Aluminum was evaporated through a metal mask on the resulting film and dissipation factor measurement at 1MHz was made using an HP 4275A LCR meter.

2. Mass Spectroscopic Analysis of Outgassing Products

DuPont PI2562 was applied to ultra clean T092 headers and subjected to 30 min. bakes at temperatures in the range 150-450°C. Metal cans were welded onto the headers in a N_2 atmosphere and the resulting hermetic packages were subjected to 30 min. bakes at the pre-bake temperature plus 50°C. The samples were loaded into the vacuum chamber, punched through in turn and the can contents analyzed at 10^{-5} torr and 150°C by an EAI quadrupole residual gas analyzer capable of resolving mass units in the range 0-150 amu.

3. Water Absorption

Water absorption by DuPont PI2545, PI2555, PI2562, PSH61453, PI2566 and Hitachi PIQ-13 was measured by weight increase as a function of soak time in distilled water at room temperature.

4. Surface I-V Characterization

Silane coated Si wafers were spin coated with DuPont PI2555, PI2566, PSH61453 or Hitachi PIQ-13, fully cured and subjected to evaporation of 4000Å of pure aluminum. Standard lithography and wet etch techniques were used to obtain a dot ring pattern where the dot area was $1.11 \times 10^{-2} cm^2$ and the gap between ring and dot was $2.22 \times 10^{-2} cm$. Following dehydration bake, the time zero surface I-V characteristics were determined using a Kepco 500V bipolar power supply and a Keithley 616 Digital electrometer. Thereafter, the wafers were soaked 1 hour in DI water at room temperature and the measurement of surface I-V characteristics was repeated.

752

5. C-V Characterization

MIS structures consisting of 8-12Ωcm, phosphorus doped Si, 100Å of thermally grown MOS quality SiO_2, 0.8-4.8μm of polyimide (DuPont PI2545, PI2555, PI2562, PSH61453 and Hitachi PIQ-13) and 4000Å of pure evaporated aluminum were subjected to CV analysis using the HP4275A LCR meter and a Fluke 8502A multimeter. A "standard" room temperature trace was performed at 100KHz from -100V to +100V using a ramp rate of 1V/sec. The MIS dot structure resting on a Temptronic TP 350A thermally controlled chuck was stressed at -5V, 200°C for 20 min. and cooled to room temperature under bias. The bias was removed and the trace under "standard" conditions was repeated. The flat band voltage shift, ΔV_{FB} was measured and compared with $\Delta V_{saturation}$ calculated from classic polarization theory.

6. Humidity Effects on Polarization

C-V analysis was performed on MIS structures which consisted of 4-7Ωcm p-type Si substrates, 1000Å of thermally grown MOS qual- ity SiO_2, 1000Å of CVD Si_3N_4, 1.5μm of PIQ-13 and 1.5μm of sputter- ed Al-Cu-Si. A MDC computerized semiconductor measurement system consisted of a Boonton 72B capacitance meter, an HP7010B XY record- er, current monitors, and power supplies supported by a 8080S microcomputer system. Standard unstressed C-V traces were made at room temperature over a voltage range of -75 to +25V at 1 MHz. Stressing was done at 225°C under a +5 volt bias for times ranging from 2 minutes to 20 minutes. The MIS structure was brought to room temperature under bias and the ΔV_{FB} was measured as a function of stress time. The time constant of polarization, τ, was calcu- lated from the slope of the $\ln \Delta V_{FB}$ versus time curves for MIS structures exposed to various humidity ambients. 100% humidity ambient was achieved by placing wafers in an air tight wafer box with water. The water at no time was in contact with the polyimide. All other relative humidities were obtained by exposing wafers to the clean room ambient. Typically, control wafers were stored at 200°C in dry N_2.

RESULTS & DISCUSSION

1. Kinetics of Cure

At least three events occur during the cure of a condensation type polyimide: these include covalent bond formation on ring closure of the polyamic acid to yield a polyimide; concomitant release of water of imidization; and release of solvent and water, components of the starting polyamic acid solution. One criterion to follow these events of cure that has been reported is dissipa- tion factor, D[6].

If D at 1 MHz is measured for a polyimide film baked for

varying times at specific temperatures a family of curves of D versus time results as shown in Figure 1 for DuPont PI2562. If, further, the minimum D value is used as a criterion of optimum cure, it is evident that the critical time and temperature for curing a 1.6μm thick PI2562 film is ∿20 min at 350–400°C. This result has been obtained for three different condensation type polyimides and reflects the need for high temperatures and relatively short times to effect optimum cure. The identical result is obtained for a 5 mil thick PI2562 film suggesting that D is not a sensitive indicator of diffusion controlled events such as solvent and water release. This hypothesis is supported by the results of mass spectroscopic analysis of outgassing products from PI2562 cured in hermetic packages containing N_2. Considerable outgassing of CO_2 believed to derive from contaminant isoimide or monomer acid anhydride[9] and H_2O occurs in the range 350–450°C. This is evident in Figure 2 where the mass spectroscopic peak heights of NMP^{99}, CO_2^{44} and H_2O^{18} relative to N_2^{29} are presented as a function of temperature for PI2562. D(1MHz) is apparently insensitive to such outgassing events since D_{min} occurs in the range 350–400°C and remains invariant throughout the temperature range to 450°C. What, then, does D measure? A possible explanation is that D reflects ring closure and, in fact, more sensitively than does IR absorption at any of the typical imide absorption bands (for example $1380cm^{-1}$) since no further increase in IR absorption occurs beyond 300°C[10]. D probably also reflects the presence of solvent and water in the early stages of cure when both are present in high concentration. Thus, from D measurements it appears that conventional condensation type polyimides require high temperatures to achieve complete cure. Conditions such as 2–20 min at 350–400°C are typical for the polyimides PI2562, PI2555, and PSH61453 investigated. However, an even more sensitive indicator of cure events is the mass spectroscopic identification of outgassing products. This latter technique confirms the high temperature requirement and indicates even higher temperatures such as 450°C may be necessary to effect complete removal of H–bonded water. One other consideration for optimized cure is the requirement for a temperature ramp to prevent excessive outgassing and consequent blow holes.

Having achieved complete cure of a condensation type polyimide, it is then possible to address two potential reliability issues of polyimide films, namely, water absorption and polarization effects on application of an external electric field.

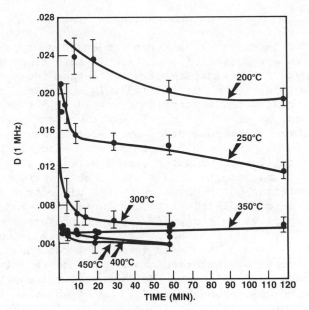

Figure 1. Dissipation factor (D, 1MHz) versus time (min) for DuPont PI2562, thickness 1.6μm.

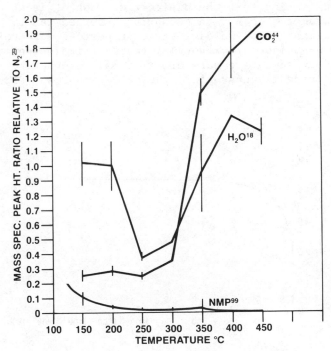

Figure 2. Mass spectroscopic peak height ratio for NMP^{99}, CO_2^{44} and H_2O^{18} versus temperature for PI2562.

2. Water Absorption/Permeation

The weight increase of numerous condensation type polyimides was measured as a function of soak time in room temperature water. A typical curve of the kinetics of water uptake for a particularly hygroscopic polyimide, DuPont PSH61453, is shown in Figure 3. The data indicate that water absorption is rapid with half of the maximum possible uptake of 4.2% occurring in the first several minutes. The water uptake by numerous other polyimide films spin coated to a thickness of 1-3μm could not be measured reproducibly by this technique because the levels of water uptake were too low.

An attempt was made to find a more sensitive test of absorbed water by correlating water uptake with the dielectric properties (dissipation factor and dielectric constant) for DuPont PSH61453. As shown in Figure 4, no change in D beyond the error of measurement was seen for 0.3-0.6 mg absorbed water, levels consistent with water saturated films of standard interlevel thickness, 1.2μm. However, with continued water absorption by the relatively thick PSH film, D increased 41% above the time zero value after a 24 hour soak period corresponding to 1.65 mg of absorbed water. This result points out the relative insensitivity of D (1MHz) to low levels of water absorption consistent with the observations made during the determination of cure kinetics. Furthermore, throughout the absorbed water range studied, no change in capacitance at 1MHz was noted suggesting that the dielectric constant is an even less sensitive parameter than D in measuring absorbed water.

While weight gain and dielectric properties were relatively insensitive parameters, surface I-V characteristics, on the other hand, were very sensitive parameters of moisture uptake. Figure 5 shows the surface I-V characteristic of PSH61453 and PI2566 films

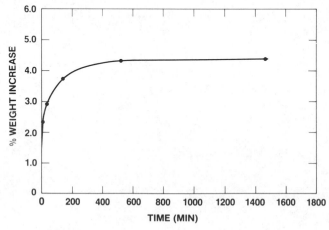

Figure 3. % weight increase versus soak time for PSH 61453, thickness = 3.6μm.

before and after 1 hour room temperature water soak. Of the two, only the PSH61453 film showed significantly increased surface leakage current following the soak period. Furthermore, the film showed a tendency towards dielectric breakdown at a surface field $>0.8 \times 10^4 V/cm$. A comparison of the surface leakage current at a surface field of $0.45 \times 10^4 V/cm$ is presented in Table I for four different polyimides.

Table I. Surface Leakage Current Following 1 Hour Room Temperature Water Soak.

Polyimide	I(amps) @ $0.45 \times 10^4 V/cm$
PSH61453	1×10^{-8}
PI2555	1×10^{-12}
PIQ-13	1×10^{-12}
PI2566	1×10^{-13}

It is obvious that considerable difference in leakage current exists between a highly hygroscopic film like PSH61453 and a relatively non-hygroscopic film such as PI2566. While little difference was evident between PIQ-13 and PI2555 films at a surface field of $0.45 \times 10^4 V/cm$, the surface leakage current for PIQ-13 continued to rise to 10^{-4} amps at $2.25 \times 10^4 V/cm$ whereas PI2555 surface leakage was maintained at $\sim 10^{-12}$ amps for a surface field of $2.25 \times 10^4 V/cm$.

At least for the films exhibiting the extremes of surface leakage following water soak, it was observed that the more hygroscopic film types were also the more readily polarizable and in fact, water does contribute to polarization effects as will be discussed.

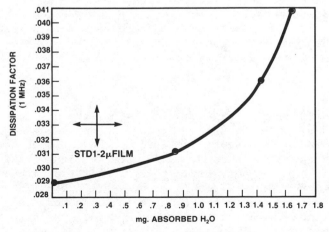

Figure 4. Dissipation factor (D, 1MHz) versus mg. absorbed H_2O for PSH61453.

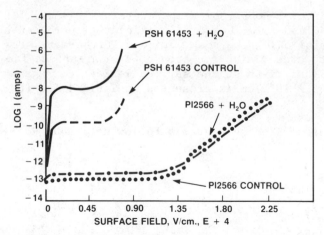

Figure 5. Surface I-V characteristics for PSH61453 and PI2566 with and without absorbed H_2O.

3. Polarization Effects versus Polyimide Chemistry

It has been noted by Brown[8] that charge effects in polyimide can cause inversion of underlying silicon and consequently undesired communication between doped areas. An effort was made to distinguish among the various commerically available polyimides with respect to the degree of polarizability at -5V and 200°C stress. Consequently, MIS (metal-polyimide-SiO_2-Si) structures were fabricated and stressed for 20 min. The C-V profile of the stressed structure was taken from -100 to +100V at a ramp rate of 1.0V/sec and the ΔV_{FB} was measured relative to V_{FB} of the non-stressed C-V profile. Table II presents ΔV_{FB} measured for each polyimide studied. The results can be contrasted with the $\Delta V_{saturation}$ calculated from classic polarization theory[11].

$$\Delta V_{sat} = - (k_o x_p / k_p x_o) V_p$$

V_p is the polarizing voltage, k_o and k_p are the dielectric constants of oxide and polyimide respectively and x_o and x_p are the thicknesses of oxide and polyimide respectively. The ratio of the ΔV_{FB} measured and the calculated ΔV_{sat} is a reflection of the relative polarizability of the polyimides tested as is the parameter ΔV_{FB} measured/μm. From Table II it is clear that PI2566, in addition to being the least hygroscopic polyimide investigated, is also the least polarizable. DuPont PI2545 and PSH61453 are both rapid polarizers and the other polyimides fall intermediate between these two extremes.

Table II. Measured and Calculated ΔV_{FB} For Various Polyimides.

Polyimide	Thickness (μm)	ΔV_{FB} Meas. (Volts)	ΔV_{sat} (Volts)	ΔV_{FB} Meas /ΔV_{sat}	ΔV_{FB} Meas/μm (Volts/μm)
PL2566	4.76	4	216	.018	0.840
PI2562	0.79	18	35.9	.501	22.8
PIQ-13	1.15	34	52.3	.650	29.6
PI2555 40% PSH	1.12	40	50.9	.794	36.1
61453	1.95	78	88.6	.880	40.0
PI2545	1.68	80	76.4	1.00	47.6

An attempt was made to measure the time constant of polarization. Figure 6 shows the C-V trace under -5V, 200°C stress for
times varying from 0-15 min for PSH61453. The undulating C-V profiles relfect the fact that complicated polarization events are
proceeding concurrently in this rapidly polarizing material and it
is thus difficult to measure a single meaningful time constant.
Figure 7 shows the comparable family of C-V traces for PI2566 also
stressed at -5V and 200°C though now for 15-40 min. The family of
curves exhibits uniform, conventional C-V behavior. A plot of
$\ln(\frac{k_o x_p}{k_p x_o} V_p - \Delta V)$ versus time as shown in Figure 8 gives a straight

line with correlation coefficient 0.999, a result consistent with
a single polarization event. The time constant τ was determined
from the slope of the curve in Figure 8 and the equation:

$$\Delta V = \frac{k_o x_p}{k_p x_o} V_p (I-e^{-t/\tau}) \quad (4)$$

The calculated τ for PI2566 was 500 min. The activation energy calculated from the Arrhenius plot of Figure 9 for the polarization
induced by the aforementioned stress conditions is 0.64 ev, a value
which is consistent with electronic as opposed to mobile ion effects.
Since polarization like behavior has been observed for virtually all polyimides investigated, it is tempting to speculate that
there is a component of polarizability inherent in the polyimide
backbone, for example, in the carbonyl group. It is easily argued
that the degree of polarizability is a function of polyimide structure. More rigid structures would impede polarization and thus τ
would be large. On the other hand, more flexible polymer chains
would allow more facile polarization and hence, smaller τ values.
Additionally, the polarization is not symmetric; negative polarizing voltages drive the C-V characteristic further positive than do
the comparable positive polarizing voltages in the negative direction. This asymmetry is reflected in the different activation energies for shifts induced by V_p's of opposite sign, i.e., 0.64 ev for
V_p = -5 and 1.32 ev for V_p = +5V for PI2566 films.

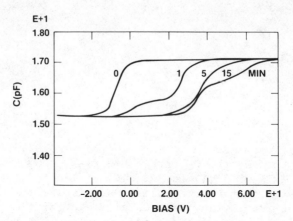

Figure 6. C–V characteristics for PSH61453 stressed at 200°C, –5V for 0, 1, 5, and 15 minutes.

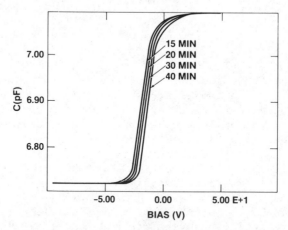

Figure 7. C–V characteristics for PI2566 stressed at 200°C, –5V for 15, 20, 30 and 40 minutes.

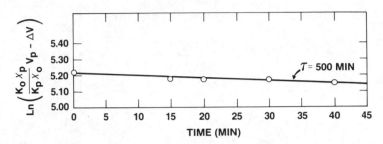

Figure 8.   LnΔV versus time for PI2566.

4.   Polarization Effects Due to Absorbed Water

As a result of the apparent correlation of water uptake and polarizability for PSH61453 and PI2566, an attempt was made to study the specific contribution of water to polarization effects in polyimide films. Consequently, MIS structures using PIQ or PI2566 were "soaked" for varying times in either a clean room environment where the relative humidity varied from 17 to 35% or a 100% relative humidity environment. The structures were stressed at -5V and 225°C following humidity exposure and C-V profiles were measured as a function of stress time. Figure 10 is an example of the results for an MIS structure using PIQ and SiO_2 where the stress times were 0-16 min. The figure illustrates the typical finding that initially a rapid shift in V_{FB} occurs which correlates directly with the relative humidity of the soak period. A term ΔV_{H_2O} is defined to be the ΔV_{FB} that arises between the zero stress C-V profile for a dehydrated film (the dotted line in Figure 10) and the zero stress C-V profile for a water "soaked" film. This ΔV_{H_2O} occurs almost instantaneously upon temperature and bias stressing and disappears following a 200°C, 30 min. dehydration bake in dry N_2. The inter- pretation of such a rapid water induced shift is that it is "free" water trapped in pores and voids in the microstructure of the poly- imide. In such a form, the water could polarize instantaneously and would be expelled as water vapor. After ΔV_{H_2O} occurs, a family of C-V curves results with continued stressing from which ΔV_{FB} values can be measured and used to calculate a time constant of polarization, τ. If such a time constant is measured as a function

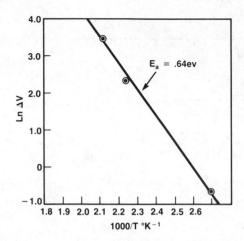

Figure 9. LnΔV versus 1/T for PI2566.

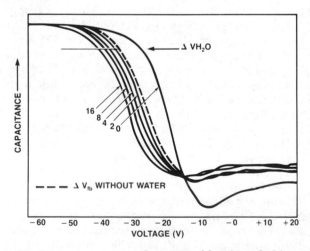

Figure 10. C-V characteristic for PIQ-13-SiO$_2$ following 22 hours exposure to 17% RH and stressing for 0-16 min at 225°C and +5V.

of exposure time in a 100% R.H. ambient, τ is seen to increase as
shown in Figure 11 for PIQ MIS structures. Samples which exhibited
increasing τ with exposure time at 100% RH did not revert to the
original τ values following 200°C dehydration bake in dry N_2 atmosphere for as long as 72 hours. They did, however revert to original
τ values following 450°C dehydration bake in dry N_2 for 30 min.
Such high temperature dehydration conditions are consistent with
the release of H bonded water. Thus, the interpretation of increasing τ with exposure to sufficiently high humidity ambients is the
increasing constraints that H bonded H_2O molecules impose on polarization of H_2O molecules and/or the carbonyl of the polyimide backbone.

If similar MIS structures of PIQ are exposed to a clean room
ambient where the relative humidity varied from 17-35%, the τ
measured after the initial ΔV_{H_2O} shift exhibited the behavior
illustrated in Figure 12. At 17% RH, τ decreases and such samples
showed complete reversibility of τ following a 200°C, 30 min dehydration bake in dry N_2. This behavior leads to a definition of yet
a third type of water, so-called "unassociated molecular" water.
However, it would still be influenced by the surrounding polymer
chains. The decrease in dielectric relaxation time would be a
result of the addition of component relaxation times as follows:

$$\frac{1}{\tau_{tot}} = \frac{V_{H_2O}}{\tau_{H_2O}} + \frac{V_{PI}}{\tau_{PI}}$$

where τ_{tot} is the total, effective relaxation time, τ_{H_2O} is the
relaxation of the molecular water, and τ_{PI} is the relaxation time
of the dry polyimide. V_{H_2O} and V_{PI} are the volume fractions of H_2O
and polyimide respectively. It is reasonable to assume that the
water dipoles would polarize more quickly than the backbone polyimide dipoles. Thus, as ΔV_{H_2O} increases, τ_{tot} would decrease.
Again with reference to Figure 12, as relative humidity increases,
τ increases presumably in response to the effects of H bonding as
just discussed.

Thus, the polarization results are consistent with the existence of three types of water: free water which gives rise to ΔV_{H_2O}
and which is readily removed by low temperature (200°C) dehydration
bake; unassociated molecular water which gives rise to a τ that
reverts to pre-soak τ values following low temperature dehydration
bake; and finally, H bonded water that gives rise to a τ which
reverts to pre-soak values only on high temperature (450°C) dehydration bake. While free water is absorbed throughout the range 17-
100% RH, the occurrence of molecular or H bonded water is dependent
on relative humidity. The preponderance of absorbed water is H-
bonded when the ambient humidity is $\geq$ 35%.

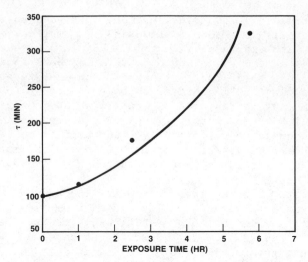

Figure 11. Polarization time constant τ versus exposure time at 100% RH for PIQ-13-SiO$_2$.

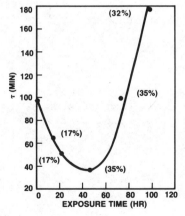

Figure 12. Polarization time constant τ versus exposure time in a clean room ambient for PIQ-13-SiO$_2$.

SUMMARY

Two major determinants of the reliability of polyimide in its use as an interlayer dielectric and passivant in integrated circuits are electric polarizability and water absorption. Polarization behavior was investigated using standard CV techniques. Data generally exhibited multiple time dependent polarization events, as well as anisotropic polarization for positive and negative biases in most polyimides tested. However, ΔV_{sat} and τ measurements showed large differences between different types of polyimides. It is concluded that degree of polarizability is highly dependent upon polyimide structure; more rigid structures with side chains and crosslinks impede polarization and the consequent result is a larger τ value.

Water absorption was determined in several different ways. While dissipation factor measurements were shown to be less sensitive to water uptake, surface I-V characteristics were very sensitive to water absorption. In conjunction with water weight gain measurements, it was shown that highly hygroscopic polyimide films also exhibit larger surface leakage currents at constant voltages. It is interesting to note that polyimide film hygroscopicity correlates well with electric polarizability, indicating that water-absorption may be highly structure dependent as well. Standard CV measurements for polyimide exposed to ambients of differing humidities provided evidence for structural models of water in polyimide. These include free water trapped in micropores which exhibits instantaneous polarization and large voltage shifts, molecular water which is in the polyimide causing an increase in τ with water absorption, and molecular water "unassociated" with any specific dipoles in the polyimide which is manifested by a decrease in τ with exposure time.

Additionally, the kinetics of cure was studied using dissipation factor measurement and mass spectroscopy of the outgassing products for various times and temperatures. Dissipation factor, although sensitive to imidization in films, was not sensitive to diffusion controlled events such as outgassing. Mass spectroscopy indicated that a high temperature requirement of 400-450°C is necessary for complete cure.

ACKNOWLEDGEMENTS

The careful technical assistance of Kathleen Ginn and Dan McGuire is gratefully acknowledged. Special thanks to Michele Soltero for typing the manuscript.

REFERENCES

1. K. Sato, S. Harada, A. Saiki, T. Kimura, T. Okubo, K. Mukai, IEEE Trans. On Parts, Hybrids & Packaging,PHP-9,No.3,175(1973).
2. S. Miller, Circuits Manufacturing, p.39 April (1977).
3. R. Larsen, IBM J. Res. Develop. 24, No. 3, 268 (1980).
4. L. Rothman, J. Electrochem. Soc., 127, 2216 (1980).
5. G. Samuelson, Organic Coatings & Plastics Preprints, 43, No. 2, 446 (1982).
6. A. Gregoritsch, Rel. Physics, 14th Annual Proceedings, p.228 (1976).
7. K. Mukai, A. Saiki, K. Yamanaka, S. Harada and S. Shoji, IEEE J. Solid State Circuits, SC-13, No. 4, 462 (1978).
8. G. Brown, Rel. Physics, 19th Annual Proceedings, p.282 (1981).
9. J. Craig, (1982) personal communication.
10. G. Samuelson and K. Ginn, Electrochemical Society Extended Abstracts, No. 241, p. 578, Denver, (1981).
11. E.H. Snow and M.E. Dumesnil, J. Appl. Phys. 37, 2123 (1966).

POLYIMIDE PLANARIZATION IN INTEGRATED CIRCUITS

D. R. Day, D. Ridley, J. Mario, and S. D. Senturia

Department of Electrical Engineering and Computer Science
and Center for Materials Science and Engineering
Massachusetts Institute of Technology
Cambridge, Massachusetts 02139

Among the many properties that make polyimides (PI)
attractive as inter-level insulators in integrated
circuits is their ability to planarize topographical
features. This paper examines the role of many of the
factors that affect planarization, such as the PI type,
percent solids, molecular weight, film thickness, and
number of layers. A study of the rate of solvent loss and
reflow characteristics of PI films during drying revealed
that a critical point is reached where flow stops and
rate of solvent evaporation decreases. A model has been
developed which assumes that surface tension in combination with flow keeps the film planar until this critical
point, after which the film shrinks due to solvent loss
and can no longer flow, leading to a loss of the perfectly planar surface. This model is shown to account for
the major effects observed. The model implies that for a
given step height and film thickness, the degree of
planarization depends primarily on the percent solids at
the critical point. This, in turn, is a function of PI
chemistry and of molecular weight, but is independent of
the percent solids of the initial solution, as observed.
The model also serves as a guide for achieving improved
planarization with multiple layers.

INTRODUCTION

A major advantage of using polyimides (PI) as an inter-level insulator in integrated circuits is their ability to planarize underlying topographical features. Ideal planarization would yield a perfectly flat surface, regardless of the complexity of the underlying topography, but this is not achieved in practice. This paper addresses many of the features that affect planarization, including PI type, percent solids, molecular weight, film thickness, and number of layers.

Following Rothman[1], the degree of planarization (DOP) is defined as the fractional reduction in step height achieved by a coating of PI. Figure 1 illustrates the relevant variables, and is drawn for the case in which the coating is thicker than the step height to be covered. Just after spin coating, and during that portion of the drying where surface tension plus flow keeps the film surface flat, the condition illustrated in Figure 1a holds. The total initial film thickness is t_{pi}, and h_i is the height of the step to be covered (equal in this illustration to the thickness of an assumed metal feature t_m). As long as the film is flat, the thickness of the film over the metal feature, t'_{pi} takes on a value such that

$$t_{pi} = t'_{pi} + t_m \qquad (1)$$

After drying (and subsequent imidization), the condition illus-

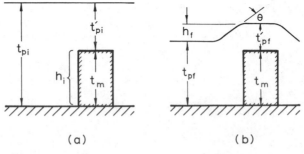

(a)   (b)

Figure 1. Schematic diagram of a) flat film before the critical point and b) non-planar film after shrinkage.

trated in Figure 1b holds. Flow ceases before the film is completely dry, and film shrinkage during drying leads to a nonplanar surface. The degree of planarization (DOP) is defined as

$$DOP = 1 - h_f/h_i \qquad (2)$$

where h_f is the final step height. From Figure 1b, it is seen that

$$DOP = 1 - (t'_{pf} + t_m - t_{pf})/t_m \qquad (3)$$

where t_{pf} and t'_{pf} are the final film thicknesses corresponding to the initial thicknesses t_{pi} and t'_{pi}.

Equation (2) defines how planarization is measured experimentally, and Equation (3) relates it to the geometrical properties of the film. Further insight is gained if one recognizes that since film shrinkage is responsible for the difference between Figures 1a and 1b, then

$$t_{pf} = t_{pi}(1-R) \qquad (4)$$

and

$$t'_{pf} = t'_{pi}(1-R) \qquad (5)$$

where R is the fractional reduction in thickness due to shrinkage. Substitution of Equations (4) and (5) into Equation (3), and making use of the equality expressed in Equation (1), we find that

$$DOP = 1-R \qquad (6)$$

This says that loss of perfect planarization arises from whatever shrinkage occurs after the film can no longer flow quickly enough to keep up with the drying (and imidization) process, an observation that is important to the intepretation of results presented in this paper.

Figure 1b also illustrates a second variable used to characterize planarization, the maximum slope θ on the shoulder covering the step. Measurements of this quantity as a function of film thickness yield useful insights into potential failure mechanisms when attempting to cover relatively large steps with PI.

EXPERIMENTAL

A series of polyimide resins were used, including several of DuPont's commercial grade materials (PI-2545, 2555, and 2562) and several samples fabricated by DuPont as higher or lower molecular weight versions of these resins. NMP was the solvent for all solutions supplied by DuPont, and was also the diluent used when thinning solutions.

Substrates used for the experiments consisted of patterned aluminum lines on oxidized 2" silicon wafers. Aluminum was evaporated either by electron-beam or filament evaporation, with thicknesses ranging from 0.5 to 1.5 µm, and was patterned with conventional photolithography. The line widths varied from 2.5 to 25 µm at spacings of 50 µm.

Films were spun on using a conventional photoresist spinner. Following spinning, films were dried at 90°C in air, a temperature low enough to permit solvent removal without significant imidization having taken place. This permitted us to separate the effects of shrinkage due to solvent loss from subsequent shrinkage during imidization. Films were then cured at 200°C for 1 hour to imidize, followed by a final bake at between 350 and 400°C in air for 30 minutes. Since dilution and spin speed both affect film thickness, a set of thickness characterization measurements was made for each PI tested. Figure 2 illustrates the variation of final film thickness with spin speed and dilution for PI-2545.

Some films were dried at room temperature instead of 90°C to facilitate monitoring of weight loss due to solvent evaporation (with a standard Mettler balance), and to permit examination of flow characteristics. No differences in planarization characteristics were observed between films dried at the two temperatures.

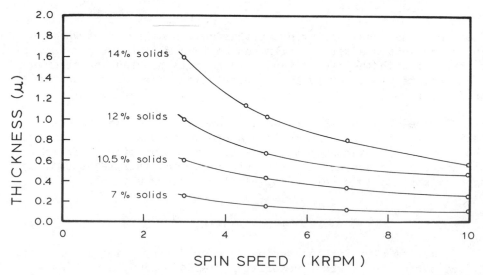

Figure 2. Cured film thickness versus spin speed for various dilutions of PI-2545

770

Measurements of step heights before and after application of the PI films were made with a Dektak surface profilometer, using a tip with a 12 μm radius. These data were used in Equation (2) to determine DOP. Since surface profile measurements were not accurate enough to permit direct measurement of the slope, a photoresist mask was applied, the PI was selectively etched off in an oxygen plasma, and the slope at the edge of the masked PI was measured from SEM photographs.

RESULTS AND DISCUSSION

DOP of Dried Films

This section reports the effects of the drying process on planarization. The first variable examined was the weight of freshly spun films as they dried in air at room temperature. Figure 3 shows the data expressed in terms of percent solids, S, versus time for the three commercial PI's studied. Note that at time zero in the Figure, the percent solids is 5-15% higher than that of the initial solution, reflecting solvent loss during the spinning process. The arrows in Figure 3 indicate the "critical point" in the drying process, as determined from the loss of reflow following a

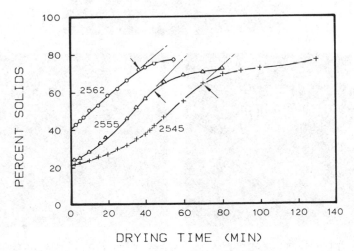

Figure 3. Percent solids increase during drying at room temperature (arrows indicate critical point as determined from reflow expts).

scratch (see following paragraph). The percent solids at the critical point, S_c, is in the range 60-75%, and is listed for each PI in Table I. Following the critical point, the rate of solvent loss decreases markedly, until the film is finally dried, at which point we assume for modelling purposes that the percent solids is 100%, even though some residual solvent may be present. Once the film is dried, both thickness and planarization measurements can be made. These results are presented after the following discussion of reflow properties.

In parallel with the solvent loss study described above, similarly prepared wafers were scratched with a fine needle at various times in the drying process, and subsequently examined in a microscope to determine the degree of flow back into the scratches. Figure 4 shows a sequence of optical micrographs taken of a dried PI-2555 film which was scratched at various times near the critical point. It is seen that the 30 and 35 min scratches fill in completely, the 40 min scratches show little flow, and by 50 min, the film was not even penetrated. The transition from reflow to no

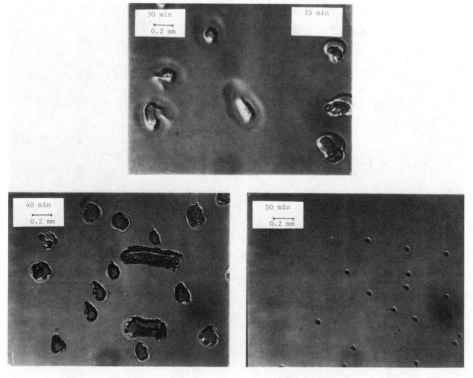

Figure 4. Optical micrographs of dried PI-2555 films scratched at various times during drying (time values indicate time film was scratched after spin deposition).

reflow is quite abrupt, occuring within a space of about two minutes at room temperature, and it appears that after this critical point, flow ceases, and the loss of planarization due to shrinkage by further solvent loss begins.

The DOP due to drying alone was measured for each commercial PI for approximately 2 μm films over features of height 0.8 μm. The results are presented as the "observed" column in Table I. The calculated DOP values were obtained from a simple model, explained below, which is based on shrinkage following the critical point.

Table I. DOP Results for Non-imidized Films Dried at Room Temperature.

PI Type	S_i initial	S_c critical	d_i solution	d_f dried	DOP observed	calculated
PI-2555	.18	.60	1.05	1.41	.48	.52
PI-2545	.14	.64	1.04	1.47	.49	.56
PI-2562	.27	.74	1.07	1.53	.69	.66

The model with which we calculated the DOP in Table I presumes that the film remains perfectly planar until the critical point, after which uniform shrinkage occurs. In general,

$$w_p = w_t S \qquad (7)$$

where w_p is the weight of the polymer in the film, w_t is the total weight of the film including solvent, and S is the percent solids (expressed as a fraction). In terms of the density of the film d and its area A, this becomes

$$w_p = d t_p S A \qquad (8)$$

where t_p is the PI thickness. Using the subscript c to denote "the critical point", and f to denote "final", and since the weight of polymer is unchanged before and after drying,

$$d_f t_{pf} S_f = d_c t_{pc} S_c \qquad (9)$$

where the final percent solids S_f is assumed below to be unity (100%). The film shrinkage R is given by

$$R = 1 - t_{pf}/t_{pc} \qquad (10)$$

which, in conjunction with Equation (9) yields

$$DOP = d_c S_c / d_f \tag{11}$$

The final density d_f and the critical percent solids S_c are both measured quantities, but the density at the critical point d_c cannot be readily measured without monitoring the thickness of the drying film. Instead, we have estimated d_c by assuming a linear variation of density with percent solids from the initial solution) value to the final value of 100%. That is, we assume

$$d_c = \left(\frac{S_c - S_i}{1 - S_i}\right)\left(d_f - d_i\right) + d_i \tag{12}$$

where d_i is the density of the solution, and S_i the corresponding percent solids. Substituting Equation (12) into Equation (11) yields for the DOP

$$DOP = S_c \left[\left(\frac{S_c - S_i}{1 - S_i}\right)\left(1 - \frac{d_i}{d_f}\right) + \frac{d_i}{d_f}\right] \tag{13}$$

This Equation was used to obtain the calculated values in Table I. In spite of the many assumptions and experimental variations that occur in measurements of this type, the agreement is within 7%, which is encouraging. Moreover, this model suggests (1) that the DOP should be independent of the percent solids of the initial solution, since until the critical point, the film remains planar, and (2) that the DOP should be independent of step height provided the PI film is thicker than the step. Results addressing these issues are presented in the following section.

DOP of Fully Cured Films

The DOP obtained with fully cured films is the composite effect of two processes, drying and curing. Following drying, the effect of imidization and associated water loss adds to shrinkage, hence reducing DOP below that obtained from drying alone. The amount of shrinkage due to imidization is between 20 and 50%, depending on the material, and in every case, we have found that the measured film shrinkage between dried and cured films matches exactly with the decrease in measured DOP due to the curing step. Thus, DOP results on fully cured films systematically reflect the sequential shrinkage steps.

Rothman has suggested a correlation between percent solids of the inital solution and DOP[1]. In fact, there is an apparent correlation, as illustrated in Figure 5, which combines our data with

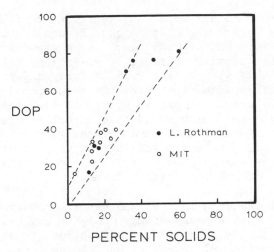

Figure 5. Degree of planarization as a function of percent solids of original solutions.

that of Rothman. However, the correlation is somewhat misleading. For a given PI material, there is an extremely small range over which percent solids can be varied by dilution and still yield a viscosity acceptable for spin casting of films. For example, films of 0.6 μm thickness can be obtained from PI-2545 with no dilution (14% solids) down to 3:1 dilution (10.5% solids), and over this range, we observe no significant variation in DOP. In order to change percent solids substantially, one must modify the system chemistry, for example, by changing molecular weight, as discussed below.

Molecular weight plays a major role in the DOP. For example, DuPont's PI-2545, with 14% solids, normally planarizes 25-30% for a 1.5 μm film over a 1 μm step. A specially prepared high molecular weight version of PI-2545, with only 5% solids at equivalent viscosity, only planarizes 17%. Similarly, the DOP for commercial grade PI-2555, with 18% solids, is 30-35%, while that of a specially prepared lower molecular weight version was 35-40%. One can speculate that since the solution percent solids with which Rothman achieved 70-80% planarizations were in the range 35-60%, the corresponding PI formulations were probably of low molecular weight.

Film thickness is not a primary determinant of DOP for films thicker than the underlying step height. Figure 6 shows DOP obtained for PI-2562 and PI-2545 over 1.5 μm steps as functions of final PI film thickness. Even films of PI-2562 as thick as 5.5 μm still planarize only 55%. This result is fully understandable in terms of the shrinkage model of planarization. The decrease in DOP at lower film thicknesses shown in Figure 6 is expected. Ideally, it would occur at that fully cured film thickness for which the corresponding critical point film thickness is just equal to the step height. In fact, we see a fall off at slightly thicker films, suggesting that wetting of the underlying step by the PI film contributes to loss of DOP for thinner films.

The general features of Figure 6 are also observed for different PI's over a wide range of step heights and film thicknesses. In Figure 7 are plotted DOP data as a function of the ratio t_{pf}/t_m, where t_{pf} is the final PI thickness and t_m the metal thickness for various metal thicknesses. Within the experimental scatter, several trends are visible. First, for $t_{pf}/t_m > 1$, DOP is independent of film thickness, and increases slightly with increasing step height. Second, variations in PI chemistry (for example, between PI-2545 and PI-2562), are more important than any geometrical feature in determining DOP. This is consistent with the idea that percent solids at the critical point and subsequent shrinkage during drying and curing control planarization.

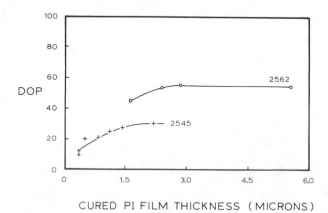

Figure 6. Degree of planarization as a function of film thickness over a 1.5 μm aluminum step.

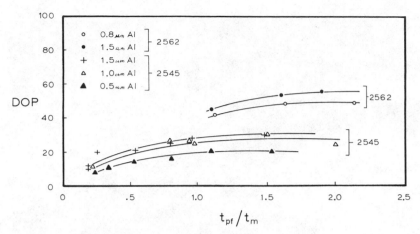

Figure 7. Degree of planarization versus t_{pf}/t_m (final PI film thickness/ metal step height)

Slope Measurements

Measurements of the maximum slope of the cured PI film over an underlying step were made from SEM photographs, several of which are shown in Figure 8. The sequence shows the cross section of PI-2545 over an aluminum step with t_{pf}/t_m ratios varying from 1.95 down to 0.59. As the film gets thinner, the maximum slope increases markedly. Figure 9 shows measured results for both PI-2545 and PI-2562. There is little difference in slope between the two materials.

In Figures 8c and 8d, where the PI thickness is less than the step height, noticeable thinning of the film over the corner can be seen. Such features can be reliability hazards, both in terms of the mechanical integrity of the film, and the results of high electric fields that can occur at the corners of conductors. Where there is a second level of metal above the PI, increased possibility of interlevel breakdown at the step would be expected. The conclusion is that considerable caution should be used in designs where step heights approach the thickness of the PI layer to be used.

DOP with Multiple Layers

DOP can be greatly improved with the use of multiple layers. The step height to be planarized is reduced with each subsequent

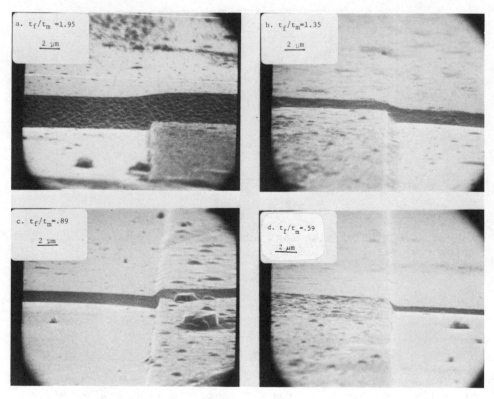

Figure 8. SEM micrographs of various PI film thicknesses on aluminum steps (PI-2545).

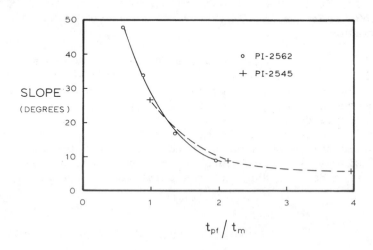

Figure 9. Slope (θ) of PI over step edge. (see Figure 1b for definition of variables)

layer. If layer thicknesses are selected in the range where DOP is independent of thickness, then the DOP achievable after n layers, DOP_n, should be related to the DOP of an individual layer, DOP_1, by

$$DOP_n = 1 - (1-DOP_1)^n \qquad (14)$$

Experimental results agree very well with Equation (14), as shown in Table II.

Table II. DOP With Multiple Identical Layers.

t_{pf}/tm	Number of PI Layers							
	1		2		3		4	
	Obs.	Calc.	Obs.	Calc.	Obs.	Calc.	Obs.	Calc.
2.1	48%	-	74%	73%	88%	86%	-	-
1.6	49%	-	71%	74%	86%	87%	91%	93%
1.2	38%	-	65%	62%	80%	76%	-	-

There is a problem with the use of multiple identical layers to improve DOP, namely, that to achieve high DOP, the total build up of material can become quite large. The total build up to achieve a given DOP can be optimized by taking advantage of the fact that DOP is independent of t_{pf}/h_i, where h_i is the height of the underlying step, for $t_{pf}/h_i>1$. If one keeps t_{pf}/h_i constant for each successive layer, then the overall DOP after n layers will still be given by Equation (14), and the overall t_p/t_m ratio will be

$$\frac{t_p}{t_m} = \sum_{j=1}^{n} \frac{t_{pf}}{h_i}(1 - DOP_1)^{j-1} \qquad (15)$$

Figure 10 shows DOP_n versus t_p/t_m as a function of n and DOP_1 assuming that for each individual layer, t_{pf} is adjusted so that $t_{pf}/h_i=1$. For DOP_1 of 0.2, a DOP_n of 0.8 can be achieved for a t_p/t_m ratio of 4 using 8 layers, while for DOP_1 of 0.5, DOP_n of 0.88 can be achieved for t_p/t_m of only 1.8 and using only three layers.

The validity of Equation (15) was tested both with PI-2545 and PI-2562 keeping t_{pf}/h_i equal to 1 for each layer. The data are shown in Figure 10. The PI-2545 data follow Equation (15) for a DOP value of 0.27, achieving a DOP_n of 0.73 after five layers, while the PI-2562 data correspond to DOP_1 of 0.43, achieving DOP_n of 0.80 after only three layers.

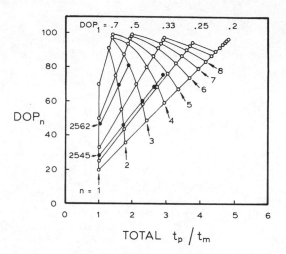

Figure 10. Theoretical DOP data from equation 15 (open circles) and experimental DOP data (closed circles) for multilayer PI film deposition.

SUMMARY

Degree of planarization was shown to depend on molecular weight of the polyamic acid precursor or chemistry and not to depend on percent solids of the starting solution (for a given molecular weight) or film thickness. This behavior can be understood with a simple film shrinkage model where only the percent solids of a drying film at the critical point (solidification point) is important. Utilizing the film shrinkage model, a multiple coat method for obtaining a high degree of planarization with minimum overall film thickness was developed.

780

ACKNOWLEDGMENTS

This work was supported in part by E.I. du Pont de Nemours and Company, which also supplied the polyimide materials. Fabrication of samples and profile measurements were made in the Microelectronics Laboratory of the MIT Center for Materials Science and Engineering, which is supported in part by the National Science Foundation under Contract DMR 78-24185. SEM photographs were made in the Electron Microscopy Laboratory of the Center, which is similarly supported. The authors wish to thank Y. K. Lee, J. Craig, and E. Yuan, all of du Pont, and T. O. Herndon and R. L. Burke of the MIT Lincoln Laboratory, for many helpful discussions.

REFERENCES

1. L. B. Rothman, "Properties of Thin Polyimide Films", J. Electrochem. Soc., <u>127</u>, 2216 (1980).

PLANARIZATION ENHANCEMENT OF POLYIMIDES BY DYNAMIC CURING

AND THE EFFECT OF MULTIPLE COATING

C. C. Chao and W. V. Wang

Hewlett Packard Laboratories
Palo Alto, California   94304

Polyimides possess a number of properties important for multilevel metallization. Successful development of multilayer metal structures depends crucially on
ability of dielectric planarization. A novel dynamic
curing technique has been demonstrated to result in as
much as 150% increase in polyimide planarization. This
technique may possibly be applied to planarization of
other materials with vehicles. Effect of multiple
coating on the resultant degree of planarization has
been studied using a mathematical model fitted with
experimental data.

I. INTRODUCTION

Development of VLSI requires multilevel metallization on both integrated circuits and higher levels of interconnection. Polyimides have been used as intermetal dielectric in muli-layer metal structures because of their thermal stability, chemical resistance, and dielectric properties.[1-2] A most desirable condition for multilevel metallization would be the ability to planarize the intermetal dielectric over an underlying topography in order to prevent reduction of subsequent lithographic resolution.

As shown in Figure 1, the degree of planarization P of a planarizing coating over an underlying step can be defined as:

$$P = 1 - \frac{t_1}{t_0} , \qquad (1)$$

where t_1 is the height of the resultant step in the planarizing surface, and t_0 is that of the underlying step. The fact that melting temperatures of commercial polyimides with high thermal stability are not identifiable below temperatures of decomposition[3] disallows flow after imidization. Moreover, planarization of polyimides is predominantly limited by the volume shrinkage due to outgassing of solvent and H_2O in soft baking and condensation reaction during imidization, respectively. Nonetheless, by increasing flow of the material while maintaining a high solids content, planarization of polyimides can be improved. One approach is to optimize the spin deposition condition. As spin speed increases, the rate of solvent evaporation increases and, therefore, a decreased volume shrinkage would result during cure. Also, as a consequence of the increased spin speed, the film spun would be more conformal to the underlying geometries. A relative maximum degree of planarization for a desired film thickness may be achieved by adjusting spin speed and solids content. Another approach, from the materials' point of view, would be to increase flow prior to imidization by synthesizing materials of low molecular weight, but one would possibly have to sacrifice the desirable characteristics in thermal stability, mechanical integrity, and chemical resistance. However, the purpose of this study is to develop a dynamic curing technique (DCT) to promote further leveling of polyimide processed with any viscosity and spin speed without compromising its performance in physical properties.

II.   EXPERIMENTAL

Polyimides used in this study were obtained in solution form of polyamic acid from Du Pont and Hitachi. Thin films of polyimide with up to 10μm thickness were spun on substrates with metal patterns by adjusting spin speed and dilution ratio. A two step B-stage was applied at 80°C followed by 120°C for 15 minutes each before a 90 minute cure at 310°C. The degree of planarization is determined from measurements of SEM and Alpha Step profilometer according to its definition in Figure 1. The dynamic curing technique is performed by employing low pressure or ultrasonic vibration during polyimide curing. Multiple layers of polyimide may be coated over substrates to increase the degree of planarization, where each layer is fully cured prior to the subsequent coating.

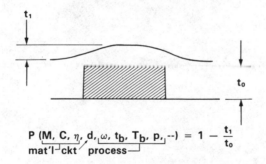

$$P\ (\underbrace{M,\ C,\ \eta,\ d,}_{\text{mat'l—ckt}}\ \underbrace{\omega,\ t_b,\ T_b,\ p,}_{\text{process}}\ \text{--})\ =\ 1\ -\ \frac{t_1}{t_0}$$

WHERE   M = MOL WT OF PLANARIZING MATERIAL
              C = SOLIDS CONTENT OF PLANARIZING MATERIAL
              η = VISCOSITY OF PLANARIZING MATERIAL
              d = CIRCUIT ELEMENT GEOMETRIES
              ω = SPIN SPEED
              t_b = BAKING TIME
              T_b= BAKING TEMPERATURE
              p = PRESSURE

Figure 1.   Definition of degree of planarization and variables in planarization.

III. RESULTS AND DISCUSSION

A. Low Pressure Dynamic Curing Technique (DCT)

It has been reported[4] that the degree of polyimide planarization decreases rapidly as a function of linewidth and spacing of
the underlying geometries. To show the effect of low pressure
dynamic curing technique, metal lines of rather large widths and
spaces were used. Figure 2 shows the planarization of a single
layer of polyimide over a 4.3µm thick 470µm wide metal line with
280µm spacing at its step after complete cure using this technique.
Degrees of planarization of several polyimide materials are listed
in Table I. Under the same spinning and curing conditions, the
improvement of planarization by low pressure DCT is evident.

Figure 2. Planarization of a single layer of polyimide over a
4.3 µm metal step after complete cure.

Low pressure dynamic curing technique employs vacuum treatment to increase the driving force for leveling as well as the
mobility of polymer molecules when flowing. The surface tension
of viscous polyamic acid-polyimide fluid induces flow of molecules resulting in surface leveling. In view of the Gibbs equation of adsorption[5], the surface tension would decrease with an
increased physical adsorption of ambient gas molecules. The initial curvature of polyimide surface immediately after spinning
is determined, in part, by the aspect ratios of the underlying
circuit geometries and the previously defined variables of material viscosity and spin speed. For an optimum condition of
viscosity and spin speed, and a given underlying geometry, planarization can be further improved with an increased surface tension. This can be obtained by the removal of the adsorbed gas
molecules using a low pressure treatment. Another effect of vacuum treatment is the increase in molecular mobility of the polymer as a result of the enhancement of the outgassing rate. This
is due to an increased rate of local free volume change giving
rise to an increased rate of molecular disorder. As this outgassing rate increases, it is believed that a reversed effect
can take place where the rate of viscosity increase for the
polyamic acid-polyimide fluid is higher than that for molecular
rearrangement. Therefore, it is expected that the outgassing
rate or the shrinkage rate of polyimide during cure can be optimized to achieve a relative maximum in planarity.

Table I. Comparison of Degree of Planarization of Polyimide
Films over Metal Lines 470μm Wide with 280μm Spacing, with and
without Low Pressure Dynamic Curing Technique (DCT). Polyimides
A, C, and E are Du Pont PI2560 of various dilutions and B is
PI2555. Polyimide D is PIQ3200 by Hitachi.

Polyimide	Degree of Planarization after Complete Cure		Enhancement in Degree of Planarization
	w/o DCT	with DCT	
A	10%	25%	150%
B	12%	22%	83%
C	14%	29%	107%
D	15%	21%	40%
E	21%	36%	71%

B. Effect of Vibration

Vibration, in theory, can be another driving force for an increased mobility resulting in better planarization. It may also be possible for the planarizing material to undergo localized heating due to absorption of kinetic energy. As a result, imidization and solvent evaporation induced agitation may take place simultaneously at a controlled rate of shrinkage by adjusting frequency and amplitude of vibration. An attempt to improve polyimide planarization by applying ultrasonic vibration at 40 kHz before curing has yielded little enhancement. Further study is necessary to better understand the effect of vibration on planarization. However, it is reasonable to expect an improved planarization condition by the use of a combined ultrasonic and low pressure DCT.

C. Multiple Coating

The resultant degree of planarization of n layers of planarizing films prepared identically, as shown in Figure 3, can be written as:

$$P_n = 1 - \frac{t_n}{t_0} \tag{2}$$

$$= 1 - \frac{t_n}{t_{n-1}} \cdot \frac{t_{n-1}}{t_{n-2}} \cdots \frac{t_1}{t_0} .$$

Hence,

$$P_n = 1 - \prod_{k=1}^{n} \gamma_k , \tag{3}$$

where a memory factor, γ_k, can be defined as

$$\gamma_k = \frac{t_k}{t_{k-1}} . \tag{4}$$

Preliminary experimental results indicate that planarization of polyimide does not seem to depend on the height of the underlying geometries when line thickness to spacing ratios are small. Therefore, for small aspect ratios,

$$\gamma_k = \gamma_1 \qquad\qquad \text{for } k = 2,3, \cdots, n. \tag{5}$$

That is,

$$P_n = 1 - (1-P_1)^n . \tag{6}$$

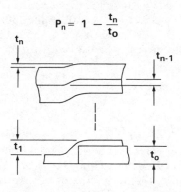

$$P_n = 1 - \frac{t_n}{t_o}$$

Figure 3.   Planarization by multiple coating.

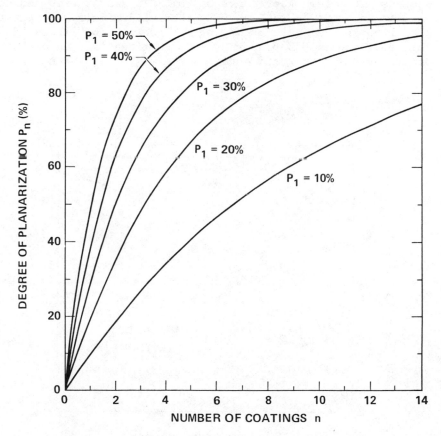

Figure 4.   Degree of planarization as a function of number of coatings.

Figure 4 shows degree of planarization as a function of number of layers n for P_1 of 10%, 20%, 30%, 40%, and 50%. From Equation (6) it is easy to determine the number of planarizing layers required to achieve a final degree of planarization P, for a given P_1, as follows:

$$n = \frac{\log\ (1-P)}{\log\ (1-P_1)}\ . \qquad (7)$$

Figure 5 shows the effect of multiple coating on planarization. The hollow circles are conventional experimental data using a polyimide with 19% solids and viscosity of 12 poises on a substrate with 4.3μm thick 470μm wide metal lines and 280μm spaces. It should be noted that these data fit quite well on the theoretical curve based on Equation (6) which is plotted with the data points. However, by applying low pressure DCT with all materials and process parameters kept constant, a 67% planarization enhancement can be achieved as indicated by the solid circle. By extrapolating both sets of data according to Equation (6), it would be possible to reach an 80% planarity in 6 coatings instead of 10 without using DCT. As noted earlier, the effect of proces-

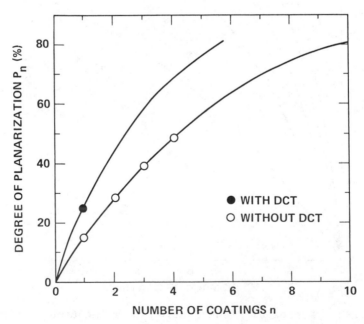

Figure 5. Degree of planarization of multiple coatings using a polyimide of 12 poise viscosity with and without DCT as function of number of coatings.

sing conditions on planarization depends on degrees of enhancement of molecular mobility due to increase of outgassing rate, and of surface tension due to pressure differential at the interface. Therefore, the degree of planarization enhancement due to vacuum DCT is in general a function of materials and processing parameters. Figure 6 shows planarization of multiple coatings using a polyimide of 15 poise viscosity with and without DCT. Over geometries of 4.3μm thick 20μm wide lines with 70μm spaces, 84% planarity is obtainable in four coats by use of DCT. Figure 7 shows an example of an 84% planarity resulting from four coats of polyimide over 4.3μm thick 60μm wide lines with 260μm spaces, using low pressure DCT.

D. Applications

Dynamic curing technique has been used to show significant improvement in polyimide leveling. In addition, DCT may find its applications in planarization of other vehicled materials such as photoresists, thick film inks, and spin-on-oxide in the process of densification.

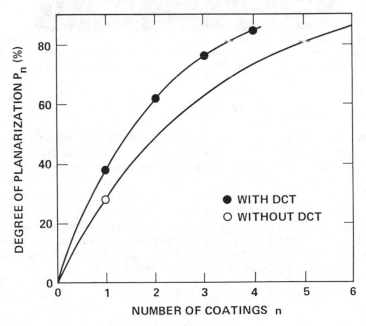

Figure 6. Degree of planarization of multiple coatings using a polyimide of 15 poise viscosity with and without DCT as function of number of coatings.

Figure 7. Planarization of four-layer polyimide over 4.3μm thick 60μm wide lines with 260μm spacing after complete cure.

IV. CONCLUSIONS

The concept of dynamic curing technique with its physical mechanisms has been introduced. Substantial improvement in planarization of polyimide has been demonstrated by means of low pressure DCT. Further planarization enhancement may possibly be achieved by combining effects of vibration with low pressure in the process of dynamic curing of polyimide. It is expected that this technique should be applicable in the planarization of other materials such as photoresists, thick films, and spin-on-glass. A mathematical model has been developed to predict the final degree of planarization as a function of number of coatings.

ACKNOWLEDGEMENTS

The authors wish to acknowledge P. S. King and J. E. Brunetti for metallization and etching work, and P. G. Mosher for her assistance in SEM sample preparation. They are grateful to R. E. Thorne for his continued support and guidance.

REFERENCES

1. K. Sato, S. Harada, A. Saiki, T. Kimura, T. Okubo and K. Mukai, IEEE Transactions on PHP: PHP-9, 176 (1973).
2. J. C. Yen, Abstract No. 170, p. 444, The Electrochem. Soc. Extended Abstracts, Dallas, Texas (1975).
3. Du Pont Pyralin Product Specifications; E. L. Yuan, Marshall R & D Labs, Du Pont, Philadelphia, Pa., private communication, 1982.
4. L. B. Rothman, J. Electrochem. Soc., 127, 2216 (1980).
5. R. Aveyard and D. A. Haydon, "An Introduction to the Principles of Surface Chemistry", Cambridge University Press, 1973.

TECHNIQUES FOR IMPROVING PACKING DENSITY IN A POLYIMIDE MULTILEVEL METALLIZATION PROCESS

S. J. Rhodes

Plessey Research (Caswell) Limited
Allen Clark Research Centre
Caswell, Towcester, Northants., UK

A two level metallization process using polyimide as an interlevel dielectric has been successfully run in a production status for over two years. The vias are formed using resist masking and oxygen plasma etching. This method is used for metallization layouts based on a 12μm grid. To reduce this grid, changes in technology are involved for via formation. In particular consideration is given to the processing of non-nested vias. A technique using a chromium etch barrier is described to overcome problems of first level metal attack whilst etching the second level metallization. A computer simulation of step coverage from a magnetron sputter deposition has been used to determine the step coverage in a non-nested via. This shows a potential problem particularly when the via sides are close to the first level metal edges.

INTRODUCTION

A double level metallization process has existed in a production status at Plessey for over two years. The process has become successful due to the incorporation of two recently established techniques, namely polyimide to provide the interlevel isolation and magnetron sputtering to deposit the silicon copper doped aluminium films. The via formation technique employed uses a thick positive resist as a mask on top of a fully cured polyimide, the resist thickness being substantially greater than that of the polyimide. The polyimide is then etched in an oxygen plasma creating the via. The resist mask is also etched in the same operation, the etch rates being very similar for the two materials, but some resist remains in the field regions due to its additional thickness. The details of this process have been reported in a previous paper[1]. It is capable of forming vias down to 4µm x 4µm at a metal pitch of 12µm utilizing conventional projection photolithography with a 3µm resolution and was designed for use on high speed bipolar gate arrays mainly employing ECL circuitry.

This basic process is now being applied to both Bipolar and CMOS logic arrays. The overall circuit packing density in gate arrays is now generally accepted as being metallization limited[2]. As computer aided design is used to determine the layout of the metallization for gate arrays, the interconnect is normally routed on a grid such that no constraints need be considered for via placement within this grid. The pitch of the grid is a convenient measure of the metallization packing density and is the one referred to throughout this paper. A series of trial layouts has been made to predict the effects of metal pitch reduction and increases in number of metal levels on the number of gates that can be expected on a 250 mil x 250 mil chip. The results of this exercise are shown in Figures 1 and 2 both for CMOS and oxide isolated I^2L. The conclusion that can be drawn from this is that both the number of metal levels and size reduction are important. To increase the number of metal levels is relatively straightforward when using polyimide as an interlevel dielectric due to its good step coverage properties. Typically third and subsequent layers follow the same processing sequence using similar design rules as the second level metal. Consideration needs to be given to the question of 'stacked' (or coincident) vias - but even 'staircasing' vias does not usually result in a serious design constraint. The more fundamental problems of reducing the metal pitch are now considered in the rest of the paper and can be seen to influence the double level metal processing technology.

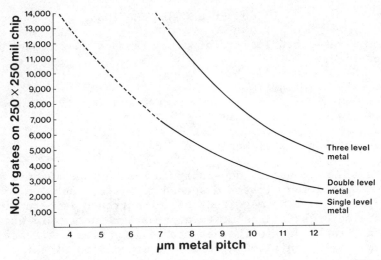

Figure 1. Estimated packing density on CMOS gate array designs.

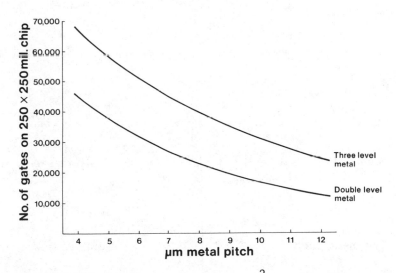

Figure 2. Estimated packing density on I^2L oxide isolated gate array designs.

LAYOUT RULE CONSIDERATIONS

The established 12µm metal pitch is made up from the following (see Figure 3).

1. Via size (4µm)

2. Via surround (2½µm)

3. Metal to metal separation (3µm)

The via surround is necessary to ensure that the via is retained within the first and second metallization, i.e. that the via is 'nested'. The magnitude of the surround is determined by the addition of an alignment tolerance, a reduction in width of the metal tracks due to the wet etched aluminium delineation process and finally an allowance for the oversizing of the via during its formation.

Let us consider each of these three terms. Firstly the via size could be reduced to below 4 x 4µm but it must be realized that for each 1µm reduction in metal pitch there is a considerable loss in via contact area. Also the lateral etching of the resist mask

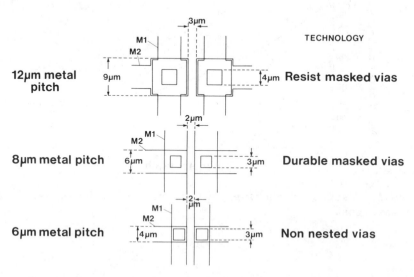

Figure 3.   Two level metal layout rules and related technology.

becomes significant in creating small vias and the via print must be undersized to allow for it. As a thick resist is also used, resolution better than 3μ is difficult to obtain. A different method is therefore needed to produce smaller vias.

Secondly, the via surround can be reduced by minimizing the lateral etching of the metallization, either by using thinner metal or anisotropic plasma etching to define the levels of metal.

Finally, as photolithographic techniques are improved, the metal to metal spacing can be reduced. It is beyond the scope of this paper to discuss this subject but a resolution of 2μm is assumed giving a consequent spacing equal to this value.

DURABLE MASK PROCESSING

If the resist mask used in the basic process is replaced by a 'durable' non-erodible mask then the final size of the via can be reduced. The procedure would be to deposit the masking layer over the polyimide, photoengrave this layer and then plasma etch the polyimide through this mask. Finally it is necessary to remove the masking layer. The advantages of this technique are that it is now no longer necessary to use a thick resist for the via definition. Also the enlargement of the via during the plasma etching of the via is reduced, thus maintaining the original mask dimensions. The via feature then can be reduced into the 2μm region using this method, the main drawback being the more vertical via profiles caused by this technique. In some cases a slight re-entrant profile occurs due to a decrease of the lateral etch rate directly underneath the masking film.

Using this process, it is feasible to design on a metal pitch in the region of 8μm. However, to go significantly beyond this figure it is seen that a distinct change in process technology is required so as to allow 'non-nesting' of the via which reduces the via surround to zero if necessary. The overall packing density is then limited by either a simple metal or via definition pitch, whichever is greater[3,4]. A summary of the layout rules and related technology is given in Figure 3.

THE NON-NESTED VIA

The reduction of the via surround to zero (or even to a negative value, where the via is larger than the metallization) gives rise to two problems in practice. The first difficulty occurs when the via is larger than the first level metal track. A trough or moat is then formed between the via side and the metal edge (see Figure 4). The second level metal then has to cover this

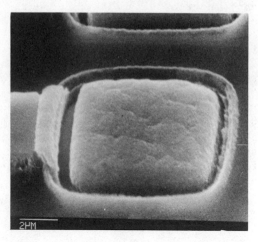

Figure 4.   S.E.M. of non-nested via chain showing moat around the via.

topography and can give very limited step coverage particularly at the base of the first level metal edge.

The second difficulty is likely to be more catastrophic in practice.   This is when the second metal no longer covers the via and leaves part of the first level metal exposed which will then be attacked during the etching of the second metal.[5]

To achieve a solution to this problem we either have to avoid etching the second level metal by using a lift off technique or provide an etch barrier between the two levels of aluminium.   A third alternative would be to control the overetch during the second metal delineation but experience has shown that this is not viable on a production line.

A test mask was designed to evaluate the performance of non-nested processing.   The structures consist of via chains using strip vias so that the metal step coverage can easily be observed in an S.E.M. (see Figure 5).   The chains are progressively non nested, firstly for first metal only, then for second metal only and lastly for both levels of metal together.

Figure 5. S.E.M. showing test structure using strip vias.

LIFT-OFF TECHNIQUES

Metallization deposited by sputtering is not particularly suitable for use in lift off processes as 'walls' are formed due to the deposition of material on the sides of the spacer layer. However, if this spacer layer is allowed to remain, an adverse topography is avoided. A method that can be used with sputtered metal is to use a thin resist coating over the 'spacer' polyimide layer which is subsequently covered with a metal 'definition' mask such as pure aluminium. The required features are then photoengraved on this masking layer and the composite resist and polyimide film ashed through in a single operation. Next the metal layer is deposited and the unwanted material floated off by dissolving the resist in a suitable solvent. A small wall equal to the thickness of the resist layer is produced but if the resist layer is made thin, then this drawback is not significant. If this process is used for second level metal then the spacer layer ashing needs to be carefully controlled to prevent undue thinning of the interlevel insulating polyimide. For this reason, this method was abandoned in favour of the etch barrier approach.

ETCH BARRIER TECHNIQUE

The original concept of this process was to provide a chromium

metal layer beneath the second level of metallization to act as
an etch stop, the chromium then being etched with good selectivity
to the underlying exposed aluminium, as reported by J R Kitcher[5].
The second level metal was etched conventionally and the chromium
layer was plasma etched in a CCl_4/O_2 plasma. It was found necessary
to remove the silicon debris from the wet etching or avoid the
debris by using pure aluminium otherwise the chromium plasma etching
was found to be unsuccessful. However, experimental results showed
that the non-nested or exposed area of the first level metal was
still attacked to some extent giving a high resistance in the test
chains.

To overcome this problem, it was necessary to also put chromium over the first level metal. This introduced further difficulties, however, as an undercut profile was produced on the edges of
the first level metal which resulted in discontinuities at this
point. A lift off first level metal process was then investigated
(similar to that described earlier) but after vias processing, the
metal 'walling' was found to be excessive due to the overetch
required at the via stage necessary to clear the polyimide over the
metal. This overetch would also etch the polyimide spacer layer
causing the sputter 'wall' to form (see Figure 6).

Figure 6.   S.E.M. showing first level metal 'walling'.

A solution was arrived at by using an identical lift-off procedure to form the vias, the spacer layer also acting as the interlevel dielectric.[6] In this process a metal layer is deposited within the via, and as this deposition is 'topped' with chromium, the required protection of the first level metal is achieved.

COMPUTER SIMULATION OF STEP COVERAGE

A computer program was developed to simulate sputter deposited metallization from a model reported by Blech [7]. However, this was modified to a cosine distribution giving a vertical side wall coverage of $1/\pi$ times the planar metal thickness, which is in good agreement with S.E.M.s from our equipment (CVC 601). This simulation was then used to look at metal coverage in $2\frac{1}{2}\mu m$ vias (see Figures 7, 8 and 9). Figure 9 is similar to the re-entrant profile mentioned earlier in durable mask processing. The program was also used to look at the non nested step coverage problems associated with the moat around the first level metal (see Figures 10, 11, 12 and 13). It is interesting to note the deterioration observed when the via edge to metal spacing is reduced. This indicates that current densities will be seriously limited when using this technology.

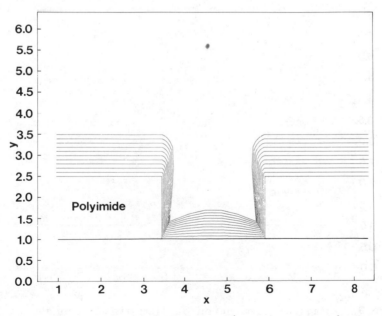

Figure 7.   Computer simulation of $2\frac{1}{2}\mu m$ via - vertical walls.

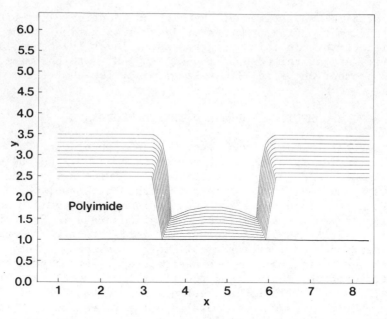

Figure 8.   Computer simulation of $2\frac{1}{2}\mu$m via – sloped walls.

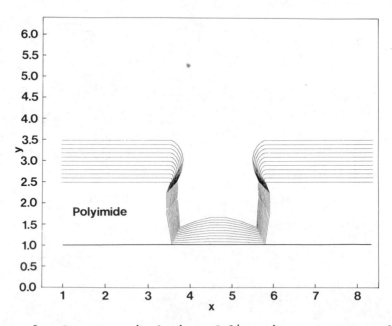

Figure 9.   Computer simulation of $2\frac{1}{2}\mu$m vias – concave walls.

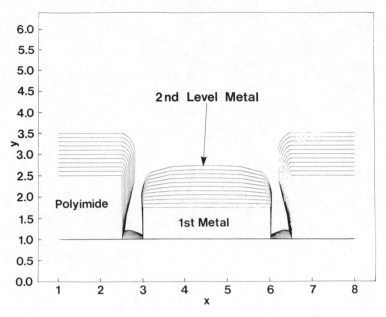

Figure 10. Computer simulation of non-nested via - wide gap without slope.

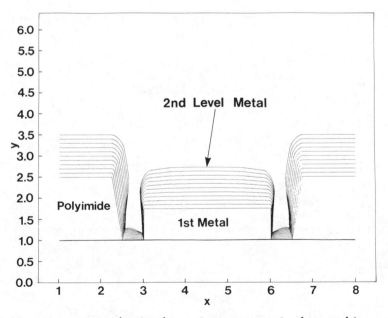

Figure 11. Computer simulation of non-nested via - wide gap with slope.

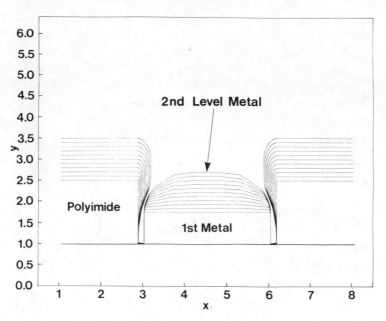

Figure 12. Computer simulation of non-nested via - narrow gap without slope.

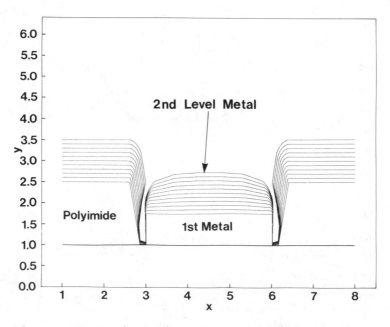

Figure 13. Computer simulation of non-nested via - narrow gap with slope.

CONCLUSIONS

Technological changes involved in adapting a polyimide double level metal process to increase metallization packing density have been described. The relative advantages of reduced metal pitch and increased levels of interconnect have been shown in terms of gate density. A solution to the problem encountered with non-nested vias has been presented, but computer simulations have shown a significant step coverage problem which presumably can only be eased by via and metal edge bevelling.

REFERENCES

1. S.J. Rhodes, Semiconductor International, 4, No. 3, 65 (March 1981).
2. J.W. Tomkins and S. Hollock, Proceedings of the 1982 Custom Integrated Circuits Conference, p. 276, IEEE, Rochester, N.Y., May 1982
3. J.R. Kitcher, J. Vac. Sci. Technol., 16(6), 2030 (1979).
4. A.M. Wilson and P.B. Ghate, in "Proc. Semiconductor Silicon 1977", H. R. Huff and E. Sirtl, Editors, p. 1047, Electro-chemical Society, Princeton, N.J., 1977.
5. J.R. Kitcher, Extended Abstracts of the Electrochemical Society, Volume 80-2, Abstract No.329, p. 851, 1980.
6. L.B. Rothman, Extended Abstracts of the Electrochemical Society, Volume 81-2, Abstract No. 243, p. 582, 1981.
7. I.A. Blech, Thin Solid Films, 6, 113 (1970).

INTER-METAL POLYIMIDE VIA CONDITIONING AND PLASMA ETCHING
TECHNIQUES*

T.O. Herndon, R.L. Burke, W.J. Landoch

M.I.T. Lincoln Laboratory

244 Wood Street, Lexington, MA. 02173

Polyimide's advantages as an intermetal insulator for VLSI has stimulated the study of factors which affect via dimensions and contact resistance.

Attempts to simplify the etching schedule for producing good via contact to increase the etch rate causes several problems. The standard schedule etches the P.I. and the photoresist at 1,000Å/min, but drops to 250Å/min. for the final 50 watt, 50 millitorr condition. At 300 watts, 50 millitorr, the polyimide etches at 3,000Å/min., but the photoresist etches at 4,300Å/min. A second undesirable effect of the increased power is that the etching of the P.I. and photoresist becomes isotropic due to sputtering. The via dimensions enlarge and considerable residue is left in the vias.

Photoresist is limited as a mask for oxygen plasma etching since it erodes at least as fast as the P.I. Aluminum nonerodible masks for P.I. have been demonstrated. Spin-on glass is an easily applied and removed noneroding mask leaving residue-free vias after etch.

---

*This work was sponsored by the Department of the Air Force. The U.S. Government assumes no responsibility for the information presented.

Surface preparation of metal exposed in the P.I. vias is essential prior to second level metal deposition. Buffered HF is effective and Shipley MF 312 also works. In situ argon sputter cleaning produces the lowest contact resistance.

Large unpatterned areas of metal on P.I. may bubble during sinter. This can be eliminated without affecting contact resistance by incremental heat treatment as a part of the sinter step.

INTRODUCTION

The advantages of polyimide as a possible inter-metal insulation between the multilevel interconnect required for VLSI have been much discussed. These include its ability to level underlying topography, and provide satisfactory high temperature characteristics, excellent insulating properties, essentially unlimited thickness potential and reasonable etching characteristics, both by means of wet or dry (plasma) chemistry. However, polyimide (P.I.) is known to have a number of potential weaknesses which must be corrected if it is to be useful for future VLSI applications. Since we have been working with P.I.[*] as an inter-metal insulator at Lincoln Laboratory for the past four years, we have learned a number of things regarding the application and subsequent processing which will be discussed here under the following headings: 1) Power and pressure considerations in plasma etching of vias in P.I.; 2) Properties of resists for plasma etching of vias in P.I.; 3) Treatment of the metal exposed at the bottom of the via prior to deposition of top level metal to assure good electrical contact; 4) Water absorption.

I. ETCHING POWER-PRESSURE CONSIDERATIONS
[Experimental]

Our previous paper on this subject details the P.I. application techniques and plasma etching apparatus employed.[1] The process currently used for applying P.I. involves heating the P.I. before and during spinning to assure that it becomes a dry film on the spinner to prevent random dewet areas or "fisheyes" in the finished film. In general, only one or two such "fisheyes" occur per three inch wafer, but since we are currently involved with die sizes in excess of 1.75" square, even this low incidence

---

[*]Dupont PI2555

of holes or thin places in the P.I. is not tolerable. Also, either Shipley AZ 1300 or 1400 series photoresist is used and processed normally to achieve a 2μ thickness for P.I. etching.

The plasma etching of vias is done using oxygen in a parallel plate reactor, and Fig. 1 is a montage of SEM's of etched vias, with the photoresist still in place. The pressure at which these vias were etched is shown on the left hand side and the power levels are shown at the bottom of the picture. It is quite clear from these images that via sidewall anisotropy, as well as photoresist attack and redeposit on the via bottom are drastically affected by plasma power/pressure conditions. The upper right hand images show considerable penetration and thinning of the photoresist, and a great deal of particulate matter is apparent at the bottom of the via. This is believed to be redeposited resist, as it can be removed by further etching in oxygen at lower power levels. It is thought that these conditions arise from increased ion bombardment occuring at the high power (voltage) conditions which physically sputters off the resist and redeposits it on the via bottom. This sputtering is also responsible for the high resist/P.I. etch rate ratio, where the resist etches at 4,300Å/min., while the P.I. etches at 3000Å/min. The sputtered resist surface is roughened, and the resist-P.I. sidewalls are eroded isotropically. The lower left hand picture shows essentially the opposite result with anisotropic etching, non-eroded photoresist surface and no visible deposit on the metal at the bottom of the via. In this case, the resist-P.I. etch rate ratio is 1:1 at 1000Å/min. At other power/pressure conditions, intermediate etching results are obtained.

Figure 2 is the same series of SEM's as before, only the photoresist has been removed by stripping in acetone. Here again, one notices the high power, low pressure case having the more isotropic, grooved sides and the low power/high pressure etched vias being nearly anisotropic and smooth.

I. (Results)

Figure 3 shows etch rate curves at three levels vs plasma power and pressure conditions. Clearly, one does not want to erode the resist, as this will etch the P.I. which could lead to shorts. Further, the redeposit at high power, and low pressure conditions has been found to cause poor intermetal contact. Therefore, Fig. 3 is divided to indicate the general regimes of power and pressure which produced acceptable resist-P.I. etch ratios and no redeposit, and those plasma conditions which are unacceptable in these respects (the unshaded area). Since it is desirable to etch at reasonable rates, our current procedure is to etch most, or all, of the way through the P.I. at 200 watts and

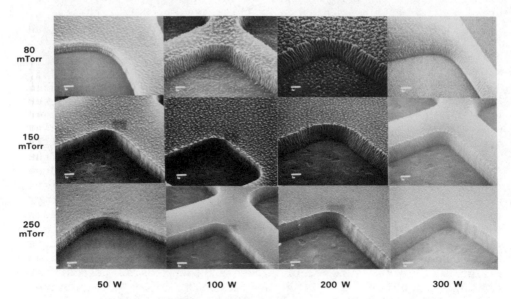

Figure 1. Effects of power and pressure on sidewall anisotropy and redeposit (polyimide and photoresist).

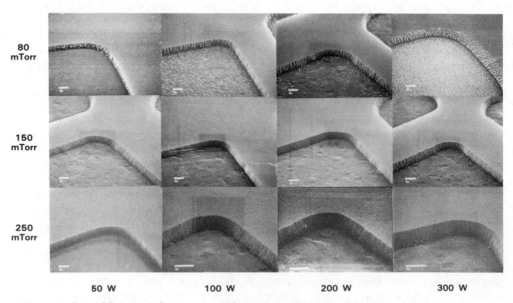

Figure 2. Effects of power and pressure on sidewall anisotropy and redeposit (polyimide with photoresist stripped).

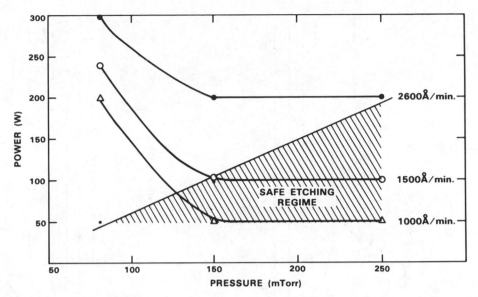

Figure 3. P.I. constant etch rate curves defining minimum redeposit conditions (safe etching regime).

250 millitorr. We then reduce the pressure/power to 50 watts and 80 millitorr and overetch under these conditions for "invisible shield" removal as described elsewhere.[2,3]

II. RESIST CONSIDERATIONS
(Experimental)

A number of complex considerations are involved in the plasma etching of P.I. which has been patterned with photoresist. First, since the resist etches at the same rate or faster than the P.I., there is a limit to the maximum thickness to which the P.I. can be applied. If the P.I. and photoresist films are uniform in thickness, the P.I. could theoretically be nearly as thick as the resist. In this case the resist thickness would be determined by the minimum sized via required. However, neither photoresist nor P.I. is conformal (uniform thickness over underlying features) nor do they level perfectly as shown by Day[4] and Rothman.[5] The thickness of both depends on the covered feature's height, dimension, vertical sidewall steepness, separation from neighbors and whether it is a hole or hump. To illustrate this, Fig. 4 shows a cross section of a VLSI design where the feature heights and proximity to each other lead to problems. Starting at the bottom of the upper view, one sees an SiO_2 layer, with a polysilicon conductor lying directly adjacent to a contact cut over the p^+ region. The PolySi is partially covered by phosphosilicate glass and then a metal conductor covers both and connects the PolySi to the p^+ region. This design is intended to save real estate by bunching

813

BEFORE PLASMA ETCH

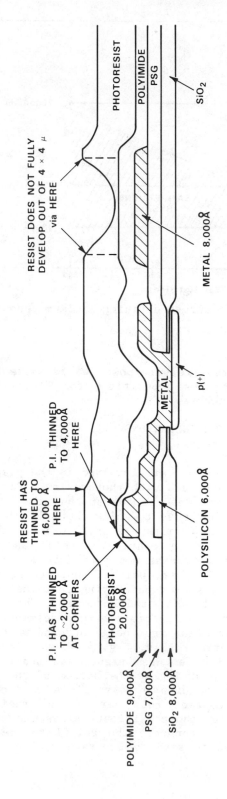

RESIST DOES NOT FULLY DEVELOP OUT OF 4 × 4 μ via HERE

PHOTORESIST

POLYIMIDE

PSG

SiO$_2$

METAL 8,000Å

p(+)

RESIST HAS THINNED TO 16,000 Å HERE

P.I. THINNED TO 4,000Å HERE

P.I. HAS THINNED TO ~2,000 Å AT CORNERS

PHOTORESIST 20,000Å

POLYSILICON 6,000Å

POLYIMIDE 9,000Å

PSG 7,000Å

SiO$_2$ 8,000Å

METAL

AFTER PLASMA ETCH

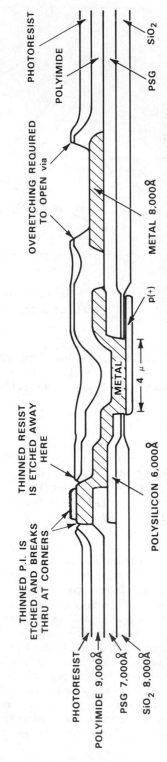

OVERETCHING REQUIRED TO OPEN via

PHOTORESIST

POLYIMIDE

PSG

SiO$_2$

METAL 8,000Å

p(+)

4 μ

THINNED RESIST IS ETCHED AWAY HERE

THINNED P.I. IS ETCHED AND BREAKS THRU AT CORNERS

POLYSILICON 6,000Å

METAL

PHOTORESIST

POLYIMIDE 9,000Å

PSG 7,000Å

SiO$_2$ 8,000Å

Figure 4. Effects of underlying features on P.I./photoresist thickness and etching.

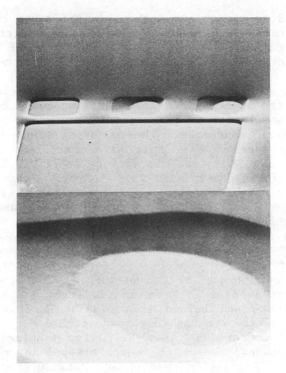

Figure 5. SEM of partially opened vias in P.I. caused by improper photoresist exposure.

contact and PolySi conductors closely so that interconnect metal consumes the minimum area. However, the PolySi-PSG-Metal stack builds to a 12,000 - 14,000Å height, where the greatest feature height elsewhere is 8,000Å. Since the design calls for first to second level metal vias in the P.I. of 4 microns square, conservative design means photoresist more than 2μm thick cannot be used with Direct-Step on Wafer exposure. Assuming that the P.I. cannot be more than one-half of the photoresist thickness, this sets the P.I. thickness at 1 micron. However, due to process variabilities, it measures 9,000Å after full cure. The resist has thinned over the top of the protrusion to approximately 16,000Å, and the P.I. has thinned to something like 4,500Å on the flat, but down to perhaps 2,000Å over the plasma etched metal corners. To the right of the contact, we see where the photoresist has been exposed and developed, but the 4 micron wide via has not developed fully or cleanly. This condition may occur in 1% of the vias, but in a VLSI circuit, this is not tolerable. Thus, the resist-P.I. must be overetched to open these vias adequately and assure good via contact between first and second level metal. Assuming an overetch by 25% in time to clean out anomalous vias with plasma etch conditions which are not optimal results in the resist being consumed at a greater rate than the P.I. These "worst case" conditions lead to the problems shown in the lower cross section. The via certainly opens, however it still has fillets of P.I. at the metal surface, reducing somewhat the area of the opening. However, the photoresist is completely stripped in the area on top of the PolySi-PSG-Metal stack, the P.I. attacked, thinning and roughening its surface on the flat region and breaking completely through it in the vicinity of the underlying metal corners. Thus if the second metal were to pass over this region, it would short to the first level conductor. The following series of SEM photos will give some idea of what this situation looks like in real life. Fig. 5 is a SEM of a pair of vias in P.I.. The upper view shows three 4 micron square vias in P.I. lying along the side of a 20 micron via. The two right hand vias are clearly not fully opened, and the lower closeup of one of these shows that the opening is perhaps only one half the proper area. This is where overetching is required. Fig. 6 shows SEM views of a wafer cleaved through a thinned area in the P.I. The upper view shows the outline of a metal conductor underlying the field of P.I. coming down from the top of the picture at the upper right. A roughened, squarish area is seen at its edge, and the PolySi is protruding beneath and beyond the metal. The lower view is a closeup where one can see the roughened, etched P.I. and how much thinner it is, compared to that P.I. flap running to the right edge of the picture. The metal is the stepped layer just beneath the P.I.. Fig. 7 is looking down on the P.I. surface, and one can see the faint outline of underlying metal, a heavily overetched via which has become nearly round, and the roughened surface of the P.I.. This is where it is highest and has been attacked by the oxygen plasma as a result of resist thinning and subsequent removal because of inadequate thickness. Notice some whitish spots in the rough areas. The lower picture clearly shows a pair of white protrusions growing out of the P.I. in the plasma attacked area adjacent to a dent in the P.I.. Fig 8 is a view of these protrusions at a steeper angle, clearly showing that they stick up from the P.I. surface. In the lower picture, the sample was immersed in dilute phosphoric acid for 1 minute which removed the protrusions to reveal holes in the P.I. where it has been etched completely through at the underlying metal corners. The composition of these protrusions is not known, but it is thought that they arise from interaction between the underlying AlSiCu and one of the wet· chemicals encountered in resist removal, rinsing, or via preparation steps. Figure 9 shows second level metal going to vias in the P.I. surface. Also, there are two rough squares visible, where the P.I. was attacked all over the top of the underlying PolySi-PSG-Metal mountain. The lower view is a closeup of the attacked area near the metal line. If the P.I. is removed with oxygen plasma to clearly reveal what is going on in these attacked areas, the conditions illustrated in Fig. 10 are seen. In the upper view, running left and right across the middle of the

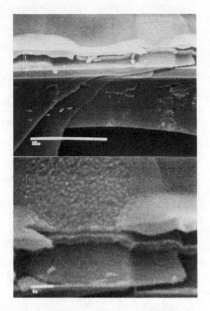

Figure 6. SEM of P.I. cross section attacked during overetching because of topography-induced photoresist thinning.

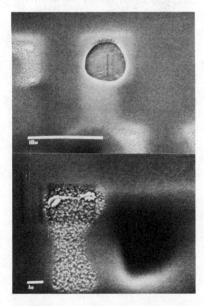

Figure 7.   SEM of P.I. attacked during overetching.

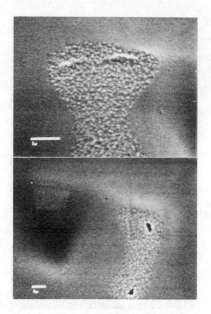

Figure 8. SEM of attacked P.I. with chemically caused growths in holes in P.I.

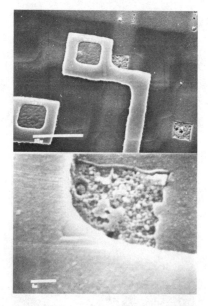

Figure 9. SEM of P.I. with second level metal, showing local overetch attack.

picture is a first level metal conductor which terminates in a contact cut on the left. In the center, running vertically is a PolySi line, and to its right, running parallel to it is a second level metal conductor. The second level metal conductor is rough where it crosses the first level conductor, because the P.I. and resist between them were higher and thinner at this crossing. The resist was etched away, the P.I. surface attacked and roughened, and the metal remaining replicates this roughness in the underlying P.I.. Next to that conductor, one can see where the first level metal was attacked during second level metal etching. The P.I. that had been over the highest crossing (first metal over PolySi) had been so badly attacked that it was not simply roughened, but nearly completely removed. The lower view is a closeup of the worst case described in the earlier cutaway drawing. Here the second metal runs vertically and left of center, lying on P.I. which has been etched off the horizontal first level metal that is crossing a vertical PolySi line (partly obscured by the top metal). The roughening of half the width of the top metal where it overlies the PolySi and all its width where it overlies the first metal, shows that the underlying P.I. was attacked in all of these areas. Further, two pits are visible in the second level metal where the underlying P.I. is gone and the second metal shorts through to the first metal. Also, one can see where the P.I. broke down near the corner of the first metal along the edge of the PolySi line and the first metal has been attacked through the P.I.

The primary points here are that the close packing desired in VLSI can lead to considerable vertical excursions in the surface by the time first level metal has been patterned. Polyimide levels these excursions somewhat, but thins at the top and corners of the highest features—as does the photoresist. This can result in the erosion of resist in the oxygen plasma etching of the P.I. and could cause problems even if no erosion took place, since the P.I. might be too thin to satisfactorily insulate first from second level metal, depending on the voltage difference existing across such thinned areas.

It appears as though a nonerodible mask will be essential to permit the necessary freedom of design for VLSI. We have investigated two approaches in this regard. In the first, a 3,000Å layer of positive photoresist is spun on the P.I. and softbaked. Then 200-300Å of aluminum is sputter deposited, and finally the normal 1 micron photoresist layer is spun and patterned. The aluminum is wet etched and the P.I. is then etched with oxygen plasma. The imaging resist is completely etched off during the oxygen plasma etching of the P.I. and the aluminum serves as a nonerodible mask. The P.I. can be overetched without significant undercut and the aluminum mask can then be stripped by dipping the wafer in acetone which dissolves the 3,000Å resist layer beneath the aluminum mask. Figure 11 shows etched P.I. lines with aluminum mask layer and underlying photoresist liftoff layer still in place. In the upper picture one can see debris between the P.I. conductors, which was the result of etching at high power and low pressure (350W and 80 mT) in an attempt to erode the mask. The debris can be removed with the usual 50 Watt, 80 millitorr step. The lower picture shows the 400Å aluminum mask overhanging the P.I. The 3,000Å thick photoresist lift-off layer can be seen as the nearly vertical sided part of the cross section beneath the metal cap. In this case, the P.I. was overetched by 50% in time, again to test for aluminum mask degradation. Undercutting is apparent, but the metal mask held up well, and serves to provide excellent protection to the P.I. during the oxygen plasma etching step.

The metal mask approach has the serious limitation of complexity, and rework means one must strip back to the P.I. and carry out two photoresist coating steps plus a metallization step. Spin-on glass was therefore investigated as an easily applied and removed noneroding mask. A material called "Silicafilm" made by Emulsitone Co., Whippany, New Jersey, is spun on to fully cured P.I. at 10,000 rpm and baked for 15 minutes at 200°C. This produces a pinhole free glass layer approximately 1,300Å thick. AZ resist is applied in the normal 1 micron thickness, exposed and developed using standard techniques. The glass is then plasma etched in CF_4 + oxygen at 250 millitorr and 50 watts power in a parallel plate reactor. The etch rate of the glass and photoresist is approximately 500Å/minute and the etch rate of the underlying P.I. is less than 25Å/minute. We have not been able to pattern the glass with HF, as the acid undercuts and lifts the glass layer beneath the imaging photoresist. Following the glass definition, the P.I. can be etched in oxygen plasma, during which time the imaging photoresist is completely removed. The spin-on glass can be stripped either with CF_4 + oxygen plasma or in 100:1 HF. Figure 12 shows a portion of a via etched with the nonerodible glass mask still in place. The upper image shows that the etching is reasonably uniform and nearly anisotropic, while in the lower picture, the edge of the spin-on glass can be seen just at the top of the etched P.I. edge. The spin-on glass has the merits of easy application and removal as well as its complete resistance to oxygen plasma which permits etching thick P.I. sections as well as the absence of residue in the via after etching regardless of plasma power. We are continuing to study the performance of spin-on glass as a noneroding mask material in terms of its freedom from pinholing, surface integrity and adhesion.

Figure 10. SEM of polysilicon - 1st metal - 2nd metal with P.I. removed.

Figure 11. Cross section of SEM's of nonerodible aluminum mask on P.I.

III. VIA PREPARATION
(Experimental)

The most serious problem in using polyimide for multilayer metal is the surface preparation of the first level metal at the bottom of the P.I. via immediately prior to second level metal deposition to produce consistently low via resistance. The use of buffered HF dip for this purpose has been described[6]. However, some applications cannot tolerate HF and we have found that a 90 second dip in a 1:1 mix of Shipley MF 312 and water is one possible substitute for HF to effect aluminum oxide removal. The alternative of a separate argon sputter cleaning step at 20 mTorr after resist removal, while usually producing low via contact resistance, results in very poor adhesion of metal to P.I. SEM analysis indicates that the argon sputter process breaks up the P.I. surface into fine, loosely bound particles which leads to such poor adhesion that the top metal may fall off nearly spontaneously. If separate argon sputter cleaning is used, it must be prior to resist removal to prevent this situation. In situ sputter cleaning at 1 mTorr immediately prior to metal deposition does not degrade adhesion and produces very good via contact. If a minimum loss of 30-60Å of exposed material on the wafer surfaces can be tolerated, in situ sputter cleaning is the superior technique. At this time there is no clear understanding of the effect of metal composition or possible P.I./metal chemistry on via yield statistics. The previously mentioned factors of getting all of the P.I. out of the via during plasma etching and making certain that wet processes associated with resist removal steps do not produce contamination must all be considered, understood, and controlled.

III. (Results)

The electrical tests employed to evaluate via resistance involve applying 10 mV to the test via and if its resistance is less than 20Ω, then 10mA is forced through the via and its resistance is measured by a 4-point probe. Via chains are measured at 1mV per via. This voltage limit is to reduce the possibility of breaking down a thin insulating layer that may exist between the first and second level metal in the single via tests. Several curious results have been observed. For instance, if the completed wafers are sintered in the usual fashion in a 425°C forming gas atmosphere, the via contact resistance between first and second level metal is sometimes higher than before sinter or the vias open up completely, an entirely opposite result to what is expected. It was established by independent measurement that the metal was not discontinuous due to step coverage problems. Experiments in sintering the vias for 30 minute intervals at 125°, 300° and 425°C and checking the via resistance after each sinter showed that the via resistance after 125°C was the same as with no sinter. However, the via resistance went up after the 300°C sinter step, sometimes by several hundred percent, and then came back down to a value below the "no sinter" resistance following the 425°C sinter. This occurs in essentially all vias which are "good". The "bad" vias simply stay at a high or infinite resistance after the 300°C sinter. This effect is observed regardless of the via preparation treatment, so long as the "before sinter" resistance is relatively low. If the vias are open at this point, they usually stay open during the sinter process. However, if first level metal is not sintered prior to the PI and second metal steps, the via resistance goes down at 300°C and continues to drop at 425°C. For 4x4 micron square vias, "glinting" in BHF prior to metallization produces resistances of 0.2 – 1.0 ohms/via, while Shipley MF 312 treated vias have resistances lying in the 0.5 – 1.5 ohms/via range. In situ sputter cleaning immediately prior to metallization results in resistances of 0.01 – 0.08 ohms/via and less scatter than with the wet processes.

Another factor affecting via resistance which we have observed is that use of P.I. which is older than 1.5 to 2 years produces high to infinite via resistance in nearly every case. This occures even though it appears visually to perform normally during application and etching procedures.

IV. WATER ABSORPTION
(Experimental)

A final general topic concerning P.I. is water absorption. Bubbling of the overlying metal was observed during the 425°C sinter step in the large (100 mm^2) unpatterned areas which remained at the outer edges of DSW exposed wafers. This was due to outgassing of the P.I. probably caused by absorbed liquid from either wet photoresist removal steps or from HF dip and rinse procedures. It was found that baking the P.I. at 300-400°C for half an hour prior to application of metal eliminates this bubbling. However, it was also discovered that baking the P.I. at temperatures above 150°C prior to metallization leads to very poor via contacts. Auger evidence suggests that outgassing from the via sidewalls or surface of the P.I. deposits a carbonaceous, insulating material on the bottom of the via. Bubbling can be eliminated without this deleterious premetallization bake by increasing the final sinter temperature after second metal definition in 15 minute increments at 125° – 300° and 425°C. Figure 13 shows this effect. The wafers in these two pictures were coated with P.I. and aluminum by using our normal process. The wafer in the upper view was placed in forming gas at 425°C after metallization, and one can see the vast number of bubbles

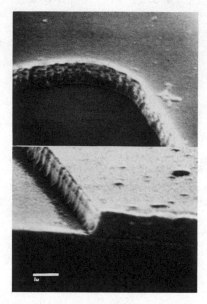

Figure 12. Cross section of SEM's of nonerodible spin-on glass mask on P.I.

Figure 13. Photograph of heat-induced bubbling of aluminum on P.I.

this produced in the overlying metal. The wafer in the lower picture was sintered in steps as described above up through 425°C and one can see only a few bubbles at the very edges near the flat and the cross. The fact that bubbling of the metal never occurs where it has been patterned into fine geometries, but only in large solid metal areas indicates that the P.I. is outgassing and that the lateral mobility of such outgassing is high enough that the gas can escape sideways from under all normal IC geometries. Further, it is believed that the stepped bake cycle permits the gasses to form slowly and diffuse through the overlying aluminum layer as compared to the nearly explosive effect of cycling the P.I. to 425°C within seconds. P.I. which has been in situ sputter cleaned immediately prior to metallization produces much less bubbling in the metal. Further evidence of water absorption is given by the observation that a 1 micron thick P.I. layer may increase 25% in weight after being soaked in water, and decrease by the same amount after a 300°C bake. D.R. Day has noted that there are five sites at each repeat unit which can accommodate a water molecule. Should the top layers of the 1 micron thick P.I. take up water in this amount, the weight increase of 25% would be expected. However, this uptake varies since the kinetics of absorption and desorption may be so rapid that the baseline is uncertain.

CONCLUSION

For VLSI, multilevel interconnect is essential. The intermetal insulation must level well to permit high resolution lithography, have excellent insulating characteristics and be easily processed and reworked. Finally, the yield and reliability of this system must be as high or higher than in LSI applications. Polyimide is an interesting candidate but a considerable amount of work remains to be done to understand its chemical and physical interactions with its on-wafer surroundings before this material can be seriously considered as an intermetal insulator.

ACKNOWLEGEMENTS

The authors wish to thank D.R. Day and S.D. Senturia of the Department of Electrical Engineering and Computer Science, and Center for Materials Science and Engineering at M.I.T. for their many helpful discussions. Also, we wish to thank our co-workers, Bimal Mathur and Robert Frank for the hours they spent at the SEM and in reducing probe test data as well as for their helpful discussions on all aspects of polyimide.

REFERENCES

1. T. O. Herndon and R. L. Burke, "Inter-Metal Polyimide Insulation for VLSI", paper presented at the Kodak '79 Microelectronics Seminar, New Orleans, LA, Oct. 1979.

2. D. R. Day, R. Whitten, S. D. Senturia, "Invisible Shield in Polyimide Vias-An Auger Analysis", paper presented at the Electrochemical Society Meeting, Denver, CO, October 1981.

3. T. O. Herndon, R. L. Burke, J. A Yasaitis, "Use of Polyimide in VLSI Fabrication", paper presented at the Electrochemical Society Meeting, Denver, CO, October 1981.

4. D. R. Day, D. Ridley, J. Mario and S. D. Senturia, "Polyimide Planarization in Integrated Circuits", these proceedings, Vol. 2, pp. 767-782.

5. L.B. Rothman, "Properties of Thin Polyimide Films", J. Electrochem. Soc., 127, 2216 (1980).

FINE PATTERN PROCESSES FOR POLYIMIDE INSULATION

Atsushi Saiki*, Kiichiro Mukai*, Takashi Nishida*
Hiroshi Suzuki** and Daisuke Makino**
*Central Research Laboratory, Hitachi Ltd., Kokubunji
Tokyo 185, Japan
**Yamazaki Factory, Hitachi Chemical Company Ltd.
Hitachi-shi, Ibaragi 317, Japan

Several etching processes for polyimide insulation are characterized and viewed from their applicability to a high density metallization. Both wet and dry etching processes are described.

An alcohol solution of tetramethyl ammonium hydroxide (TMA alcohol) is proposed as the most suitable etchant for high packing-density metallization with polyimide insulation. This etchant can form fine via holes as small as 3x3 µm with a 60 degree taper slope, and barely corrodes aluminum.

On the other hand, a TMA-aqueous solution is shown to have no potential for future etching processes.

Looking into the future, a hydrazine hydrate and ethylenediamine mixture can form 3x3 µm via holes but their 45 degree taper slope is a limitation in applicability to high density metallization applications.

Oxygen plasma is also shown to be poor for via hole size controllability. On the other hand, it is shown that oxygen sputtering can lead to precise via holes which have a slope perpendicular to the substrate. Full utilization of the sputtering process will thus require new technology.

INTRODUCTION

Polyimide has been increasingly used as a passivant and intermetal insulator for ICs and LSIs[1,2,3,4]. This is because of its planarizing ability, low process cost, high device performance and reliability, and high production yield. As the demand for higher packing density has intensified, fine via hole formation has become a requirement of polyimide film.

Via holes in polyimide film can be formed by both dry and wet processes. For dry processing, there are two methods: oxygen plasma and oxygen sputtering. Dry processes require an additional inorganic masking material such as molybdenum film besides the photoresist. As for wet processes, two etchants are currently used, i.e., tetramethyl ammonium hydroxide (TMA) and hydrazine hydrate. The TMA etchant is used as an aqueous solution for semicured (B-stage) polyimide film with a positive photoresist. Hydrazine hydrate is used for fully imidized polyimide with a negative photoresist.

Each etching process has its advantages and disadvantages. This paper discusses each process and proposes a new etching process based on a TMA etchant.

EXPERIMENTS

Polyimide Film   Polyimide isoindoloquinazolinedione (PIQ®) was used as a polyimide type resin. A PIQ prepolymer solution of 1.1 Pa·s viscosity was spun on a silicon wafer and then cured. A 3000 rpm spin speed produced a 2 μm thick PIQ film. Curing conditions were properly chosen according to the etchants, as described later.

TMA Based Etching   Two kinds of solutions were investigated. One was an aqueous solution well known as a developer for positive photoresists. The PIQ prepolymer was coated on a wafer and baked at 150°C for 60 minutes. Then AZ-1350J, as a positive photoresist, was applied and exposed to ultraviolet light. While the photoresist was developed with the use of a TMA-aqueous solution, the PIQ film was simultaneously etched. TMA concentration used in this experiment was 5 %.

The other TMA etchant is new. The solvent is an alcohol solution instead of water. The main feature of this etchant is that the dissolution of the 150°C baked PIQ is much smoother and uniform than in the case of aqueous solution, thus ensuring improvement via hole size controllability. In addition, this etchant attacks the aluminum to a lesser degree than the aqueous solution. Etching time was 40 to 60 seconds at 30°C for 2 μm PIQ.

In this experiment, 25% TMA-75% methanol solution was used.

After via holes were etched, PIQ film and the photoresist were baked at 200°C for 20 minutes and then the photoresist was removed by a photoresist stripper, such as J-100.

Hydrazine Based Etchant   This consists of hydrazine hydrate and ethylenediamine. After PIQ film was fully cured at over 200°C, a negative photoresist was applied and developed. Then, the film was etched. With the etchant temperature at 30°C, etching time was 10 minutes for 2 μm PIQ film.

Oxygen Plasma Etching   A barrel type etcher was used. After PIQ film was formed on a silicon wafer, 200 nm thick molybdenum was sputtered onto it. Then the molybdenum film was photoetched and the PIQ was etched using oxygen plasma. Oxygen pressure was 38 mPa (5 Torr) and RF power was 300 W.

A 5 minute continuous running operation was sufficient to etch 2 μm thick PIQ film. However, when the operation was performed in successive timed intervals, reproducibility was not obtained. In the actual experiments, etching was carried out by repeating two minute etching operation with a two minute break between each operation until completion. In this case, total etching time was typically 8 minutes.

Oxygen Sputter Etching   A parallel plate type etcher was used. As with plasma etching, molybdenum was used as the etching mask. Etching was carried out at a 380 μPa (50 mTorr) oxygen pressure and 100 W power. Typical etching time was 30 minutes for 2 μm PIQ.

RESULTS AND DISCUSSIONS

TMA Based Etchant   Etching characteristics of both TMA-aqueous and alcohol solutions were evaluated by examining side etch width. Here, side etch width is defined as the pattern shift from the photoresist edge, as is shown in Figure 1. Side etch widths are shown as a function of etching time in the figure. When PIQ film was etched with a TMA-aqueous solution, etching proceeded according to via hole size, as shown in curves a, b and c. That is, the larger the via hole, the shorter the time for proper etching. Therefore, all the via holes were not properly etched at the same time. At 120 seconds, a 10x10 μm via hole is properly etched, but larger via holes were overetched, while smaller via holes were underetched. Via holes smaller than 5x5 μm were not etched even after 180 seconds etching.

Curve d in Figure 1 is for TMA-alcohol. Side etch width

increases with etching time, and does not depend on via hole size.
This is a great improvement and promises uniform and reproducible
etching processes. In adittion, TMA-alcohol can form via hole as
small as 3x3 µm as well as patterns as large as 50x50 µm within
the same etching period.

Figure 2 shows the via holes formed by TMA-aqueous solution.
After 100 seconds etching, 7x7 µm and 10x10 µm via holes are
underetched, while 16x16 um via holes are slightly overetched. At
120 seconds, 7x7 µm via holes are still underetched, 10x10 um via
holes are properly etched and 16x16 µm via holes are considerably
overetched.

Figure 3 shows the via holes formed by TMA-alcohol solution.
After 40 seconds etching, 3x3 to 10x10 µm via holes are all just
etched. At 60 seconds, the side etch width for all the via holes
is, independent of via hole size, the same.

SEMs of the via holes are shown in Figure 4. The left side
of the figure shows via holes for the TMA-aqueous case. A 5x5 µm
via hole is not completely etched, while 7x7 µm via hole is
underetched. These via holes are round in shape and isotropically
etched.

On the other hand, for TMA-alcohol case, formation of square
shaped via holes from 3x3 to 10x10 µm with steep tapered slope is
illustrated on the right side of the figure. The tapering slope
is typically 60 degrees. This taper angle satisfies both
requirements: good step coverage and high density multilevel
metallization.

One more reason why a TMA-alcohol solution is superior to
aqueous solution is that the former causes less corrosion of
aluminum metallization. Figure 5 shows aluminum corrosion with a
TMA solution. The aluminum etching rate with an alcohol solution
is one fifth that with an aqueous solution. During 60 seconds
etching by the alcohol solution, aluminum etching is less than 20
nm, while it is 100 nm with the aqueous solution.

The etching mechanism for the TMA-etchant may be explained as
follows. At first, TMA attacks PIQ imide bonds and subsequently the reaction products dissolve into the solution.
The dissolution rate is thought to depend on via hole size and the
solubility of the reaction products in the solution. Alcohol may
dissolve the reaction product more easily than water does. This
explains why etching properties of TMA-alcohol are more uniform
than is the case for aqueous solution.

The rate at which TMA etches aluminum is thought to depend on
the hydroxyl ion concentration. The degree of electrolytic

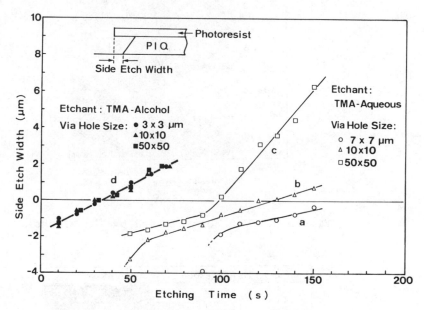

Figure 1. Side etch width vs. etching time.

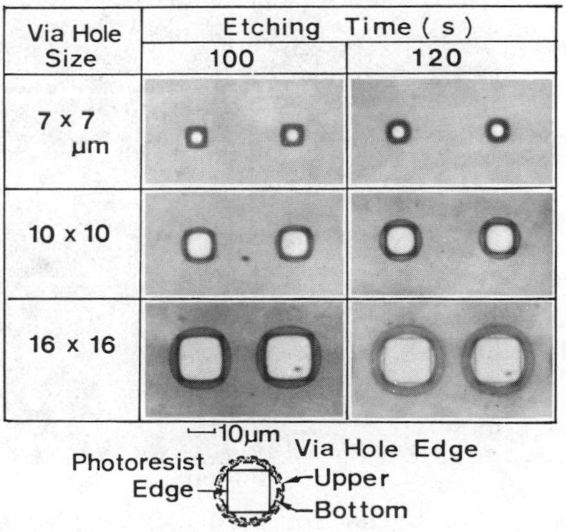

Via Hole	Etching Time (s)	
Size	100	120
7 x 7 μm		
10 x 10		
16 x 16		

⊢—10μm

Via Hole Edge

Photoresist Edge — Upper — Bottom

Figure 2.   Via holes formed with TMA-aqueous solution.

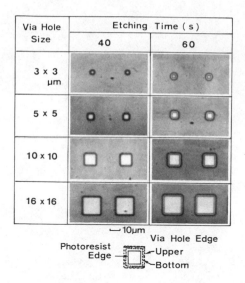

Figure 3. Via holes formed by TMA-alcohol.

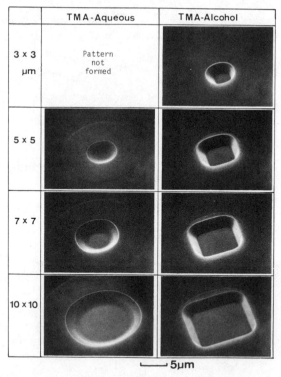

Figure 4. SEM's of via holes formed by TMA-aqueous and alcohol solutions.

dissolution of TMA in alcohol is thought to be smaller than in the aqueous solution, thus resulting in a smaller hydroxyl ion concentration in alcohol solution. Therefore, aluminum is barely etched by TMA-alcohol as compared to the TMA-aqueous case.

What has been presented up to now verifies that a TMA-alcohol etchant offers uniform etching process and fine via holes as small as 3x3 μm with steep taper slopes. The proper etching time does not depend on via hole size. In addition, this etchant barely corrodes aluminum.

Hydrazine Based Etchant   When PIQ is etched by hydrazine hydrate, the film swells at the beginning of etching, and a few minutes later, is dissolved into the etchant. This is shown in Figure 6. However, the mixed solution of hydrazine hydrate and ethylenediamine reduces the swelling. The greater the ethylenediamine content, the smaller the swelling. This is also shown in Figure 7. A hydrazine hydrate etchant contains 60% ethylenediamine does not cause swelling.

Figure 7 shows the via holes formed by various ethylenediamine contents. These figures show that with an increase in the ethylenediamine content, the swelling decreases. Above 60% ethylenediamine the swelling is not observed, and this is accord with the results shown in Figure 6. Even a 5x5 μm via hole can be formed with the required shape.

Figure 8 shows a 3 x 3 μm via hole covered with second level aluminum metallization. The taper angle is about 45 degrees, and offers good step coverage with the second level metallizations.

The etching mechanism was investigated as follows. A PIQ film was immersed in a hydrazine hydrate solution, followed by immersion in ethylenediamine. Figure 9 shows the film thickness change in ethylenediamine for PIQ film previously immersed in a hydrazine hydrate for 4 minutes. It was found that the ethylenediamine dissolves only the reacted portion of PIQ film. The etch rate of reacted PIQ in ethylenediamine is about 2 μm/min. Figure 10 shows the PIQ residual thickness after immersion into ethylenediamine as a function of previous immersion time in hydrazine hydrate. The "reaction rate" of PIQ with hydrazine hydrate is about 0.5 μm/min.

Using these results, the etching mechanism of hydrazine hydrate and ethylenediamine mixed solutions can be explained as follows. First, hydrazine reacts with PIQ at a "hydrazidation rate" of 0.5 μm/min. Then, ethylenediamine dissolves the hydrazided portion at a dissolution rate of 2 μm/min, which is four times faster than the hydrazidation. Thus with the mixed solution, dissolution proceeds more smoothly and swelling does not occur.

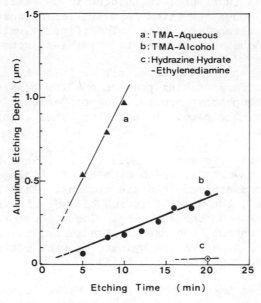

Figure 5.   Aluminum corrosion due to polyimide etchants.

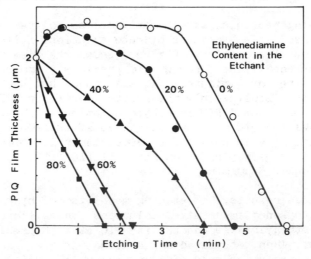

Figure 6.   PIQ film thickness in various hydrazine hydrate-
ethylenediamine etchants (PIQ cure: 220°C for 1 hr.).

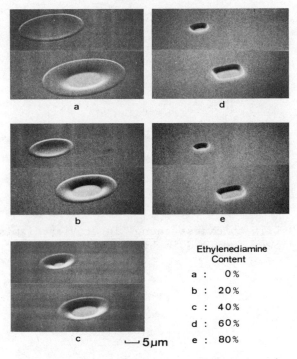

Ethylenediamine
Content

a : 0%
b : 20%
c : 40%
d : 60%
e : 80%

⟶5µm

Figure 7. Via holes formed by hydrazine hydrate and ethylenediamine etchants (upper via hole: 5x5 µm; lower via hole: 7x7 µm).

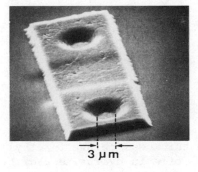

3 µm

Figure 8. 3 x 3 µm via holes covered by the second level metallization.

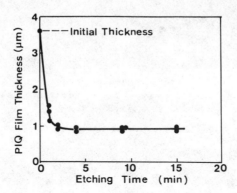

Figure 9. PIQ film thickness change in ethylenediamine. The film had previously been immersed in hydrazine hydrate for 4 min.

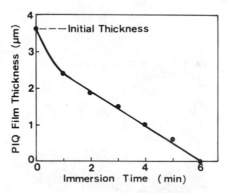

Figure 10. PIQ film thickness change in ethylenediamine as function of previous immersion time in hydrazine hydrate.

This makes it possible to form via holes as small as 3x3 µm in 2 µm thick PIQ film.

Oxygen Plasma Etching   Side etch width rapidly increases with each time during oxygen plasma etching. This is shown in curve a of Figure 11. Even when etching time was suitable for proper etching, that is when the via hole bottom surface just appeared, the side etch width was about 1.5 µm. Therefore, a 2.5x2.5 µm via hole widened to a diameter of about 5 µm.

On the contrary, when PIQ film was etched by oxygen sputtering, the side etch rate was very small, as is shown in curve b. The side etch rate when oxygen sputtering is employed is about two orders smaller than in the oxygen plasma case. Therefore, oxygen sputtering is considerably superior to oxygen plasma in terms of via hole size controllability.

Via holes formed by oxygen plasma and sputtering are shown in Figure 12. Oxygen plasma produces a via hole taper of 45 degrees. This offers good step coverage for the second level metallization. However, severe undercutting can be seen, indicating poor controllability. In contrast, the via hole taper formed as a results of oxygen sputtering is perpendicular to the substrate surface. The via hole size is almost the same as the window size formed in molybdenum masking film. This is suitable for high density metallization and precise via hole size controlling. However, step coverage at the via hole is very poor, resulting in an open circuit in the second metallization. There was no via hole size dependence on side etch rate for both plasma and sputter etching.

SUMMARY

Five etching processes that have been described are compared and summarized in Table I. A comparison of these processes was made considering fine patternability, via hole size controllability and step coverage in overlaying metallization. The potential for high density metallization is also discussed.

A TMA-alcohol solution shows excellent characteristics with respect to all the items above. This etchant can form fine via holes as small as 3x3 µm. There is no via hole size dependence on etching time, thus ensuring good controllability. Step coverage is good because of the 60 degree taper, which is also suitable for high density metallization.

A TMA-aqueous etchant provides little advantage except for step coveerage. Thus it has little potential for future via hole processes.

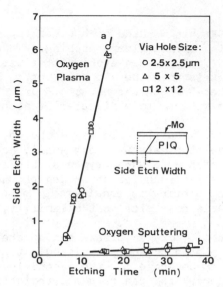

Figure 11. Side etch width vs. etching time.

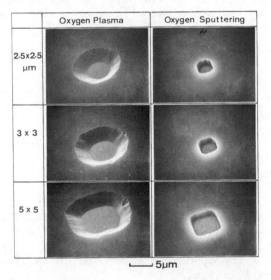

Figure 12. SEM's of via holes formed by oxygen plasma and sputtering.

A hydrazine hydrate-ethylenediamine mixed etchant, on the other hand, shows required fine patternability, controllability and step coverage. However, its potential for high density metallization is somewhat lower than that of TMA/alcohol because of a gentle via hole slope.

Oxygen plasma and sputtering complement each other. In the future, oxygen sputtering will undoubtedly be put to practical use because of its superior potential for high density metallization. However, step coverage problems need to be solved, for example, by developing a method of filling via holes with conductive materials.

Table I. Comparison of Several Etching Processes.

	TMA-Alcohol	TMA-Aqueous	Hydrazine Ethylene-diamine	Oxygen Plasma	Oxygen Sputter-ing
Fine Pattern-ability	good	poor	good	poor	good
Via hole Control-lability	good	poor	good	poor	good
Step Coverage	good	good	good	good	poor

REFERENCES

1. K. Sato, S. Harada, S. Saiki, T. Kimura, T. Okubo and K. Mukai, IEEE Trans. Parts, Hybrid and Packaging, PHP-9, 176 (Sep. 1973).
2. K. Mukai, A. Saiki, K. Yamanaka, S. Harada and S. Shoji, IEEE J, Solid-State Circuits, SC-13, 462 (Aug. 1978).
3. P. Shah, D. Laks and A. Wilson, IEDM Technical Digest, 465 (Dec. 1979).
4. R. A. Larsen, IBM J. Res. Develop., 24(3), 268(May 1980).

REACTIVE ION BEAM ETCHING (RIBE) OF POLYIMIDE

J. Tkach

Commonwealth Scientific Corporation
500 Pendleton Street
Alexandria, Virginia   22314

A broad-beam Kaufman-type (Millatron^{TM}) ion source has been used to demonstrate a practical method of etching microstructures into a polyimide layer, 1-2 microns deep. Control of the wall profiles has been achieved.

Contact resistance measurements made between the exposed underlying layers and sputter deposited aluminum pads showed the contact was highly conductive. This indicates the etch process left little, if any, residual contamination on the via bottoms.

INTRODUCTION

Etching polyimide in a precise manner is of significant importance to the microelectronic industry because of the need to clear via holes through polyimide used in multilayer integrated circuit devices. The vias are then metallized, forming electrical contacts between the upper and lower devices. Some of the problems encountered in realizing the via hole by more conventional methods (i.e. wet chemical and direct plasma etching) are: 1) profile control of the sides of the via (the goal is to produce vertical or tapered sidewalls) which affect the continuity of the subsequently deposited metallizations (usually aluminum) and 2) the cleanliness of the bottom of the via which exposes the underlying device and which must be electrically and mechanically free of debris in order not to reduce the contact resistance. The work resulting from our Reactive Ion Beam Etching of the vias demonstrates controllability of the wall profiles and cleanliness of the via bottom.

EXPERIMENTAL

Detailed descriptions of Kaufman-type ion sources are given elsewhere[1], but the principles of the equipment used are as follows:

A collimated ion beam is extracted from a plasma discharge confined within the ion source. In the source, electrons emitted from a cathode travel toward the anode surface and create ionized species from the gas admitted into the source via collisional effects. A variable axial magnetic field is supplied by an electromagnet surrounding the source to increase the electron path length to the anode.

Typical cathode to anode (discharge) voltages range from 40 to 70 volts, while the gas pressure in the source is on the order of 4 to 5 x 10^{-4} Torr. The front or "downstream" end of the source is comprised of a dual grid assembly, also known as the ion optics.[2] The grids are comprised of thin plates of molybdenum or pyrolytic graphite on the order of 1 mm thick. Holes (2 mm diameter) are drilled into the material through which ion "beamlets" are accelerated. Spacings between these parallel grids are also typically 2 mm. The inner grid is at a positive voltage with respect to ground, with the voltage determining the ion energy in eV. The downstream grid is biased negatively and influences the ion current and the amount of beam divergence.

842

Since the ion beam has a net positive charge, it is necessary to neutralize it to avoid surface charge build-up of the substrate as well as to decrease the level of beam divergence caused by electrostatic repulsion. A filament is provided for emitting the neutralizing electrons and is placed in the beam, close to the source for efficient operation.

The neutralizer filament temperature is controlled by a feedback signal from the sample fixture. The process of neutralizing is macroscopic and takes place at the sample surface exposed to the beam, as opposed to an atomic or molecular recombination during the beam flight.

Our samples were prepared on 4" diameter silicon wafers. A layer of Hitachi PIQ^{TM}, approximately 1.5 microns thick was applied, on top of which was a patterned silicon oxide or silicon nitride and photoresist combination mask. The wafer was Drichuked* (using an elastomeric pad) onto a rotating platen and heat sink. The ion beam uniformly covered the wafer diameter and the angle of beam incidence was adjustable between normal and 45 degrees. Typical beam parameters explored were ion energies of 600-900 eV and current densities between 0.5 and 2 mA/cm^2 at the sample surface.

Gas mixtures of oxygen and argon were introduced into the ion source, using mass flow controllers. The vacuum system used cryogenic pumping and maintained a total process pressure of 1 to 3 x 10^{-4} Torr. The interior of the 8" diameter ion source contained a graphite lining to enhance stable operation with reactive gases. Oxide reactions and polymeric deposits from reactive gases in an unlined, aluminum anode source caused a decrease in the source operating lifetime.

RESULTS

It was found that resist thickness, ion gas mixtures, ion beam parameters and etch times were influential in controlling wall profiles in the polyimide. Etching rates ranged between 2,000 and 2,500 Å/minute, and were measured by dividing the etched step height (as measured on a surface profilometer) by the etching time. The SEM photographs (Figures 1 and 2) show the polyimide sidewall geometries achieved. (Photographs of more sloping-tapered wall profiles were also produced but were considered proprietary and were not available for submission.) Selectivities of etch rate of the polyimide over the underlying aluminum layer averaged between 30:1 and 40:1. Modest over-etching insured low electrical resistance contacts between the exposed layer and

---

*"Drichuk" is a trademark of Commonwealth Scientific Corporation.

Figure 1.   SEM of polyimide corner pattern.   Ion current
density was 2 mA/cm^2 and ion energy was 800V.

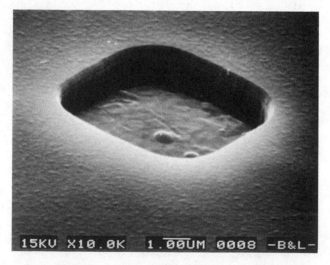

Figure 2.   SEM of polyimide via hole.   Same operating
conditions as Figure 1.

subsequently deposited metallizations. The over-etching with pure argon (i.e. ion milling) removes residual polyimide debris and oxides that are insulating. Resistance checks were made by probes attached to several via contact pairs isolated from each other. The measured resistance was similar to that of one continuous line of aluminum of the same effective length as the contact path. This value of resistance was below 10 ohms in the samples checked.

DISCUSSION

It is believed that in RIBE, the polyimide removal mechanism consists of a combination of directional physical sputter erosion and a chemical formation of volatile or easily sputtered material formed of polyimide and reactive gas constituents.

A more detailed discussion of the physical sputtering process in ion beam milling and the effects of masking on etch profiles are discussed in References 3 and 4, while general treatments of RIBE processing are given in References 5 and 6. RIBE of polyimides has also been previously explored by other workers (References 7 and 8). Etching rates and the selectivity of this former work are similar to our findings. This previous work, however, did not indicate a tapering profile of etch into the polyimide, but that a "highly anisotropic"[7] etch was found. Our ability to reproduce a tapered profile is significant in allowing an electrically continuous and reliable metallizing contact from the sides to the bottom of the via.

A comparison of RIBE to plasma or RIE (Reactive Ion Etching) processing of polyimides indicates the following fundamental differences:

1. In RIBE the samples are located outside of the plasma discharge area and thus are not subject to possible direct secondary electron impingement. Furthermore, an arbitrary angle of the sample with respect to the beam can be chosen.

2. The processing pressures in RIBE are on the order of 1×10^{-4} Torr, as opposed to several mTorr in plasma and RIE, so that effects such as redeposition of the underlying metallization surface by random scattering during the over-etching can be neglected.

3. The beam energy and current density are directly read and can, to a good degree, be independently varied in RIBE.

CONCLUSION

The etching rates (2500 Å/min.) and control of geometry (enabled by varying the process parameters) coupled with practical equipment considerations, indicate RIBE is a viable means of etching polyimide structures for microelectronic applications.

REFERENCES

1. H. R. Kaufman and R. S. Robinson, "Ion Source Design for Industrial Applications", in Proceedings of the International Electric Propulsion Conference, AIAA-81-0668, (1981). Available from American Institute of Aeronautics and Astronautics, New York, New York 10104.
2. H. R. Kaufman, J. J. Cuomo and J.M.E. Harper, Part I, J. Vac. Sci. Technol., 21(3), 725 (1982).
3. H. L. Garvin, Solid State Technol., 31 (Nov. 1973).
4. P. G. Gloersen, Solid State Technol., 68 (April 1976).
5. T. M. Mayer, R. A. Barker and L. J. Whitman, J. Vac. Sci. Tech., 18(2), 349 (March 1981).
6. B. A. Heath, Solid State Technol., 75 (Oct. 1981).
7. P. D. DeGraff and D. C. Flanders, J. Vac. Sci. Technol., 16(6), 1906 (1979).
8. H. R. Kaufman, J. J. Cuomo and J. M. E. Harper, J. Vac. Sci., Technol. Part II, 21(3), 737 (1982).

POLYIMIDE SILOXANE: PROPERTIES AND CHARACTERIZATION FOR THIN FILM ELECTRONIC APPLICATIONS

G. C. Davis, B. A. Heath* and G. Gildenblat

General Electric Company
Corporate Research and Development
P.O. Box 8, Schenectady, New York 12301

Polyimide siloxane has been shown to be an excellent dielectric material for thin film applications. It is comparable to and in some cases better than Pyralin® 2555, a commercially available polyimide, in meeting the requirements for electronic application yet has the important advantage of not requiring an adhesion promoter to adhere to nearly all surfaces. Data on SiPI are presented with respect to planarization, adhesion, electrical properties, thermal properties, purity and patternability, demonstrating that SiPI has excellent potential in the field of electronics.

---

*Present Address: Inmos Corporation, P.O. Box 16000
Colorado Springs, Colorado 80935

INTRODUCTION

The use of polyimides in microelectronics, particularly as dielectric interlayers and passivants, is growing at a rapid rate.[1] At the present time, only a limited number of polyimides are available for these applications. It is our aim in this paper to present data on a silicon containing polyimide which maintains the excellent properties of polyimides for electronic applications yet in some respects is superior to commercial polyimides, particularly with respect to adhesion and resistance to water. Some comparisons to DuPont Pyralin® 2555 polyimide will be made.

A number of requirements and properties of a polymer must be met for it to be useful in a thin film electronic application. Included in this list are requirements of surface planarization, adhesion, electrical properties, thermal properties, purity and methods of patterning. An understanding of the practical aspects of preparing, applying and handling the polymer and its solutions is also necessary.

POLYIMIDE SILOXANE (SiPI) SYNTHESIS

The polymer reported here is a block copolymer of benzophenone dianhydride (BTDA) with methylene dianiline (MDA) and bis-gamma aminopropyltetramethyldisiloxane (GAPD). The polymer solution is prepared by dissolving the BTDA in a dipolar aprotic solvent, preferably N-methylpyrrolidone (NMP) followed by addition of MDA. Once the MDA has completely dissolved, the GAPD is added and the entire polymer solution is stirred overnight. Room temperature is maintained throughout the operations. The above solution gives a polyamic acid solution of extended A-B-A type copolymer. Random copolymers have also been prepared by adding the two diamines simultaneously to the dianhydride solution. The data presented in this paper are on the 70/30 MDA/GAPD block copolymer, although other diamine ratios as well as other polyimide siloxane prepared from other monomers have also been investigated. The acid amide copolymer is soluble in neat dipolar aprotic solvents, as well as dipolar aprotic solvents mixed with small amounts of aromatics (e.g. toluene) or alcohols (e.g. cellosolve). The polyamic acid can be imidized by the application of heat to give a polymer which is no longer soluble in organic solvents and is resistant to dilute acids and most other aqueous solutions. The synthesis sequence of SiPI is shown in Scheme I. The ^{1}H nmr, ^{13}C nmr, ^{29}Si nmr, and IR spectra of SiPI amic acid are shown in Figure 1.

®Pyralin is a Registered Trademark of E.I. Dupont de Nemours Company.

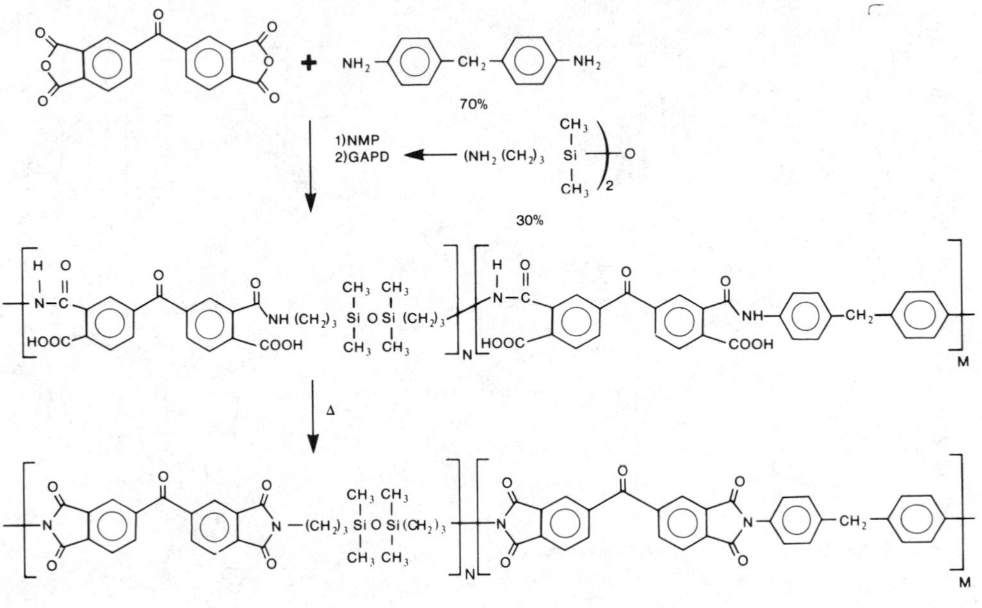

Scheme I: Preparation of SiPI.

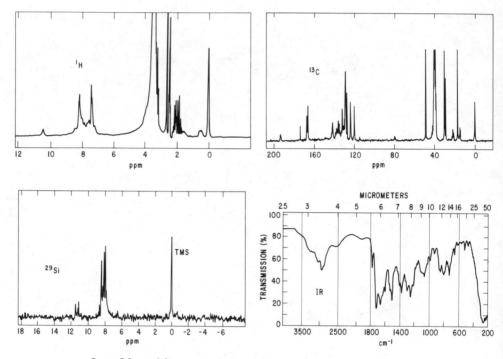

Figure 1. ^1H, ^{13}C, ^{29}Si nmrs and IR Spectra of Polyimide Siloxane (amic acid).

SOLUTION PROPERTIES OF SiPI

Polyimide siloxane solutions up to 35% solids have been prepared. Above this level, high solution viscosity makes working
with the solution difficult as a spin-on coating material. A plot
of solution viscosity versus percent solids is shown in Figure 2.
The solutions of SiPI amic acid are hygroscopic and water can lead
to a rapid reduction in solution viscosity due to hydrolysis and
chain scission. As shown in Figure 3, the viscosity loss is
dramatic at elevated solution temperatures. However, refrigerated
samples (~4°C) show little or no loss in viscosity over periods of
at least 1 year. GPC measurements made on SiPI polyamic acid
solutions show that the dispersivity increases with time, Table 1.
This broadening of the molecular weight distribution is an
indication that the SiPI is a living polymer system.

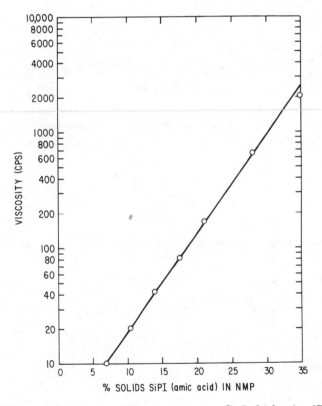

Figure 2. Plot of Solution Viscosity vs. % Solids in NMP of SiPi
(amic acid).

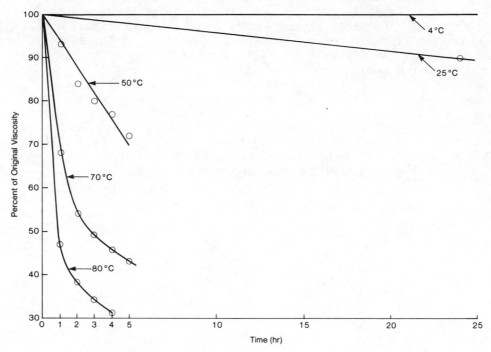

Figure 3.   Effect of Temperature on the Viscosity Loss in SiPI
(amic acid) Solutions in NMP.   Solutions Contain about
0.9% Water.

Table 1.   Changes in Dispersivity of SiPI Amic Acid Solutions as a
Function of Time.

SiPI
Age of Polymer Solution
(months)

Age of Polymer Solution (months)	Dispersivity
6	4.73
19	5.60
29	7.02

SPINCOATING OF POLYIMIDE SILOXANE SOLUTION

Films ranging from 800Å to 10μ have been successfully prepared in one application by spincoating SiPI amic acid solutions. Pinholes when observed can be successfully eliminated by double coating the polymer. Surface preparation is extremely important in the successful application of polymer films. The surface must be clean. Plasma descumming is sufficient to give a clean surface.[2] SiPI can be coated on an uncured previously deposited SiPI layer without wetting problems; however, on a cured SiPI film a plasma descum between layers is often needed. Dilution of the NMP solution of SiPI amic acid with solvents such as toluene or cellosolve has proved beneficial in improving solution surface wetting with a subsequent improvement in film quality.

After coating with SiPI amic acid, the film is dried at 100°C for 1 hour. This leaves the polymer in the acid amide form, although a considerable amount of NMP remains (Table II). To convert the acid amide to imide the film is heated at 200°C for 1 hour followed by a 300°C bake for 1 hour. Fourier transform IR measurements show that the imidization at 200°C requires less than 1 hour and at 300°C it is less than 15 minutes. The 300°C bake is needed to remove the last traces of solvent and the tightly bound water of imidization. Table II shows trapped NMP as a function of drying cycle.

Film thickness decreases of 25 to 30% have been noticed in SiPI polymer films upon curing, due primarily to the large amount of NMP in the film after 100°C drying and partially to the loss of water upon imidization. On wafers which have been patterned at acid amide stage, this decrease is seen only in the z (thickness) dimension. Other line dimensions are not affected.

Table II. NMP Remaining in SiPI Film After Various Drying Sequences.

Drying Sequence (1 hour)	% NMP in Film[a]
100°C	22
100°C, 200°C	6
100°C, 200°C, 300°C	<<1

[a] % NMP was measured by film pyrolysis at 400°C, trapping volatiles, and analysis by gas chromatography.

SURFACE PLANARIZATION BY POLYIMIDE SILOXANE

One of the major advantages of polyimides as dielectrics over inorganic materials such as SiO_2 or Si_3N_4 is their ability to planarize a wafer surface. A method to measure the degree of planarization has been developed by Rothman.[3] The illustration below defines the parameters. The degree of planarization is defined as:

$$\text{Degree of Planarization (DP)} = \left(1 - \frac{t_s}{t_L}\right) \times 100\%$$

and θ is the slope of the polyimide mound.

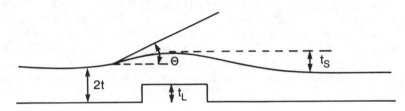

Figure 4. Structure Used to Determine Degree of Planarization of Polymer Over Isolated Step.

Using a structure close to that shown in Figure 4, it was determined that the SiPI at 27% solids in NMP gives a DP of about 50% and the angle θ of about 15°. These values compare favorably to values for commercial polyimides reported by Rothman.[3] An SEM step-coverage comparison of SiPI versus Pyralin 2555 is shown in Figure 5. SiPI is as good if not better than Pyralin in this comparison and more than adequate for microelectronic applications. Attempts to flow the polymer to improve planarization by heating at elevated temperatures (<400°C) had no affect.

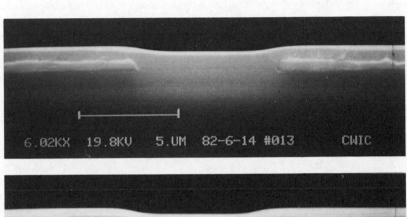

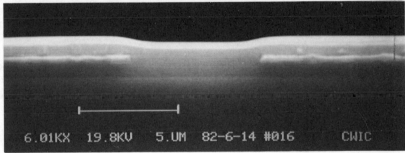

Figure 5.   SEM of Polyimide Over Aluminum Step: A. Polyimide
Siloxane; B. Pyralin 2555.

ADHESION OF POLYIMIDE SILOXANE

All commercial conventional polyimides require an adhesion promoter when used as thin films on microelectronic devices.[4a-c] The fate of the adhesion promoter during polyimide cure and subsequent high temperature processing steps is unknown and the subsequent effectiveness of the polyimide to maintain adhesion and prevent corrosion between the polyimide and the substrate surface must be questioned. Removal of adhesion promoter during polyimide wet etching has been shown to be a problem.[4d] Polyimide siloxane requires no adhesion promoter to adhere to all surfaces tested including gold. Comparisons of SiPI to Pyralin 2555 were made using the standard scribe/tape pull test. The wafers were coated with the polyimide films (1μm) and cured. Samples of Pyralin with and without the use of supplied adhesion promoter were prepared. Manufacturer's specified cure cycle, final cure at 300°C for 1 hr, was employed for Pyralin. The same cure cycle was used for SiPI. The wafers were then scribed with a cross hatch pattern giving 100 squares, 1mm on a side. Adhesion was tested by pressing 3M-710 tape onto the pattern and quickly removing the tape at an angle of 90° to the wafer surface. This was done three times on a fresh surface of tape. The wafers were tested prior to submersion into boiling water and at various times thereafter. The adhesion results on a number of surfaces is presented in Table III.

As shown in Table III, SiPI adhesion was superior to adhesion promoted Pyralin 2555 on single crystal Si and SiO_2 surfaces. Adhesion was maintained with SiPI even after three days in boiling water on these surfaces. SiPI was better than non-promoted Pyralin 2555 on a Si_3N_4 surface and both polymers showed excellent adhesion on an aluminum surface. Intercoat adhesion of SiPI is good.

Since no adhesion promoter is needed for SiPI, device fabrication is simplified by one step each time the polymer is used and the adhesion characteristics will be consistent and not dependent on an adhesion promoter.

ELECTRICAL PROPERTIES OF POLYIMIDE SILOXANE

In order to be useful as a dielectric material, some important electrical properties must be met. Some of the properties which have been determined are shown in Table IV.

Table III. Adhesion of Polyimide Siloxane and Pyralin on Various Wafer Surfaces.

| Sample | Surface | Test Results[a] | |
		No Boiling Water	1 hr Boiling Water[c]
SiPI	Si	P	P
Pyralin w/AP[b]	Si	P	F
Pyralin wo/AP	Si	P	F
SiPI	SiO_2	P	P
Pyralin w/AP	SiO_2	P	F
Pyralin wo/AP	SiO_2	P	F
SiPI	Si_3N_4	P	P
Pyralin w/AP	Si_3N_4	P	P
Pyralin wo/AP	Si_3N_4	F	F
SiPI	Al	P	P
Pyralin w/AP	Al	P	P
Pyralin wo/AP	Al	P	P
SiPI	SiPI	P	P

[a] P - passed tape pull test; F - failed tape pull test.

[b] AP - Adhesion Promoter

[c] All samples that withstood 1 hr. boiling water test withstood boiling water for at least 3 days.

Table IV. Electrical Properties of SiPI.

Property	Value
Dielectric Constant (Ref. 5)	3.0
Bulk Resistivity (Ref. 5)	10^{17} ohm-cm at 25°C
Dielectric Strength	5.5 MV/cm

Electrical characteristics of SiPI and Pyralin 2555 films were also compared. As shown in Table V, the breakdown voltage for SiPI films of thicknesses greater than 2,500Å up to at least 7,000Å scales linearly. The breakdown voltages were obtained on Si wafers by using Al gate capacitors with the polyimide as the dielectric material. The average breakdown field of 5.5 MV/cm compares favorably with that measured for Pyralin of 6.0 MV/cm and sufficient for most electrical thin film applications. In both polyimide siloxane and Pyralin 2555 hysteresis was observed in the high frequency C(V) curves of capacitor structures employing the polyimide layers. This hysteresis is explained by the presence of trapping centers in the polymer films. The effect can be suppressed by a thin layer of LPCVD SiO_2 (500Å) deposited prior to polyimide application. Figure 6 illustrates these results for polyimide siloxane. Similar results were observed in capacitors containing Pyralin 2555. In a few cases, residual hysteresis was observed even after SiO_2 deposition. The width of this hysteresis loop did not exceed 0.4V as compared with 4-8V without SiO_2.

Table V. Breakdown Voltage and Field in SiPI Films of Varying Thicknesses.

Polyimide Siloxane Thickness	Average Breakdown Voltage	Average Dielectric Strength
Å	V	MV/cm
2700	155	5.7
5000	275	5.5
7000	375	5.4

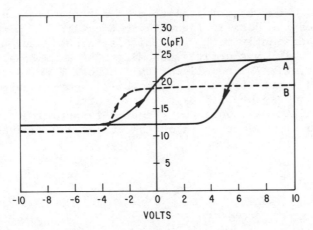

Figure 6. High Frequency (10^6Hz) C(V) Plots For: A. Al/3000Å Polyimide Siloxane/Si; and B. Al/3000Å Polyimide Siloxane/500Å SiO$_2$/Si Capacitor Structures.

Charge storage at an oxide/polyimide interface has been described by Brown[6] to be the cause of a field threshold instability. The following Equation has been derived for the threshold (or flat band) voltage shift (ΔV):

$$\Delta V = - V_s \frac{C_o}{C_p}(1 - e^{-t/\tau}) \qquad (1)$$

$$\text{where} \quad \tau = (C_o + C_p)/G_p \qquad (2)$$

V_s is the stress voltage, C_o and C_p are the capacitances of the oxide and the polyimide, respectively, G_p is the conductivity of the polyimide, and t is the stress time. The above Equation assumes a linear conduction of the polyimide and no conductivity of the oxide.

Using a SiPI containing capacitor similar to that used to obtain the C(V) curve, the flat-band voltage shift was measured after a bias temperature stress. Both positive and negative stress voltages were applied. The results of this study are shown in Figure 7. The flat-band voltage shift has the opposite polarity to that of the stress voltage, its absolute value does not depend on the sign of V_s, and it increases with the absolute value of V_s. All these observations can be predicted from

Equation (1). However, the values of τ determined from a least squares fit of the data in Figure 7 to Equation (1) were found to decrease significantly with V_S. This is inconsistent with Equation (1) which was derived assuming that G_p in Equation (2) remains constant. The non-linearity of the I-V characteristics of polyimides is a probable explanation. This non-linearity might result in a non-exponential $\Delta V(t)$ dependence. In addition to the shift of the flat-band voltage, the bias-temperature stress described above produces a distortion of the C(V) curves as shown in Figure 8.

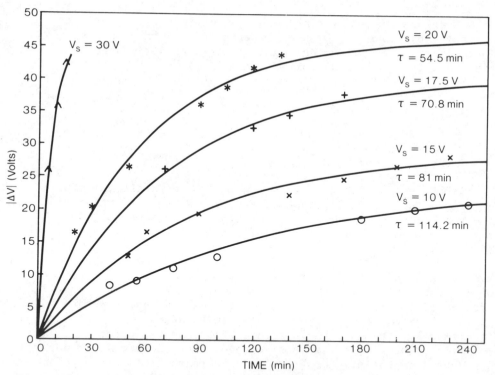

Figure 7. Absolute Value of the Flat-Band Voltage Shift As a Function of the Bias Temperature (150°C) Stress Time. The Solid Lines and the Values of τ Were Obtained by a Least Squares Fit to Equation (1).

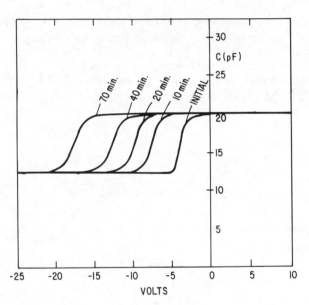

Figure 8. Shift and Distortion of the High Frequency (10^6 Hz) C(V) Curves As a Result of Bias Temperature (150°C) Stress on Capacitor Structure Similar to Sample B, Figure 6. Stress Voltage is 15V.

THERMAL PROPERTIES OF POLYIMIDE SILOXANE

The 70/30 SiPI has a glass transition temperature (T_g) of 190°C; however, crosslinking of the polymer film occurs when heated to temperatures above 300°C in air. Dimensional stability is maintained by SiPI structures when subjected to temperatures above 300°C. Thermogravimetric analysis (TGA) show that 10% weight loss (not including weight loss observed for removal of trapped solvents and water of imidization) in both N_2 and air occurs at about 500°C. The TGA analysis is shown in Figure 9. Isothermal TGA data indicates the polymer to be stable to about 450°C in N_2 and 400°C in air (Figure 10). SiPI was shown to be comparable in stability to Pyralin 2555 in air; however, stability of Pyralin 2555 as measured bv isothermal weight loss was about 50°C higher in N_2. In devices containing SiPI which were intentionally subjected to temperatures above those at which the polyimide siloxane begins to lose weight, no evidence of cracking, peeling, blistering or loss of dimensional stability was observed, although Dektak measurements and SEM pictures indicate film thinning had occurred.

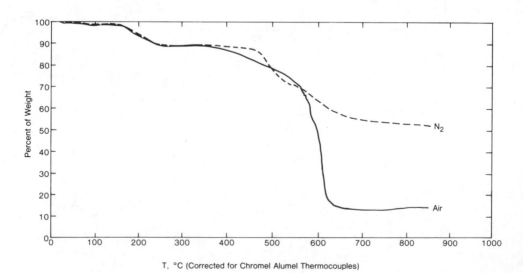

Figure 9. Thermogravimetric Analysis of SiPI (Film Cast Dried 100°C/1 hr) at 10°C/min in Air and Nitrogen.

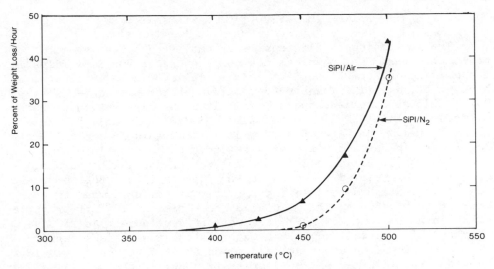

Figure 10. Isothermal TGA Results of SiPI (imide) Represented As Percent Weight Loss/Hr at Given Temperature.

PURITY OF POLYIMIDE SILOXANE

Mobile sodium has been implicated as a possible source of device instability in structures containing polyimides.[6] There is some evidence, however, that under some conditions sodium is not mobile in polyimide films.[7] Despite the arguments it is advantageous to maintain sodium levels as low as possible in the polyimide polymers. Our attempts to decrease sodium from SiPI amic acid solutions by adsorption techniques has met with limited success. It is therefore advisable to remove sodium from the monomers and solvents prior to polymer preparation. Sodium levels in monomers and solvent used for SiPI are indicated in Table VI. Sodium levels in SiPI amic acid solutions have been controlled to under 2ppm.

Table VI. Typical Sodium Levels in Monomers and Solvent Used to Prepare SiPI.

Material	Na (ppm)
BTDA	2.7
MDA	2.1
GAPD	1.9
NMP	0.2
SiPI Solution	1.7

PATTERNING OF POLYIMIDE SILOXANE

The patterning of SiPI can be accomplished by wet chemical or dry techniques. Wet chemical methods are useful for SiPI films in the acid amide form. The SiPI acid amide is soluble in aqueous base which makes it amenable to positive photoresist developing. Using Shipley base developers, a simultaneous dissolution of both exposed positive Shipley photoresist and SiPI amic acid has been accomplished. The rate of SiPI amic acid dissolution in the developer can be controlled by altering the ratio of insoluble imide to soluble amic acid in the polymer chain, and by varying the amount of trapped solvents in the polymer film. Both of these can be accomplished by changing the polymer drying temperature or time or both. For a SiPI amic acid film dried at 100°C/1 hr (this film contains about 22% NMP), the rate of development at 25°C in Shipley 351 diluted 1 to 4 with water is about six times faster than the rate of development of exposed AZ 4620 (exposure setting 125 on a Perkin Elmer 111 projection printer) in the same developer.

Direct solution development of the SiPI amic acid film can lead to undercutting in fine geometries. This undercutting can be severe if care is not taken to prevent it (see Figure 11). Better control of polyimide siloxane patterning can be obtained by reactive ion etching (RIE). RIE patterning of SiPI (imide) has been demonstrated to be useful at VLSI geometries.

Reactive ion etching experiments on polyimide siloxane, Pyralin, Shipley AZ 1470 photoresist and various other materials used in semiconductor processing were carried out in Plasmatrac 2406 single wafer etching system. Figure 12 shows the etch rates of polyimide siloxane, Pyralin and photoresist as a function of pressure in an oxygen plasma. At all pressures investigated polyimide siloxane etches more slowly than do the other materials. In addition, a Si containing residue is left behind after etching polyimide siloxane using pure oxygen.

Figure 13 shows the effect of adding CF_4 to the plasma[8] on the etch rates of these materials at a constant pressure of 50 mTorr. An increase in the etch rate of polyimide siloxane is observed as CF_4 is added to the O_2 plasma, presumably due to the formation of volatile silicon fluoride products. At CF_4 concentrations in oxygen of 20% or greater the etch rate of polyimide siloxane is comparable to resist and Pyralin under these conditions. The addition of CF_4 to the oxygen plasma reduces the amount of residue left after etching polyimide siloxane, but at 50 mTorr total pressure, some residue still remains when using a resist mask.

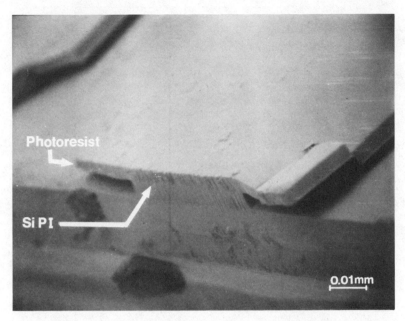

Figure 11. SEM of SiPI (amic acid) Dried at 90°C/Hr Under AZ 4620
Photoresist, Spray Developed 2 min With 4/1, Water/Shipley 351,
Demonstrating Undercut Phenomenon.

In Figure 14, both the etch rates of polyimide siloxane and
resist as well as the extent of residue formation are presented as
a function of pressure in a 20% CF_4 in O_2 plasma. No residue was
observed for SiO_2 samples at 50 mTorr. Unmasked polyimide
siloxane samples show minimal residue under these conditions. The
arrows indicate the amount of residue observed for resist masked
samples at 50, 150 and 300 mTorr. Though residue was observed on
the resist and on the etched portions of these samples at 50 mTorr,
increasing the pressure tended to eliminate the problem. When
etching polyimide siloxane with a resist mask, a pressure of 150-
200 mTorr was found to minimize both resist etch rate and residue
formation. When using an aluminum mask, a higher pressure was
required to etch polyimide siloxane cleanly, though increasing the
pressure did cause undercutting of the mask. At 400 mTorr, a
minimal residue and isotropic etching characteristics were
observed.

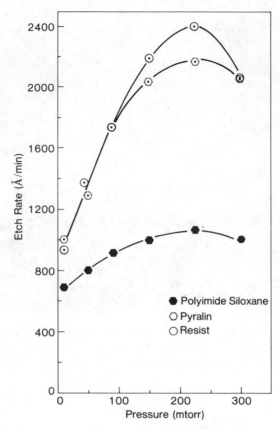

Figure 12. Etch Rate vs. Pressure for Polyimide Siloxane, Pyralin 2555, and Shipley AZ 1470 photoresist in an O_2 Plasma in the RIE Mode: Flow Rate 100 sccm, RF Power Density 0.4 W/cm^2.

SUMMARY

Polyimide siloxane has been shown to be an excellent dielectric material for thin film applications. It is comparable to and in some cases better than Pyralin® 2555 in meeting the requirements for electronic applications, yet has the important advantage of not requiring an adhesion promoter to adhere to nearly all surfaces. Data on SiPI are presented with respect to planarization, adhesion, electrical properties, thermal properties, purity and patternability, demonstrating that SiPI has excellent potential in the field of microelectronics.

866

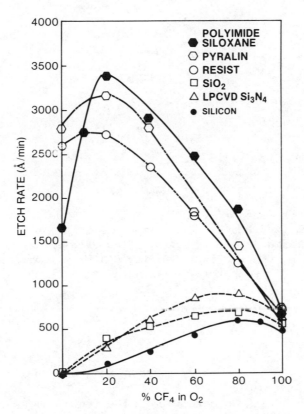

Figure 13. Etch Rates of Polyimide Siloxane, Pyralin 2555,
Shipley AZ 1470 Photoresist, SiO_2, Si_3N_4, and Single
Crystal Si in Mixtures of CF_4 in O_2 in the RIE Mode:
Pressure 50 mTorr, RF Power Density 0.4 W/cm^2.

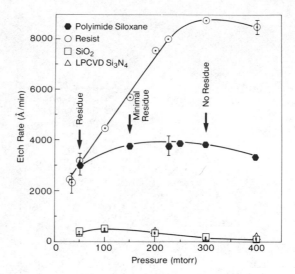

Figure 14. Etch Rate vs. Pressure For Polyimide Siloxane, Shipley AZ 1470 Photoresist, SiO_2 and Si_3N_4 in a 20% CF_4/80% O_2 Plasma in the RIE Mode: Flow Rates 20 sccm CF_4 and 80 sccm O_2, RF Power Density 0.4 W/cm^2. Arrows Indicate the Extent of the Residue After Etching Using a Resist Mask.

ACKNOWLEDGEMENTS

The authors wish to acknowledge the technical assistance of Carol Fasoldt, Bill Hennesy and George Charney in sample preparations and measurements and to George Loucks for supplies of polyimide siloxane.

REFERENCES

1. A. M. Wilson, Thin Solid Films, 83, 145 (1981).
2. D. F. O'Kane and K. L. Mittal, J. Vac. Sci. Technol. 11, 567 (1974).
3. L. B. Rothman, J. Electrochem. Soc., 127, 2216 (1980).

4a. Y. K. Lee, paper presented at the Electrochemical Society
     meeting in Minneapolis, MN, May 1981.
4b. D. Suryanarayana and K. L. Mittal, J. Appl. Polym. Sci., $\underline{29}$,
     2039 (1984).
4c. A. Saiki and S. Harada, J. Electrochem. Soc., $\underline{129}$, 2278 (1982).
4d. J. W. Lin, "Polyimide Coatings for Microelectronics with
     Applications", Continuing Education in Engineering, University
     of California, Berkeley, August 1981.
5.   A. Yerman (1976), Unpublished data.
6.   G. A. Brown, in "Polymer Materials for Electronic Applications",
     E. D. Feit and C. W. Wilkins, Editors, ACS Symposium Series
     No. 184, p. 151, American Chemical Society, Washington, D.C.
     (1982).
7.   A. J. Gregoritsch, Reliability Physics, $\underline{14}$, 228 (1976).
8.   L. Stein, U.S. Patent 4209356.

POLYIMIDE DIE ATTACH ADHESIVES FOR LSI CERAMIC PACKAGES

Justin C. Bolger

Amicon Corporation

Lexington, Massachusetts 02173

This paper presents new equations which permit calculation of thermal stresses on rectangular dies bonded with gold eutectic, glass, polyimide or epoxy adhesives. Epoxies provide lower stresses than any other die attach method. Unfortunately the most stable epoxies begin to decompose at about 350°C, making them useful for many I.C. packages but not for most hermetics.

Addition curing polyimide adhesives produce higher stresses than epoxies but much lower stresses than glass or gold eutectic. The most widely used present-day silver-filled polyimides are, however, not stable enough for use in hermetic packages which are sealed, in air, at 420-450°C. The degradation mechanism is shown, in this paper, to be due to catalytic effects at the silver-polyimide interface.

Silver-filled, condensation-curing, polyimides are more stable than silver-filled addition curing polyimides, in air, above 350°C, but are difficult to process to yield void-free bonds. The best choice depends, therefore, on whether the die attach adhesive must provide electrical as well as thermal conductivity. If back-face electrical conductivity is not needed, aluminum-filled addition curing polyimides provide important processing, heat transfer, strength, thermal shock, and reliability advantages.

I. INTRODUCTION

During the 1980's, the trend in I.C. packaging will continue in the direction of larger dies, increased power and more active elements per die. Many of these larger, higher density I.C.'s will be packaged in hermetic or ceramic packages ("CERDIP's"), in which thermal stresses will become a serious design as well as device reliability problem.

Die attach adhesives have the potential advantage of providing lower thermal stresses than other die attach methods. Epoxy adhesives do not have sufficient thermal stability to survive the glass operations in CERDIP manufacture, but unfilled polyimide resins are available which are stable in air at $450^{\circ}C$ for processing times on the order of 10 minutes. The polyimide adhesives formulated to date for die attach, however, have all contained a high loading of silver powder filler to provide thermal and electrical conductivity, and these silver-filled polyimides are not stable enough for use in most CERDIP's, for reasons explained in Section III.

These silver-filled polyimide adhesives have, rather surprisingly, become widely used in many "high reliability" plastic encapsulated I.C.'s because of the low levels of Cl^-, Na^+, K^+, NH_4^+ and other extractable ionic impurities in polyimides. Until recently, all epoxy die attach adhesives yielded very high levels, typically 500 ppm or more, of extractable Cl^- and other ions. Recently, however, new high purity silver-filled epoxies have become available (Table I) with very low extractable ionic levels and with major cost, strength, processing and cure temperature advantages over the polyimides. These low chloride epoxies are ultimately expected to displace the polyimides for use in plastic packages.

This means that the future use of polyimide die attach adhesives must be in CERDIP's, where $450^{\circ}C$ stability in air is required. For this, improved polyimide adhesives are needed which provide a better compromise between the conflicting requirements of thermal stress, thermal stability, and thermal and electrical conductivity.

II. THERMAL STRESS

Thermal stresses may be generated between die and substrate, between lids and heat sinks and at other interfaces within a ceramic package. The stresses considered in this paper are those for silicon dies bonded to a ceramic substrate with either a gold eutectic solder, a silver-filled glass frit, or epoxy or polyimide die attach adhesives. These stresses are caused by differences in expansion coefficient between die, adhesive and substrate and are generated either during cool down after the die is first bonded, or subsequently as the die temperature rises and falls during cycling.

Table I. Typical Extractable Ionic Content (PPM).
(from Ref. 1)

ADHESIVE	TYPE	SUPPLIER	Cl^-	Na^+
C-805	Epoxy	Amicon	240	50
826-1	Epoxy	Ablestik	160	20
C-850-6	Epoxy	Amicon	160	10
C-940-4	P-imide	Amicon	6	7
C-940-7	P-imide	Amicon	3	3
AB-71-1	P-imide	Ablestik	6	6
CRM-1050	P-imide	Sumitomo	5	4
CRM-1020	Epoxy	Sumitomo	8	5
EN-4000	Epoxy	Hitachi	6	2
C-860-3	Epoxy	Amicon	8	3
C-868	Epoxy	Amicon	9	7
AB-843-1	Epoxy	Ablestik	10	5

Figure 1 shows a rectangular die bonded to a substrate of thickness x_s, by an "adhesive," which can be glass, solder or a polymer, of thickness, x_a. When the die is first bonded, the adhesive layer is in a liquid state and the die is in the zero stress condition. As the bonded unit cools, the glass and solder pass through a solidification temperature, or the adhesive passes through its glass transition temperature, T_g. Continued cooling then generates the stresses shown in Figure 1. For larger dies, these stresses can lead to cracking, spalling or delamination of the die. Even if cracking does not occur, bending stresses can cause dimensional changes on the die surface which can crack passivation coatings, or change resistor or other thin film circuit values.

Many previous authors[2-3] have studied aspects of this thermal stress problem. Chen and Nelson[4] have derived equations for stress distributions in bonded materials differing in expansion coefficient and for thermal stresses at the joint edges when a thin elastic layer is bonded between two rigid, circular wafers.

Niwa et al[5] have studied the warpage of large silicon dies and full (3-inch diameter) silicon wafers, bonded to mullite or to alumina substrates and have derived equations for the maximum shear stress, σ_{max}, at the edge of a circular die or full wafer.

Lewis and Adams[6] have studied thermal stresses in dies bonded to circular metal heat sinks with epoxy or with solder and have obtained a relationship which can be written as:

$$\sigma_{max} = K\, E_a\, (T_o - T)(\alpha_s - \alpha_{si}) \tag{1}$$

where E_a is the modulus of the adhesive

T is the temperature at the interface

T_o is the solidification temperature of the solder or epoxy.

These previous stress analyses have been performed for bonds of circular geometry, which has the advantage of mathematical simplicity and is useful for stresses at heat sink interfaces. Heat sinks in most I.C. packages are invariably either circular wafers or wafers with rounded corners. But I.C. dies are either square or rectangular. The corners are right angles, and maximum thermal stresses occur at these corners. Any meaningful study of die thermal stresses must focus on the edge and corner stresses of Figure 1.

For this paper we assume that:

1. A rectangular silicon die of longest side, L, is bonded to a much larger substrate.

2. The adhesive bond thickness is much less than the thickness of either die or substrate; and the adhesive modulus is less than either silicon or the substrate.

Under these assumptions, a thin film of relatively flexible adhesive can dissipate shear stresses between the two rigid adherends shown in Figure 1. Using Krieger's square root relationship[7] between modulii and peak stress at corners, and other assumptions explained[8] in the original derivation, dimensional analysis leads to the following equation:

$$\sigma_{max} = K\,(\alpha_s - \alpha_{si})\, \Delta T \sqrt{\frac{E_a E_s L}{x_a}} \tag{2}$$

where σ_{max} = maximum stresses at die corners, psi

$\quad K$ = geometric constant. Function of die shape and amount of fillet. Dimensionless.

$\quad \Delta T$ = difference between zero stress temperature and test temperature, ^{o}C

$\quad E_a$ = adhesive tensile modulus, psi

$\quad E_s$ = substrate tensile modulus, psi

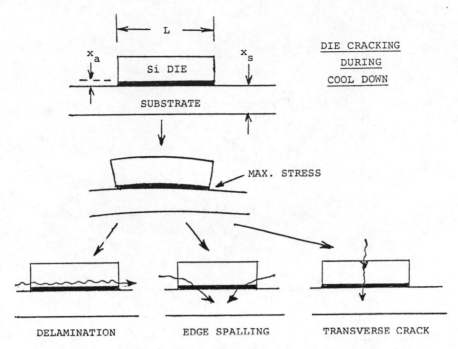

Figure 1. Thermal stresses on bonded die after cure.

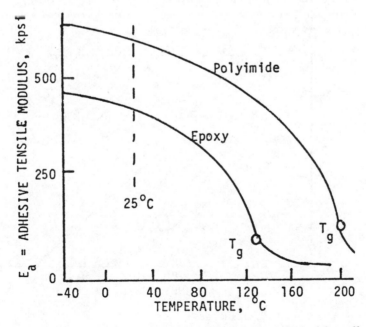

Figure 2. TMA for silver filled epoxy and polyimide adhesives.

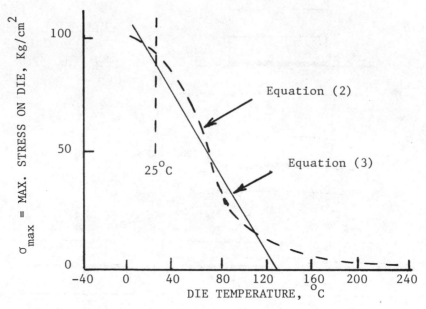

Figure 3. Equations (2) and (3) for the epoxy shown in Figure 2.

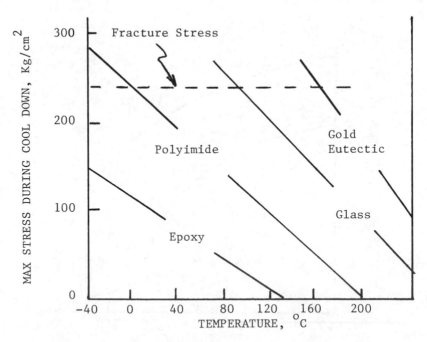

Figure 4. Maximum stresses from Equation (3).

Figure 2 shows tensile modulus, measured by TMA for two typical die attach adhesives; a silver-filled epoxy with T_g about $125^\circ C$ and a modulus at $25^\circ C$ of about 450,000 psi, and a silver-filled poly-imide of $T_g = 200^\circ C$ and a modulus of 580,000 psi. Because many of the terms in Equation (2) vary with temperature, a plot of stress versus temperature is a curve of complex shape, as shown in Figure 3.

Equation (2) can be linearized by assuming that the zero stress point is T_g for the adhesive, and by using constant $25^\circ C$ values for the adhesive and substrate moduli and for the thermal expansion coefficients, to obtain:

$$\sigma_{max} = K\,(\alpha_s - \alpha_{si})(T_o - T)\sqrt{\frac{E_a E_s L}{x_a}} \tag{3}$$

where $T_o = T_g$ for polymeric adhesives or the melting temperature for glass or eutectic

and α_s, α_{si}, E_a and E_s are values at $25^\circ C$.

Equation (3) gives the solid straight line shown in Figure 3 and provides a reasonably good approximation of the stresses predicted by Equation 2. Equation (3) gives the same linear relationship bet-ween stress and temperature as predicted by Lewis and Adams[6]. It is most useful in comparing stresses produced for a given die-substrate combination, by different adhesive types, by different die dimen-sions, or by different bond thicknesses. For the usual case of a silicon die and an alumina substrate:

$$\frac{(\sigma_{max})_1}{(\sigma_{max})_2} = \frac{T_{01}-T}{T_{02}-T}\sqrt{\frac{E_{a1}}{E_{a2}}\frac{L_1\,x_{a2}}{L_2\,x_{a1}}} \tag{4}$$

If the same adhesive is used for case 1 and case 2, the effect of die length and adhesive bond thickness would be:

$$\frac{(\sigma_{max})_1}{(\sigma_{max})_2} = \sqrt{\frac{L_1 x_{a2}}{L_2 x_{a1}}} \tag{5}$$

If die size is constant, but the adhesive type is varied:

$$\frac{(\sigma_{max})_1}{(\sigma_{max})_2} = \frac{T_{01}-T}{T_{02}-T}\sqrt{\frac{E_{a1} x_{a2}}{E_{a2} x_{a1}}} \tag{6}$$

Figure 4 shows the use of Equation(3)to compare stresses produced for the epoxy and polyimide adhesives of Figure 2, for gold
eutectic and for a glass adhesive. Highest stresses are produced
by the glass and eutectic die attach methods. The polyimide is
intermediate between the epoxy and the eutectic solder.

Very simply, these equations say that if an adhesive is expected
to provide strain relief between two rigid adherends, then the adhesive must be as flexible as possible. This means a reduced E_a and a
low transition temperature, T_g. The bond should be as thick as
possible, consistent with heat transfer requirements.

These equations also show why the thermal stress problem is
now becoming so serious. Die sizes are increasing. The L term in
Equation 3 used to be 40 to 60 mils; now dies of 400 mils or larger
are common. I.C. manufacturers are testing by thermal cycling down
to $-40^{\circ}C$. And most importantly, the die bond adhesives themselves
are becoming more rigid. New polyimides and epoxies have T_g values
of $200^{\circ}C$ or higher and are much more rigid than the early epoxies.
Glass frit adhesives are now used extensively in hermetic packages.
And in response to increased power dissipation requirements, bond
thicknesses are being minimized. Together, these changes can mean
peak stresses which are 10 to 100 times larger than the stresses on
small dies, exerted by softer adhesives, several years ago.

III. THERMAL STABILITY

Today, essentially only two types of polymeric adhesives are
used for die attach: silver-filled epoxies and silver-filled polyimides. The epoxy adhesives offer the advantages of lower cost,
higher bond strengths and simple, one-step cure to yield void-free
bonds, in addition to the lower thermal stresses noted earlier due
to their higher elongation and lower T_g relative to the polyimides.

Epoxies and polyimides are now widely used in plastic encapsulated I.C.'s and in special packages, such as pre-molded plastics,
"quasi" hermetics,and side-seam brazed packages. But unfortunately,
none of the epoxy or polyimide die attach adhesives available to
date has had sufficient thermal/oxidative stability for the most
widely used hermetic packages which require glass lid sealing, in
air, at temperatures of $410-450^{\circ}C$.

TGA measurements, in air and in N_2, can be used to estimate the
maximum temperature which an adhesive can survive before generating
sufficient gaseous decomposition products to cause pressure buildup
in a hermetic package after lid sealing.

Figure 5 shows the typical TGA curve obtained for silver-filled epoxy die attach adhesives; in this case, Amicon C-860-1. Almost all epoxies show this same behavior. Onset of decomposition depends only on temperature and is independent of whether silver or some other filler is used, or whether the TGA is run in air or in N_2. In N_2, the decomposition residue is char and filler; but in air the char ultimately oxidizes, causing a second weight loss and ultimately leaving only filler as the residue. Hence, as shown in Table II, the maximum recommended processing temperature (defined as the temperature where weight loss exceeds 2% at a heating rate of $10^{\circ}C/$ minute) is the same for epoxies in air as in N_2. As expected, the maximum processing temperature for epoxies correlates with T_g and with cross-link density. Unfortunately, even C-860-1, with a T_g of $190^{\circ}C$ and the highest processing temperature ($360^{\circ}C$) of any epoxy tested to date, is still not stable enough for use in most ceramic packages.

The polyimide die attach adhesives show an entirely different stability pattern. For the unfilled polyimide (Figure 6), thermal stability is adequate to well over $400^{\circ}C$ and just slightly better in N_2 than in air. But for the same polyimide filled with powdered silver (Figure 7), the decomposition temperature is $150-200^{\circ}C$ lower in air than in nitrogen. This effect is clearly due to the surface catalytic effect of the silver, as evidenced by the TGA curves in Figure 8, where the same polyimide resin is shown unfilled, with silver filler, and with a heat conductive aluminum powder filler. Table II shows that decomposition temperature for polyimides does not correlate with T_g; and that in nitrogen the filler effect is very slight. But for all present polyimide resins, the combination of air plus silver filler greatly reduces thermal processing stability.

The catalytic effect of silver is also shown by the die shear measurements in Figures 9 and 10. After long-term aging in air at $200^{\circ}C$ neither the aluminum-filled, nor the silver-filled, polyimide loses strength due to thermal degradation. Both show strength improvement after 1 day at $200^{\circ}C$, probably due to post cure or to additional solvent removal. But at $300^{\circ}C$, the silver-filled polyimides degrade very rapidly in strength, and the percentage of strength loss clearly increases as silver content increases. The same polyimide resin, filled with aluminum powder, loses less than 10% of its initial shear strength after the same period in air at $300^{\circ}C$.

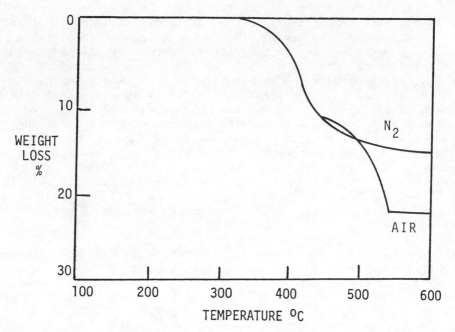

Figure 5. TGA at heating rate of 10°C/min. Silver-filled epoxy (Amicon C-860-1).

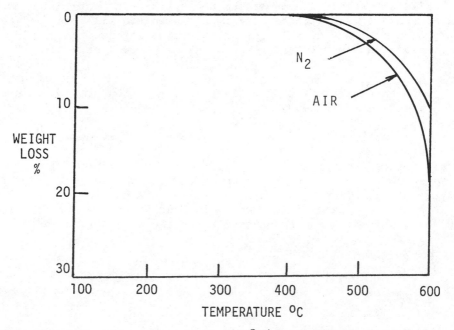

Figure 6. TGA at heating rate of 10°C/min. Unfilled polyimide.

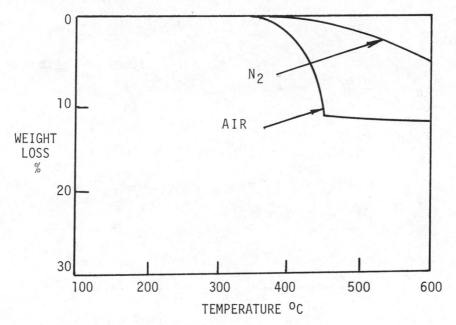

Figure 7.   TGA at heating rate of 10°C/min.   Silver-filled poly-
            imide.

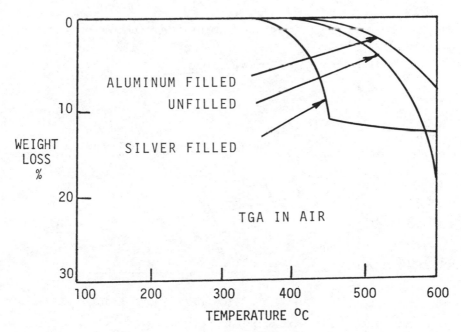

Figure 8.   TGA at heating rate of 10°C/min.   Filled and unfilled
            polyimide.

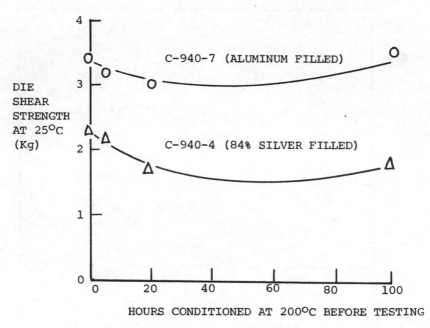

Figure 9. Die shear strength for polyimide adhesives after 200°C aging in air. Die size is 44 x 44 mils.

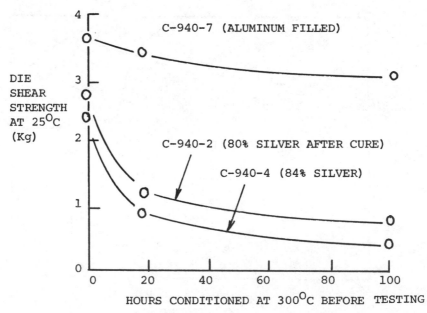

Figure 10. Die shear strength for polyimide adhesives after 300°C aging in air. Die size is 44 x 44 mils.

Table II. Thermal Stability of Die Attach Adhesives.

DIE ATTACH ADHESIVE	FILLER	T_g	TEMPERATURE AT WHICH TGA WT. LOSS > 2%	
			IN AIR	IN N2
Amicon C-805 Epoxy	Silver	90°C	275°C	275°C
Amicon C-850 Epoxy	Silver	140°C	310°C	310°C
Amicon C-860 Epoxy	Silver	190°C	360°C	360°C
Addition-cured Polyimide	Silver	200°C	320°C	460°C
Addition-cured Polyimide	None	190°C	430°C	445°C
Addition-cured Polyimide	Aluminum	205°C	450°C	470°C
Condensation-cured Polyimide	Silver	280°C	420°C	520°C

IV.   DISCUSSION OF RESULTS

The catalytic effect of the silver surface can be explained by
referring to Figure 11, which shows the structure of the oxide layer
on the surface of all glasses, ceramics, and almost all metals.
This surface model, first proposed in 1967[9], can be used to explain
how oxide surfaces interact with organic compounds by an acid-base
mechanism[10].

The surface hydroxyl groups, written as -MOH, may interact with
water or with other organic compounds via H-bond interactions where-
in the surface acts either as the acid (proton donor) or base
(proton accepter).

$$-MOH\cdots OH_2 \; \rightleftharpoons \; -MOH + H_2O \; \rightleftharpoons \; -MO\overset{H}{\diagup} \cdots HOH \qquad (7)$$

As the pH of the aqueous phase is altered by electrolyte addition,
the surface acquires an ionic charge via

$$\overset{+}{-MOH_2}\cdots OH_2 \; \overset{H^+}{\underset{\rightarrow}{\leftarrow}} \; -MOH + H_2O \; \overset{OH^-}{\underset{\leftarrow}{\rightarrow}} \; -M\bar{O}\cdots HOH \qquad (8)$$

Surface charge can be calculated from zeta potential measurements as a function of pH, using suspensions of the powdered oxide
in water. For every oxide, there exists some pH at which the number
of positive charges equals the number of negative charges and the
zeta potential is zero. This pH value is defined as the IEPS, the
isoelectric point of the surface. A low IEPS value indicates an
acidic oxide surface. A high IEPS value indicates a basic surface.

Parks[11] has shown that the IEPS for an oxide depends upon the valence (Z) and the radius (R) of the cation (Figure 12). IEPS is lowest (most acidic) for metals of valence 4 or higher (e.g., SiO_2, MnO_2, WO_2). IEPS is in the medium range for trivalent metal oxides (Z=3), such as Al_2O_3 or Fe_2O_3, and is highest for oxides of Z=2 or lower, such as MgO, ZnO, FeO and CuO. Oxides of Z=1 are the strongest bases of all, with IEPS values estimated to be above 14 for Ag_2O.

Figure 13 shows the decomposition temperature for the same polyimide resin unfilled and with seven different metals and metal oxide fillers, plotted against IEPS values taken from Figure 12. Clearly, as base strength of the filler surface increases, thermal stability of the polyimide resin decreases. This suggests the following decomposition mechanism:

$$\text{POLYIMIDE} + O_2 \xrightarrow[\Delta]{Ag} \text{DECOMPOSITION} + H_2O \qquad (9)$$

Neutral and acidic surfaces have essentially no effect on thermal/oxidative decomposition of these polyimide resins. The 30% palladium/70% silver alloy in Figure 13 appears to have almost the same base strength, and the same effect on polyimide decomposition, as pure silver. This is disappointing because the substitution of powdered Pd/Ag alloy is a very effective and (relative to the use of pure gold filler) an inexpensive way to avoid silver migration problems in I.C.'s having high internal voltage gradients.

As indicated elsewhere in these proceedings, there are two general classes of polyimide resins, i.e., high molecular weight condensation curing resins and lower molecular weight thermoplastic or addition curing oligomers. Most of the silver-filled polyimides now used for die attach are made from the thermoplastic or addition curing resins because these yield high solids adhesives with minimum outgassing and volatile evolution during cure. The polyimide shown in Figure 6 through 10 is an acetylene-terminated polyimide as described previously by Bilow[12], but the same effect of silver on decomposition in air has been observed in thermoplastic polyimides, such as the LARC TPI resins, the G.E. "ULTEM" resin and Ciba's XU-218, and also for the addition cured PMR polyimides and the Rhone-Poulenc Kerimid 601 bis-maleimide.

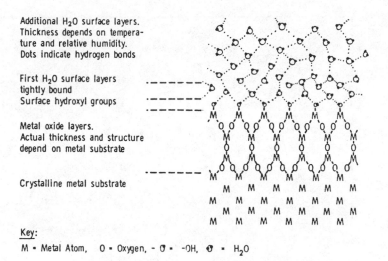

Additional H_2O surface layers.
Thickness depends on tempera-
ture and relative humidity.
Dots indicate hydrogen bonds

First H_2O surface layers
tightly bound
Surface hydroxyl groups

Metal oxide layers.
Actual thickness and structure
depend on metal substrate

Crystalline metal substrate

Key:

M ▪ Metal Atom,   O ▪ Oxygen, - ◌ ▪ -OH, ◌ ▪ H_2O

Figure 11.   Reactive groups, water and oxide layers on metal surface[9].

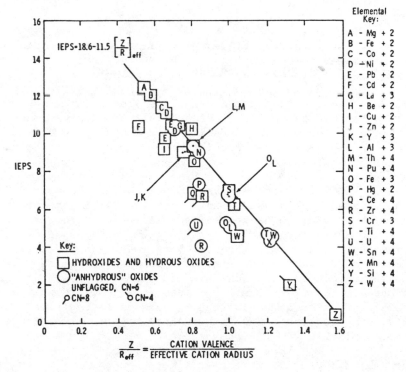

Figure 12.   Parks' equation[11] for IEPS.   Radius corrected for crystal field effects, coordination, and hydration.

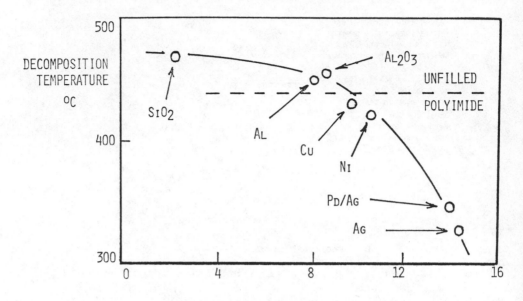

IEPS = ISOELECTRIC POINT OF SURFACE (pH UNITS)

Figure 13.   Thermal stability of polyimide vs. IEPS of filler.

The most stable silver-filled polyimides tested to date are the high molecular, condensation curing PMDA-ODA resins, described elsewhere in these proceedings. As shown in Table II, these have sufficient stability in air for use in I.C. packages sealed below about 420°C. These condensation curing polyimides are, however, extremely difficult to cure to yield void-free adhesive bonds, because of the water generated during cure and also because of the need for much higher solvent levels than are needed for the addition curing polyimides.

Clearly, silver filler has many disadvantages in polyimides. Why then is it so widely used? In theory, silver is one of the few fillers which provides <u>both</u> thermal <u>and</u> electrical conductivity. Most other metal fillers, including aluminum, have an oxide surface which prevents electron flow between particles. In practice, back-face electrical conductivity is erratic for silver-filled poly-imides. It depends on cure schedule and interfacial effects. This problem has escaped the attention it deserves, because only a small fraction of I.C. dies actually require back-face electrical con-ductivity; and even these few (e.g. RAM's) might be grounded in other, more reliable ways.

Thermal conductivity is almost always needed, but other metal fillers (notably aluminum) provide thermal conductivity equal to silver in polyimide adhesives and provide as much as twice the bond strength and shock resistance of silver fillers. These other metal fillers avoid the catalytic degradation effects of silver, yield addition cured polyimide adhesives which do not outgas at $450°C$ in air, eliminate entirely the silver migration problem, and also avoid the phase separation problems (settling during storage, resin bleed during cure) characteristic of silver-filled adhesives. Silver is also a major source of Cl^- and Na^+ contamination. The aluminum-filled polyimide in Table I (Amicon C-940-7) contains less than 5 ppm extractable Cl^- or Na^+, and still lower quantities of K^+ and NH_4^+.

Hence, the best polyimide choice depends on whether the adhesive must provide electrical as well as thermal conductivity. If so, silver-filled condensation-cured polyimides may be used; but these have processing and other (silver related) disadvantages. If electrical conductivity is not required, new aluminum-filled addition curing polyimides provide important processing, strength and reliability advantages.

V.   REFERENCES

1. "Determination of Extractable Ionic Contaminants in Cured Adhesives." TP-91, Amicon Corp., Lexington, MA, June 1983.
2. R.J. Usell and S.A. Smiley. Proceedings 31st Electronic Components Conf., pp. 65-73, IEEE/Proc. 1981.
3. W.H. Schroen, J. Spencer, J. Bryan, R. Cleveland, T. Metzgar, and D. Edwards. Ibid. pp. 82-87.
4. W.T. Chen and C.W. Nelson. IBM Res. Develop., 23 (2),179 March 1979.
5. K. Niwa, K. Hashimoto, N. Kamehara and K. Murakawa. Semi-Conductor International, 49-56, April 1980.
6. T.E. Lewis and D.L. Adams, 1982 Electronic Components Conf., San Diego, CA, pp. 166-173, May 1982.
7. R.L. Krieger. 22nd Annual SAMPE Symposium, San Diego, CA, April 1977.
8. J.C. Bolger. 14th Nat'l. SAMPE Tech. Conf., Atlanta, GA, October 1982.
9. J.C. Bolger and A.S. Michaels. in "Interface Conversion For Polymeric Coatings". P. Weiss and D. Cheevers, Editors. Chapter 1, Elsevier, New York, 1969.
10. J.C. Bolger. in "Adhesion Aspects of Polymeric Coatings," K.L. Mittal, Editor, pp. 3-18, Plenum Press, New York, April 1983.
11. G. Parks. Chem. Rev., 65, 177 (1965).
12. N. Bilow. 23rd Nat'l. SAMPE Tech. Conf., Anaheim, CA, May 1978.

THE EFFECT OF POLYIMIDE PASSIVATION ON THE ELECTROMIGRATION OF AL-CU THIN FILMS

E. S. Anolick, J. R. Lloyd, G. T. Chiu, and V. Korobov

International Business Machines Corporation

P. O. Box 390, Poughkeepsie, New York 12601

A comparison is made of the effect of two polyimide coatings and a sputtered glass coating on the electromigration of Al-Cu. The findings, both by resistance changes and by open circuit failures, indicate that the electromigration is significantly retarded by the various polyimide coatings. The data were insufficient to show any differences between the two polyimides tested. If this behavior could be projected to product-like structures in VLSI, it would possibly counter any harmful effects expected due to the reduced thermal conductivity of the polyimides when compared to most passivation glasses. The source of the difference in the behavior of polyimide and quartz is unknown but may be attributed to any one of three possible effects: the formation of a relatively defect-free coating, the penetration into the grain boundaries by a component part of the polyimide (such as hydrogen), or the blocking of surface sites which are active in the metal transport by the combination of polyimide and the adhesion promoter used.

INTRODUCTION

Electromigration or electrotransport has been a major topic of interest in the microelectronics industry ever since it was recognized as a principal failure mode in integrated circuits almost 20 years ago[1]. A number of excellent review papers have appeared in the literature[2,3,4] and the interested reader is urged to consult them for a more detailed discussion and for an extensive list of references.

Basically, electromigration is mass transport where the driving force is supplied by the momentum exchange between conduction electrons and the metal atoms making up the conductor. When an electric current is passed through the conductor, this momentum exchange becomes biased in the direction of electron flow in n-type conduction and in the opposite direction in p-type conductors[5]. In the presence of flux divergences, caused by any one of a number of conditions, the conductor will fail by open circuit by the mechanism of void formation and growth at the sites of negative divergence. Conversely, a failure may occur by a short circuit at sites of positive divergence.

It has been found that the presence of a dielectric coating on top of the metallization can enhance the reliability of the underlying conducting lands[6-10]. This effect has been attributed to the ability of the passivation to physically restrain the film by permitting high compressive stresses to develop[11,12]. These stresses retard the diffusion and should provide longer lifetimes, but there is some indication that a flawed strong passivation may, in fact, hurt reliability[9,10].

The passivation most commonly used in integrated circuits is simply silica glass (SiO_2), deposited either by chemical vapor deposition (CVD) or by sputtering from a glass target. Silica glass is strong, but is brittle, prone to pinhole formation, and expensive to deposit. By contrast, polyimide passivations are usually free from the defects that plague silica glass and are relatively inexpensive to apply. As a result, great interest has been generated in polyimide as an alternative to silica.

The purpose of this paper was to extend the study of polyimide effects to various widths of AlCu metallizations and to other vendors over and above those reported by Lloyd[9]. The level of test was reduced to a lower temperature and lower current density than used in the earlier study. It had been reported that the thermal resistivity of polyimides was higher and that testing at higher current densities could lead to excessive increments above ambient temperatures[9]. Excessive temperatures would not occur at the lower current density in cases where one surface of the stripe was in contact with a better thermal conductor, such as silicon nitride or silica glass. This study was confined to a structure of polyimide over metal over silicon nitride. There were control samples in the form of silica glass over metal over the nitride. Two polyimides were compared, designated as vendor A and vendor B.

Product studies of the application of polyimide to actual devices have been favorable relative to electromigration. No great problems have been reported with respect to the reliability of products fabricated with layers of polyimide[13].

EXPERIMENTAL

Sample Preparation

The samples were a simple 4-point probe structure as shown in Figure 1. They were all prepared together up to the point of final coating, so that all samples had the same history through metallization. The substrate used was a p-Si wafer. The quality of the substrate was immaterial in this study, as no contact was made to the Si itself. A thin layer of thermal oxide was grown on the silicon by a high temperature anneal in oxygen. Silicon nitride was deposited by chemical vapor deposition. Metal was evaporated in a vacuum to obtain a thickness of 2 microns of AlCu. The addition of copper was 4 weight percent. The final structure was defined by a lift-off process into the structure illustrated in Figure 1. Three different widths of metal were defined on the same wafer on different reproductions of the same structure. The widths were subsequently measured to be 2.25, 5.5, and 12.7 microns (see below). The defined structures were then all sintered at 425°C for 60 minutes in a nitrogen atmosphere.

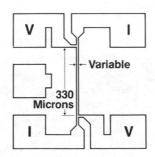

Figure 1. Drawing of the structure used to measure the electro-migration. V represents the voltage probe, I the current input.

The wafers were then separated into three groups. The first group had a glass structure sputtered onto the surface to a thickness of 3.7 microns. This structure is called quartz throughout the paper but is an amorphous silicon dioxide, not the crystalline quartz. The wafers were spin-coated with polymer A on group two and polymer B on group three. To open contacts through the passivation, a photoresist was used with exposure through a mask, development, removal of the photoresist material, and a wet etch of the exposed passivation. The quartz was removed by using a buffered hydrofluoric acid. The polyimides were removed by using potassium hydroxide. The excess photoresist was then removed and the wafers with polyimide passivation were subjected to an additional cure cycle in nitrogen, consisting of 10 minutes at 200°C, 30 minutes at 300°C, and finally 30 minutes at 400°C. The final cured thickness was 3.7 microns.

All of the wafers were then combined again and cleaned, and layers of Cr-Cu-Au were sequentially deposited through a mask onto the opened contact regions. The wafers were then diced into chips and the samples of varying widths were selected for bonding. Bonding was done to a ceramic 28-pin DIP package using aluminum wire and ultrasonic bonding.

Measurement Method

The electrical circuit is a simple voltage measurement utilizing a constant current supply and scanning device to sequentially read the voltage drop across the metal stripe. The circuit is shown schematically in Figure 2.

The device under test (D. U. T.) had the voltage probes connected to a channel on the data logger. All devices of the same nominal width were in series, there being three different power supplies for the three different widths. The voltage drop across each stripe was less than the voltage required to turn on the diode in parallel with each stripe. The voltage drop was monitored on the data logger. Data was printed out once every four hours during the test.

Resistances of the various stripes were first measured at 3×10^5 amps/cm^2. This level of current was believed to create a minimal problem due to joule heating of the stripe. Thus the resistance obtained closely represented the resistance at ambient temperatures. The widest stripe was used to estimate the resistivity of the AlCu, at room temperature, based on the resistance values and on the dimensions measured under the microscope. The thickness of the AlCu was determined by sectioning several parts and making the measurement with a scanning electron microscope. The resistivity values calculated were then used to determine the cross-sectional

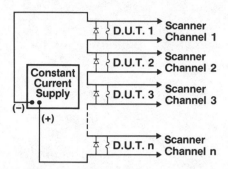

Figure 2. Schematic representation of the electrical circuit used to monitor the electromigration.

area of the various width stripes as well as to calculate the widths. The widths were determined to be 12.7, 5.5, and 2.25 microns. As a further check, sections of the narrower stripes were obtained and the average widths were measured directly. All estimates of the widths and the cross-sectional areas were within 8 percent of each other. The stripe thickness was measured to be 2.2 microns, as against the nominal value of 2 microns.

Once the cross-sectional areas were established, the thermal coefficient of resistivity could be measured. The resistance was monitored as the stripe was subjected to increasing oven temperatures up to 200°C. There were 15 stripes per width per coating. The average resistivity was calculated and plotted as a function of temperature for the different widths. Figure 3 shows how well the three different widths tracked for vendor A.

There were three ovens and three power supplies. Monitoring of the temperatures and the currents was done by the same data logger, so each reading had 135 voltage readings and 6 control readings. The ovens were operated at $200\,^{\circ}$C and the current at 1×10^{6} amps/cm^{2}. In spite of the high current density, the temperature rise in the stripes averaged only 4°C.

The samples were then placed on test. The total number of samples was initially 135, with 45 samples of each coating in each oven. There were 15 samples for each width for each coating.

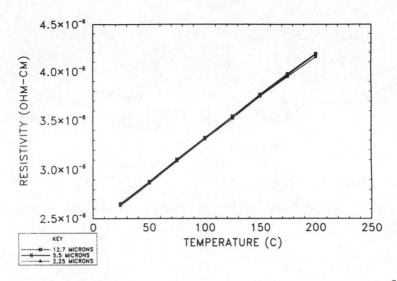

Figure 3. Variation in the resistivity as measured at $3x10^5$ amps/cm^2 with temperature for the various widths of AlCu coated with polyimide A.

First the ovens were brought up to 200°C. Then the current was brought to $1x10^6$amps/cm^2. The voltage drop on all stripes, at operating temperature and current, was approximately 0.13 volts initially. A failure was defined as a doubling of the initial voltage to 0.26 volts. Most of the failures were true open circuits or very high resistance values. When a D. U. T. failed, the diode in parallel would turn on and the reading would exceed 0.6 volts. An open bond, on the other hand, would show up as zero volts on the scanner. This test was plagued by open bonds, due to corrosion, mostly in the case of the polyimide-coated samples. There were practically no bonds lost in the samples coated with quartz.

Data

Readings were taken every 4 hours automatically. The data logger was set to flag a failure by giving a red printout. Unfailed data were recorded in black. During the first 100 hours many of the bonds opened, especially for vendor B. Most of the opens appeared to be due to local contamination. The source of the contamination was not determined (see Data Analysis and Failure Analysis).

894

The method of data analysis used was a censorship method which allowed a separation of the two types of failure[14,15]. This meant that the final numerical analysis had to be done on greatly reduced populations. Polyimide B, in fact, had only a few survivors. To determine the relative effectiveness of the polyimide coating, the resistance changes were tracked as well. Experience had shown that resistance changes were expected to occur during electromigration. The resistance changes are not regular enough to do a clean quantitative analysis. One can do qualitative analysis, however, to show that differences are occurring and that they may indeed be significant. The narrowest stripes also give clearer manifestations of steadily increasing resistance with time. Figures 4 and 5 are plots of resistance versus time for the 2.25 and the 12.7 micron stripes. The plots are for only 4 of 14 stripes in each case and for the case of quartz coating. To compare the various coatings on a normalized basis, an average effective resistivity was calculated taking into account the measured stripe dimensions and the measured variations of the resistance with time. This is not the true resistivity but rather reflects a changing resistance, due to irregular internal structural changes such as the growth and the merger of voids all normalized to fit to the same scale in the figures. Effective resistivity versus time plots are shown in Figures 6, 7, and 8.

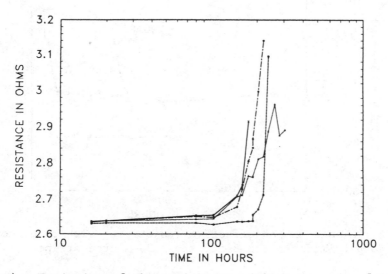

Figure 4. Variation of the resistance with stress time for four of the 2.25-micron stripes coated with quartz.

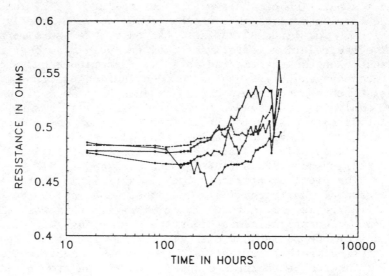

Figure 5. Variation of the resistance with stress time for four of the 12.7-micron stripes coated with quartz.

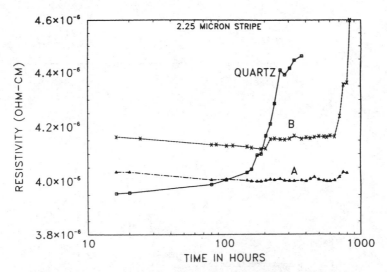

Figure 6. Change in the effective resistivity with time, comparing quartz to polyimides A and B for the 2.25-micron stripe.

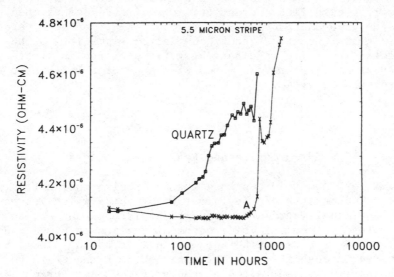

Figure 7.   Comparison of the effective resistivity change with time of quartz and polyimide A for the 5.5-micron stripe.

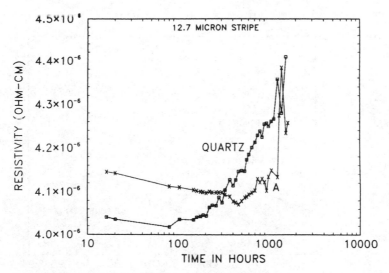

Figure 8.   Comparison of the effective resistivity change with time of quartz and polyimide A for the 12.7-micron stripe.

Where sufficient failures were obtained, a lognormal analysis was done utilizing APL programs available on IBM in-house computers. The programs take into account the loss of data due to open bonds during the test and provide an output of the time to 50 percent failure (t(50)) and the standard deviation sigma (σ) of the time-to-failure data. The resultant best fit estimate is plotted in Figure 9 along with the estimates of the percent failures obtained by censorship analysis.

RESULTS AND DISCUSSION

Void Formation and Resistance Changes

Failures, due to opens as a result of electromigration, tend to come about because of the formation of voids which subsequently grow and eventually traverse the entire width of the stripe, leading to an open circuit. Voids occur at regions of high gradient and may or may not remain at the original location. Atoms are transported through the bulk material, through the grain boundaries and possibly via vacant sites on the surface. Bulk transport is less significant than the other two possibilities because of the higher energy required for diffusion of Al atoms through the bulk structure. The grain boundary diffusion activation energy has been measured as 0.55 electron volts, while the bulk contribution

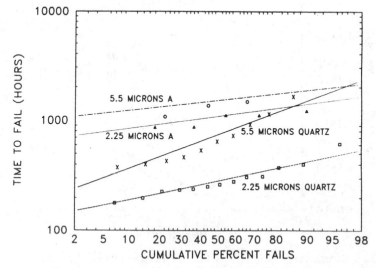

Figure 9. Lognormal estimates for the electromigration data. Comparison of polyimide A to quartz, calculated with censorship method.

898

has been reported to be 1.3 electron volts[16]. The activation
energy determined for the open stripe failure mode is in the range
0.65 to 0.75 electron volts. This is consistent with the Al/Cu
activation energy previously reported[17]. Many of the past measurements were made on stripes greater than 2 microns. Since
the grain sizes are usually smaller than 2 microns, all failures
are attributed to mostly grain boundary migration. The thinner
and narrower stripes also present a greater ratio of surface area
to volume, so surface effects can become significant. At 1 micron
and less, the stripe width becomes equivalent to the average grain
sizes in most fabrications, and transport depends mainly on the
surface and bulk properties.

If a sufficient number of voids form during a test, then the
effective cross-sectional area of the stripe is reduced and the
resistance increases. Such increases are evident in Figures 4-8.
The resistance increase is then a manifestation of the electromigration in the stripes. The narrower the stripe, the more the
effect of void volume on the resistance and also the sooner the
failure. So until the stripe widths are equivalent in size to the
grain size, the narrower stripes are expected to fail sooner and
show sharper rises in resistance.

Figure 4 shows the rise in resistance for the 2.25-micron
stripe under quartz. At less than 100 hours there is generally
little or no rise in resistance. This acts like a nucleation
period prior to the growth of sufficient void volume to impact the
value of the resistance. Most of the failures occur in these
stripes soon after the sharp rise in resistance. Figure 5 shows
the situation for the widest stripes coated with quartz. The
resistances are fluctuating widely during the entire test period.
In the 1660 hours on test, only one of the widest stripes failed
for the quartz-coated samples and none failed for the polyimide-
coated samples. That there is void formation is evident in Figure
10, which shows the region of a tested stripe near the negative
terminal. What is interesting is that some of the resistances
decreased with time and some fluctuated up and down. This suggests
that such voids as are formed may in fact also disappear or break
up.

Figure 6 shows the composite effective resistivity obtained
for quartz and the two polyimides. The quartz shows a sharp rise
after 100 hours, while polyimides A and B remain quite flat up to
about 800 hours where failures began to occur. Therefore, one must
conclude that the electromigration is somewhat inhibited by the
presence of both types of polyimides. The factor of approximately
3 times difference can project to very large differences at a use
temperature of 75°C and use current densities less than 2×10^{5}
amps/cm^2. Progressing to the next widest stripe shown in Figure 7,

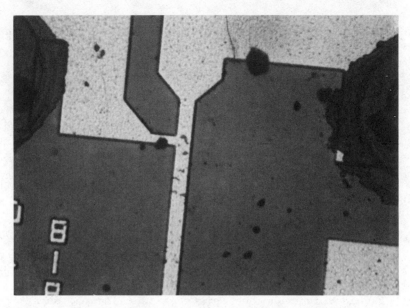

Figure 10. This shows the void formation at the negative terminal of a 12.7-micron stripe run at 200°C and 1x10^6 amps/cm.

only the A polyimide is compared to the quartz. This was because of the high losses in the bonds for polyimide B. There is still a delay in the increase in resistance for the polyimide-coated stripe, but the rise in resistance is relatively closer for the two. The two coatings, quartz and polyimide A, approach each other even more for the widest stripe, as seen in Figure 8. In the last case, the possible benefits of a polyimide coating appear to be diminishing. The widest stripe also showed a decrease in the resistance initially for both coatings. Decreases were not as evident for the 2.25 and 5.5 micron stripes. The decrease in resistance was determined as not being due to any measurement problem. The cause of the decrease is not known at this time.

Data Analysis and Failure Analysis

A lognormal analysis was done using censorship methods[14,15]. This analysis could only be applied to failed stripes defined by a doubling of the resistance or greater. All open stripes were electrically verified after the test by direct measurement. X-ray probe analysis was done on the bond regions to determine some of the causes of bond failure. Many of the failed and unfailed stripes were examined microscopically to look at void formation and try to determine the locations of the opens.

Statistically, all stripes are on test until either the stripe opens (electromigration failure) or a bond opens. As mentioned

above, the two types of failures could be distinguished electrically. Bond failures were found to be due to actual lifting or to
formation of a high resistance corrosion product at the bond. The
x-ray probe analysis revealed that both Cl and S were present in
the case of vendor B. These two types of contamination apparently
were the primary causes of excessive corrosion observed during the
test. It was not evident that the polyimide itself was the source
of the contamination.

Polyimide A showed a great deal of crazing at the end of the
test, while polyimide B did not show any structural changes at all.
The time of the occurrence of the crazing was unknown. The crazing
was greatly reduced around the region of the via hole. Also, at
the beginning of the test polyimide A had interference fringes
(see Figure 11) surrounding the metal lands. Cross-section examination revealed some possible lifting very close to the stripe
(less than 1 micron away), but not enough to account for the
fringes seen. These fringes are still visible through the crazing,
as can be seen in Figure 12. Polyimide B was clear at the start
of the test and remained clear at the end.

The statistical analysis provided both the parameters t(50)
and σ as described above. A full summary of the analysis is given
in Table I. In the same table are the ratios of surface area to
volume as stripe width varies -- the narrower the stripe, the
higher the ratio. The analysis involved 15 samples per cell.

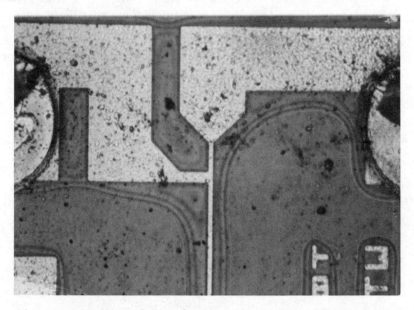

Figure 11. An example of the fringes seen around the stripes
coated with polyimide A at time zero.

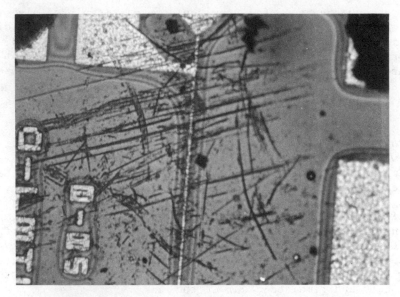

Figure 12. Polyimide A coated samples after 1660 hours at 200 °C. Note the complete breakup of the coating. The fringes are still visible through the crazing.

From both the resistance measurements and the failure data, it is evident that the differences between the polyimide and the quartz passivations are decreasing as the stripe width increases. Thus the t(50) ratios decrease between the two types of coatings as the width increases. Similarly, the time differences for the same amount of increased resistance are also decreasing with increased stripe width. This behavior is in line with the decreasing surface-to-volume ratio in Table I.

Table I. Summary of the Lognormal Data Analysis.

Stripe Width Microns	Surface/Volume Microns^{-1}	Coating	t(50) Hours	Sigma σ
2.25	1.32	Quartz	286	0.31
		Vendor A	1092	0.20
5.5	0.80	Quartz	747	0.55
		Vendor A	1418	0.20
12.7	0.59	Quartz	> 1590	--
		Vendor A	> 1660	--

CONCLUSIONS

The effect of polyimide, in general, is to decrease the tendency for electromigration damage as measured by resistance changes or by stripe opens. Eventually the polyimide-coated samples do fail, and the failures take place in a very narrow time range. This rapid failure mode after an extended time period is not understood. While one polyimide showed crazing, which may be due to some coating delamination, the other did not. Both types had failures occurring in the same time frame. The degree of electromigration failure separation in time, with coating, increases with greater surface-to-volume ratios.

There are several speculations regarding the source of the differences between the behavior of polyimide and quartz:

1) The polyimides tend to coat the metals better than the quartz, leading to a defect-free environment at the surface.

2) There is penetration of the polyimide into the grain boundaries. The presence of Cu in the grain boundaries is known to inhibit electromigration[18]. Studies have also indicated that electromigration can be inhibited by the presence of hydrogen in the grain boundaries[19].

3) The variation in behavior with increased surface-to-volume ratio may reflect a blocking of surface sites which play a role in the migration. These sites, being blocked, will not contribute to the transport of materials.

Other aspects of the use of polyimides in integrated circuits should take into account the greater thermal resistance of the polymer over many of the glass passivations. Local hot spots along a stripe coated with polyimide may not dissipate the heat as readily, leading to a premature failure in the product. Tests at more than 100 times the use power may also be misleading, due to excessive heating of the tested stripe. Considerably more study is required on the effect of polyimide on electromigration, using samples with varying topology and geometry. There is also a need for studying other possible sources of failure associated with polyimide, such as contamination and corrosion.

ACKNOWLEDGEMENTS

We are indebted to L. Timperio, S. Gehrer, and R. Edwards, the technicians who enabled us to carry this project to completion by their timely and precise efforts. The support by A. Bross,

J. Kitcher, R. J. Foley, L. Rothman, and S. Ahmed, the managers
who aided us in getting the necessary things done, is also
gratefully acknowledged.

REFERENCES

1. I. A. Blech and H. Sello, paper presented at the 5th Annual
   Symposium on Physics of Failure in Electronics, Columbus,
   Ohio, Nov. 15-17, 1966.
2. J. R. Black, Proc. IEEE, 57, 1587 (1969).
3. R. E. Hummel and H. B. Huntington (Editors), "Electro and
   Thermo Transport in Metals and Alloys," American Society
   for Metals, 1976.
4. F. M. D'Heurle and D. T. Petersen, in "Encyclopedia of
   Chemical Technology," Vol. 8, 3rd Ed., p. 763, John Wiley
   & Sons, 1979.
5. J. R. Lloyd, M. R. Polcari and G. A. MacKenzie, Appl. Phys.
   Lett., 36, 428 (1980).
6. S. M. Spitzer and S. Schwartz, IEEE Trans. on Electron Dev.,
   ED-16, 348 (1969).
7. A. J. Learn, J. Appl. Phys., 44, 1251 (1973).
8. T. Satake, K. Yokoyama, S. Shirakawa, and K. Sawaguchi, Jap.
   J. Appl.Phys., 12, 518 (1973).
9. J. R. Lloyd, Thin Solid Films, 91, 175 (1982).
10. J. R. Lloyd, Thin Solid Films, 93, 385 (1982).
11. I. A. Blech and C. Herring, Appl. Phys. Lett., 29, 131 (1976).
12. I. A. Blech and K. I. Tai, Appl. Phys. Lett., 30, 387 (1977).
13. K. Mukai, A. Saiki, K. Yamanaka, S. Harada, and S. Shoji,
    IEEE J. Solid-State Circuits, SC-13, No. 4, 462 (1978).
14. W. Nelson, J. Quality Technol., 1, 27 (1969).
15. W. Nelson, IEEE Trans. on Reliability, R-24, 230, (1975).
16. I. A. Blech, J. Appl. Phys., 47, 1203 (1976).
17. P. B. Ghate and J. C. Blair, Thin Solid Films, 55, 113 (1978).
18. F. M. D'Heurle and A. Gangulee, "Nature and Behavior of Grain
    Boundaries," H. Hu (Editor), p. 339, Plenum Press, 1972.
19. D. E. Meyer, J. Vac. Sci. Tech., 17, 322 (1980).

POLYIMIDE PATTERNS MADE DIRECTLY FROM PHOTOPOLYMERS

H. Ahne, H. Krüger, E. Pammer, and R. Rubner

Siemens AG

D-8520 Erlangen 2, West Germany

The increasing interest in microelectronics and
other application areas for polyimide-based photo-
patterned protective and insulating layers which are
dimensionally stable under heat has led to the development of polyimide precursors with photo-reactive
properties. The process of direct photo-patterning is
discussed here using the example of a polyamide carboxylic acid methacrylate ester developed by us. The
same example is used for a detailed look at aspects
of application technology and special properties of
this polyimide precursor. Further development has
since led to the possibility of producing and photo-
patterning both extremely thick and extremely thin
layers using a simple method. Two applications are
described in more detail. One is the protection of
cell arrays of memory chips against α-radiation from
the ceramic or plastic package using very thick polyimide layers, and the other is the employment of extremely thin polyimide films as alignment layers for liquid crystals in liquid crystal displays.

INTRODUCTION

Photo-patterned polyimide layers are finding increasing application in microelectronics and other fields as protective and insulating layers with dimensional stability at high temperatures. We have developed photopolymers, from which polyimide patterns can be produced directly.[1]

The production of photo-patterned polyimide layers was carried out up until now using non-photoreactive polyimide precursors with the aid of conventional photoresists. Figure 1 demonstrates the various processing steps in a conventional method (in this case using a polyimide-copolymer precursor) and in the process of direct photo-patterning with the photo-cross-linkable polyimide precursor, PVI 210, developed by us. In addition to the prior treatment of the substrate with a coupling agent, the conventional indirect photo-patterning process includes coating with polyimide precursor, thermal treatment and photoresist coating of the polyimide layer produced by thermal treatment. The photoresist is photo-patterned, and then serves as an etching mask for the underlying polyimide layer to be patterned. In order to uncover the substrate surface, the coupler must also be removed. Once this has been done, the photoresist is also removed.

The processing steps of applying and removing photo-resist and coupler and the associated drying processes can be omitted when using PVI 210 photopolymer, as PVI 210 displays photoresist properties itself and the ready-to-use solution already contains a coupling agent, for instance vinyl-tris(beta-methoxyethoxy)-silane This implies a process simplification: fewer processing steps and removal of the coupling agent during the development step. High differences in solubility are produced directly in the layer to be structured by partial photocrosslinking . Following development, this results in sharply contoured patterns with a reproducibly good resolution.

Figure 2 shows our chemical principle of direct polyimide photo-patterning using the example of our polyamide carboxylic acid methacrylate ester.

It is a general principle and applies not only to polyamide carboxylic acid ester but also to polymers with aromatic chain segments and bridges, which contain amide partial structures, the polymers carrying photoreactive groups bound esterlike adjacent to the bridges. From these polymers relief patterns can be produced by photocrosslinking and transformed into highly heat resistant relief patterns by thermal treatment.

In the following, a discussion of the properties of PVI 210 and two technical applications are presented.

906

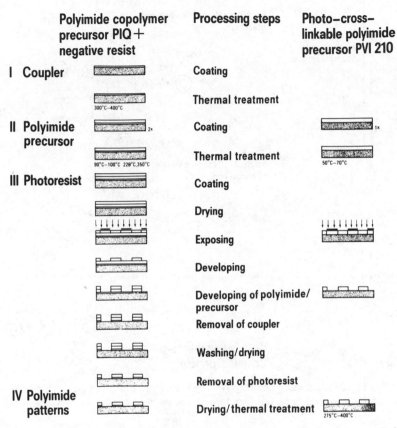

Figure 1. Process for the photolithographic production of highly
heat-resistant relief patterns.

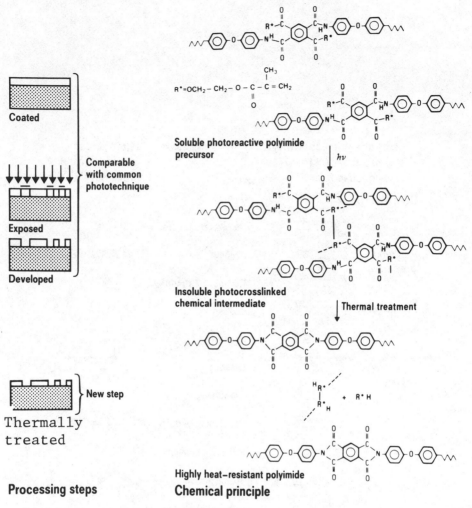

Processing steps

Chemical principle

Figure 2. Chemical principle and processing steps for direct production of polyimide relief patterns.

EXPERIMENTAL

Photo-patterning is carried out using commercial equipment in air conditioned rooms at 23 °C ± 1 °C, a relative air humidity of 45 % ± 5 % and yellow light to exclude light with a wavelength of <500 nm.

RESULTS AND DISCUSSION

Some of the properties and applications of the photopolymer PVI 210 have already been reported[2]. In the meantime, development has progressed so far that unusually thick and extremely thin layers can be produced and photo-patterned. This has led to some new technical applications.

Attention will first be given to a number of special properties of the photopolymer which is subsequently discussed from various aspects of application technology.

PROPERTIES

PURITY

The photopolymer PVI 210 is a light yellow to light brown highly purified powder for semiconductor applications with very low, of the order of magnitude of ppm or less, ionogenic metallic and non-metallic impurities.

STABILITY DURING STORAGE

The solid resin is stable for one year if stored in the dark at room temperature. It is highly soluble in polar solvents.

The stability during storage of ready-to-use solutions made from this resin, i.e., solutions with photoreactive additives and coupling agent, depends on the proportion of solid resin, the solvent, the photoreactive additives and, last but not least, on the storage temperature.

Our standard ready-to-use solutions, for instance, with approx. 37 % solid resin in N-methylpyrrolidone retain viscosity stability at room temperature for at least 25 days. These stability times are prolonged with increasing dilution or at lower temperatures, for example, to months when refrigerated.

VISCOSITY AND LAYER THICKNESS RANGE

Compared to nonphotoreactive polyamide carboxylic acid solutions, solutions of PVI 210 of the same concentration have a considerably lower viscosity. This makes it possible, for example, to use PVI 210 to produce solutions with a very high solids content and good handling properties. For example, solutions with a solid content of 45 - 50 % can be pressure-filtered through a 0.8 μm membrane filter. By using spinning techniques, layers approx. 80 μm thick can be produced from such highly concentrated solutions in just one spinning process. Unlike multiple coating, which is otherwise necessary, homogeneously thick layers with adequately small edge bead can therefore be obtained in a very simple way.

Figure 3 shows the correlation between viscosity, the spinning speed and the resulting layer thickness for a medium layer range. Thinner film thicknesses are achieved with increasing speed or decreasing viscosity.

The values refer to a spin time of 20 seconds. Longer spin times also reduce the film thickness. Extremely thin films, < 0.1 µm, can be produced from solutions with a solid resin content ≤ 10 % by using a spinning technique or the roller coating process.

EXPOSURE

The exposure times depend on the film thickness and the required imaging accuracy. With contact exposure using a 350 watt ultra high-pressure mercury vapour lamp and a radiant power of 40 mW/cm² (measured with 400 probe from the OAI Company), exposure times between 40 and 180 seconds were necessary up to now using standard photoreactive additives (Michler's ketone, maleimides). Using other sensitizers it has, in the meantime, been possible to reduce the exposure time by a factor of 10 - 20 for thickness up to approximately 7 µm³.

DEVELOPMENT

Best development results are achieved in a spray development system. The development times required are between 8 and 40 seconds depending on the film thickness. Sharply contoured patterns are obtained.

RESOLUTION

2 µm patterns are exposed in layers ≤ 2 µm at a line to space ratio of 1:1. In the case of thicker layers, an aspect ratio of 1:1 can be realized. Optimization for the individual geometries is possible and is usually necessary.
Figure 4 shows the image of a test mask with various 2 µm patterns before and after thermal conversion to polyimide.

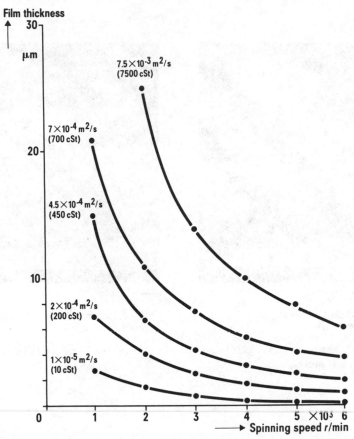

Figure 3. Correlation between viscosity, spinning speed and film thickness.

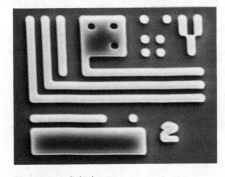

Photo–cross–linked patterns
layer thickness 2μm

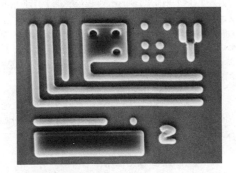

Polyimide patterns
after thermal treatment
layer thickness 1μm

Figure 4. Image of a test mask with 2 μm relief patterns on silicon nitride.

├────┤ 7 μm
Photo–cross–linked patterns
Layer thickness: 7 μm

├────┤ 7 μm
Polyimide patterns after thermal treatment
Layer thickness: 3.5 μm

Figure 5. SEM of test patterns of silicon wafers before and after thermal treatment.

CHEMICAL CONVERSION TO POLYIMIDE

Conversion of the photocrosslinked polyimide precursor to polyimide and removal of the main cleavage product component is effected within 30 minutes by thermal treatment at 275°C in nitrogen. To completely remove the cleavage products, thermal treatment at 400 °C for 1 hour is necessary. This reduces the layer thickness of the relief patterns by 50 %. The lateral shrinkage is very low.

In spite of the high proportion of volatile cleavage products, sharply contoured patterns with a very homogeneous surface are produced after thermal treatment.

Figure 5 shows SEMS's; on the left are photocrosslinked line and gap patterns and on the right the same patterns after thermal treatment at 400 °C with a layer thickness reduced by half.

912

THERMAL STABILITY

The polyimide relief patterns are stable for several hours at 400°C in nitrogen. Even brief thermal exposures up to 450°C do not lead to any adverse effects on the quality of the relief patterns.

ELECTRICAL PROPERTIES

The electrical properties have already been reported[2]. They are comparable to the properties of other polyimides used in semiconductor applications.

APPLICATIONS

Thick polyimide layers for α-radiation protection in large-scale integrated memory modules:

It has been known since 1978[4] that certain error functions of largescale integrated memory modules, so-called soft errors, are attributable to α radiation from the ceramic or filler-containing plastic package. The cause for the α radiation is the decay of radioactive trace elements (U, Th) in these component materials. The α particles penetrating the semiconductor chip statistically cause temporary changes in stored data.

It has since been possible to achieve a reduction in these soft errors by covering the integrated circuits with thick protective layers of polyimide- or silicon-based materials. The method generally used involves application and hardening of a drop on the already bonded chip. The shrinkage stresses which occur during this process can lead to breaking off of the fine contact wires, thereby impairing the bonding reliability.

We have adopted a different approach and, using photolithographic techniques, have attempted selective covering with thick polyimide layers of those parts of the chip surface sensitive to α radiation, leaving the contact points free. With the aid of the photopolymer discussed, PVI 210, it has been possible to develop a process suitable for production which is now used for effective protection of 16 and 64 K MOS RAMs produced by Siemens.

The following are the principal reasons for selection of the photolithographic process with PVI 210 for chip coating against α radiation:

- the high purity and reproducible quality of the material,
- the simple production and photo-patterning of adequately thick, well sticking layers (refer to Figure 6),

which also withstand rigorous production stress tests (cycles
between – 65 and + 150°C, 85°/85 % relative air humidity at
12 V and thermal treatment in a closed water pot to 120°C for
about 10 hours (steam pot test))
the early and low-cost covering of approx. 400 16K or 200
64K MOS RAMs on a 4-inch wafer at a time (refer to Figure 7).

Figure 6. Edge profile of 35 µm thick cell array coatings of
directly photo-patterned polyimide layers.

    In this way, an additional handling protection for subsequent
processes such as separating the chips by sawing the 4-inch wafer
and assembly in a ceramic or plastic package is achieved for the
chip surface sensitive to mechanical influences. This is favour-
able from yield and economic points of view.

    Figure 8 shows a 64 K MOS RAM module in a ceramic package
which is still open; the module has already been bonded and is
subsequently closed with a lid.

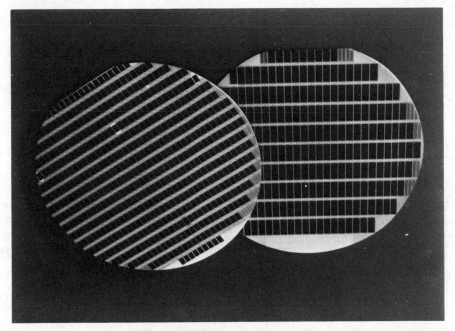

Figure 7. Photolithographically produced polyimide coatings on 16 and 64 K MOS RAMs.

Figure 8. 64 K MOS RAM in a ceramic package which is still open.

Extremely thin polyimide films as alignment layers for liquid crystal displays (LCD):

Selection of the alignment layer is of decisive importance for the production of LCDs for applications in the automotive sector. A uniform, stable, tilted layer is preferred today for LCDs.

Figure 9 shows a cross section of a twisted nematic (TN) liquid crystal layer.

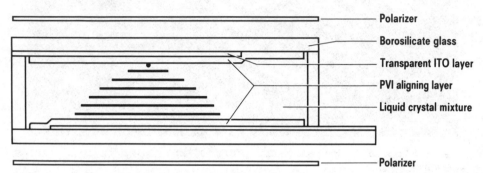

Figure 9. Cross section of a twisted nematic (TN) liquid crystal layer.

Located on the photo-patterned indium tin oxide (ITO) electrode layer is a transparent plastic orientation layer, which orients the liquid crystals parallel to the substrate surface. Parallel orientation in the required direction is achieved by rubbing this layer in that direction. The parallel alignment for a TN-LCD is twisted by 90 ° from one substrate to another. The liquid crystal layer is usually 9-10 μm.

The following series of requirements must be satisfied when selecting the plastic:

- Stability during LCD production against temperature and solvent
- Production of very thin well-bonding transparent films with a narrow layer thickness tolerance
- Tilt angle after rubbing = 2 °
- Stability under stress tests, e.g., 125 °C dry heat 1000 h, 65 °C/95 % relative air humidity 1000 h.

It has been possible to produce layers of this quality with ready-to-use PVI 210 solutions in N-methylpyrrolidone with a solid resin content ～ 10 %. The layers can be produced by using a spinning technique or the roller coating process.

In addition to the high dimensional stability at high temperature, and purity, the following features were significant for the selection of PVI 210:

- a high surface resistance,
- no contamination of the liquid crystals,
- good refractive index,
- the measured tilt angle is around 2.4°,
- high reproducibility of films < 0.1 µm,
- photo-patterning allows the polyimide layer to be designed so that contacts and adhesive borders remain free (Figure 10).

After patterning on the front and back electrode glasses, the frame is screenprinted and the two glasses are sealed together.

Figure 10. Layout of a typical LCD with directly photo-patterned polyimide alignment layer (dark areas).

These double glasses are scratched and broken into single displays, which are filled with liquid crystal materials and then closed.

OUTLOOK

We have discussed the exceptional properties of the photo-polymer PVI 210 with regard to simple production of highly heat-resistant relief patterns over a very wide layer thickness range, resulting in application in the memory and in the LCD fields. There should also mentioned the possibility of screening substrates for high temperature applications, of using the polyimide patterns as masks for evaporation-coating as well as for wet and dry etching processes in general. A more detailed report will be made elsewhere on applications beyond those briefly discussed here. We are confident that the properties mentioned here will arouse interest in many other fields.

ACKNOWLEDGEMENTS

The authors wish to thank the following: Herr Kühn and Herr Plundrich for preparing the solid resin; Frau Niederle and Herr Schmidt for the technological development.

REFERENCES

1.   R. Rubner, W. Bartel and G. Bald, "Production of Highly Heat-Resistant Film Patterns from Photoreactive Polymeric Precursors", Part 2, Polyimide Film Patterns, Siemens Research and Development Reports (Siemens Forschungs- und Entwicklungs-berichte, 5, 235 (1976).

2.   R. Rubner, H. Ahne, E. Kühn and G. Kolodziej, Photographic Sci. Eng., 23, No. 5, 303, September/October 1979.

3a.   European patent application, publication number 0047 184 A3, 1982. Applicant: E.I. du Pont de Nemours & Co., Wilmington.

3b.   Results on samples from licensees. Licensees for production and marketing are:
Ciba Geigy AG, Basel, Switzerland
Contact person: Dr. K. H. Rembold KA 5.25
E. I. du Pont de Nemours and Company, Wilmington.
Contact person: J. C. Chevrier, F. & F. Dept. Electronics
E. Merck, Darmstadt, W. Germany
Contact person: M. Vornoff, Beratung und Anwendungstechnik.

4.   H. J. Harloff, Electron. Rechenanl., 24, No. 2, 78 (1982).

NEW DEVELOPMENTS IN PHOTOSENSITIVE POLYIMIDES

H.J. Merrem, R. Klug and H. Härtner

E. Merck
Research and Development Dept.
Frankfurter Str. 250
D-6100 Darmstadt 1
Federal Republic of Germany

The use of polyimides for passivation or insulation
purposes is well established; however, one disadvantage of the products used today is the complicated
procedure to form patterns into the polyimide film. We
have developed a polyimide-precursor photoresist with
a sensitivity of 70-80 mJ/cm^2 . μm which is applied as a
common negative photoresist and gives a structured polyimide layer after a curing process. Examples of the
high resolution of 2.5 micron wide lines and spaces
in a 2.6 micron thick resist layer are given.

INTRODUCTION

Of the variety of high temperature resistant polymers, the
polyimides in particular have proved their suitability for
practical use for passivation or insulation purposes due to their
excellent properties such as chemical and mechanical resistance,
good insulation, planarizability and low defect density. One
problem concerning the application becomes evident if patterns
need to be formed into a polyimide layer, because the most
frequently used polyimide-precursors are not light sensitive
and the known light sensitive polyimide-precursors require high
exposure energies to become sufficiently crosslinked. The aim
of our research was, therefore, the development of a highly
sensitive polyimide-precursor photoresist which meets the
requirements of semiconductor production.

919

TECHNIQUES TO PATTERN POLYIMIDE LAYERS

Various techniques to pattern polyimide layers have been described including wet and dry etching methods.

Well known is the use of both negative and positive photoresists which are structured and act as masks in the etching procedure of the polyimide layer[1,2].

Van Pelt and coworkers[3] studied the application of positive photoresists and the dissolution of partially cured polyimide in alkaline solution, i.e., positive resist developer. The findings showed that the process required a very tight control to ensure consistent results. Particularly the prebake conditions to form the partially imidized polyimide had an important influence on the etch rate.

The use of tetramethylammonium acetate as etchant to achieve a better polyimide edge slope was published by Brunner and coworkers[4].

Negative photoresists can be applied in a similar way. To etch the partially cured polyimide, both hydrazine hydrate[2] and ethylenediamine[5] are used.

Dry etch techniques were investigated by Samuelson[1]. A fully cured polyimide layer was coated with plasma enhanced CVD SiO_2 or Si_xN_y. The inorganic film was structured by plasma etching and subsequently used as a mask for RIE etching of the polyimide.

Several photosensitive polyimide-precursor photoresists have been described recently. The most interesting system was published by Rubner and coworkers[6,7]. It is based on a polyamic acid esterified with photoreactive alcohols, e.g., hydroxyethyl-methacrylate. This system can be used as a normal negative photoresist resulting in a patterned polyimide after imidization.

POLYIMIDE-PRECURSOR PHOTORESIST OF HIGH SENSITIVITY

The object of our research was the development of a polyimide-precursor photoresist of higher sensitivity as compared with systems known today.

For the system discussed here we chose the best known polyimide
which is poly-/N,N´-(p,p´oxydiphenylene)pyromellitimide_/ (II)
whose excellent mechanical, thermal and electrical properties are
well established'. This polyimide is normally formed in an
imidization reaction of the polyamic acid (I) synthesized from
pyromellitic acid anhydride and p,p´-diamino-diphenylether
/Equation (1)_/.

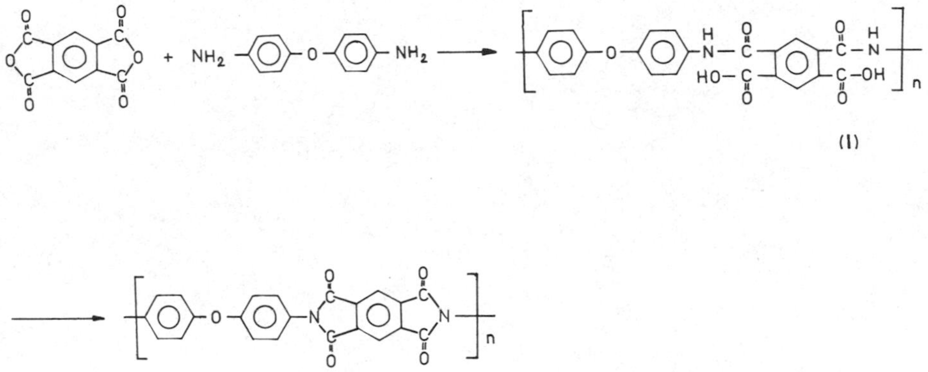

(1)

(II)   Equation 1

Similarly the polyimide (II) can be formed from polyamic acid
esters containing light-sensitive, crosslinked ester groups.

These polyamic acid esters are synthesized by the following
procedure: starting from pyrromellitic acid dianhydride the
diester was formed by adding light sensitive compounds containing
hydroxyl groups, e.g., pentaerythritriallylether /Equation (2)_/.
Subsequently, thionyl chloride was added to form the pyromellitic
acid dichloride and finally p,p´-diamino·diphenylether was
added to form the light-sensitive resin. The average chain
length can easily be controlled by varying the ratio of
pyromellitic acid dichloride and the amino compound.

A second method of increasing the sensitivity is the
addition of monomeric polyfunctional crosslinkable systems to
the polyamide acid ester, e.g. 2,4,5-trisallyloxy-1,3,5-triazin.
In order to obtain optimum sensitivity of 70-80 mJ/cm² μm it
is necessary to add a comonomer such as maleimide and a photo-
initiator, e.g., Michler´s ketone.

921

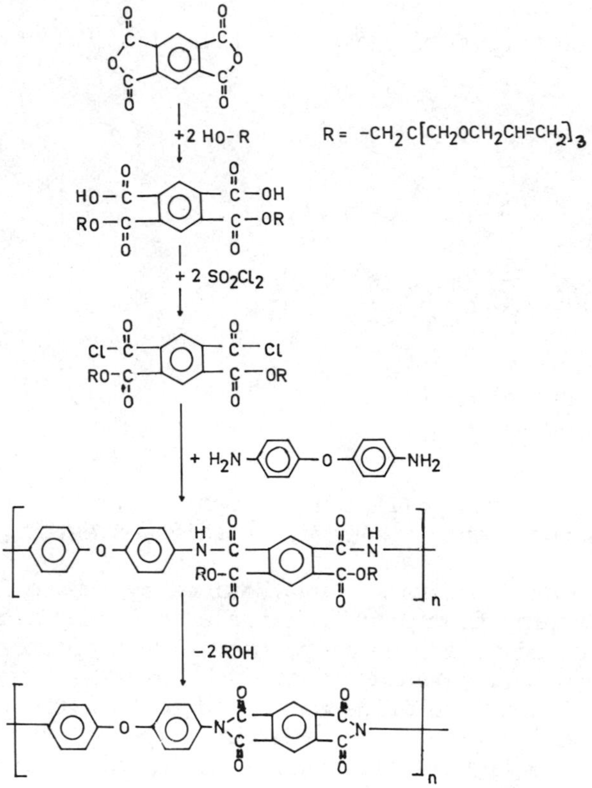

$R = -CH_2C[CH_2OCH_2CH=CH_2]_3$

+2 HO-R

HO—C   C—OH
RO—C   C—OR

+2 SO_2Cl_2

Cl—C   C—Cl
RO—C   C—OR

+ H_2N—⟨⟩—O—⟨⟩—NH_2

H O   O H
N—C   C—N
RO—C   C—OR

$\left[\qquad \right]_n$

−2 ROH

$\left[\qquad \right]_n$

Equation (2)

A comparison of the sensitivity of the system described here
with the polyimide precursor photoresist described by Rubner
and coworkers[6] is given in Table 1. Using the same photo-
initiator and comonomer the exchange of the photocross-
linkable ester group of the polyamic acid ester increases
the photosensitivity by a factor of at least five.

Table 1. Comparison of polyimide-precursor photoresists.

Light sensitive group	Substrate	Sensitivity $(mJ/cm^2 . \mu m)$
R: $-OCH_2CH_2OCC=CH_2$ ‖O, CH_3	SiO_2	650-700
R: $-OCH_2C[CH_2OCH_2CH=CH_2]_3$	SiO_2 Al	120-130 70-80

The dependence of light sensitivity of the photoresist on the
number of light sensitive groups per polymer unit of the polyamic
acid is linear in the range discussed here / Figure (1) /.

The polyimide-precursor photoresist is applied in the same way as
any common negative photoresist. The substrate to be coated has
to be dry and free of impurities to avoid adhesion problems.
Pretreatment with the well-known adhesion promotors for polyimide-
precursors[10] such as γ-aminopropyl triethoxysilane   or aluminum
chelates as well as addition of an adhesion promotor to the
photoresist solution is recommended to ensure optimum adhesion.
We have used the second method adding a vinylsilane-based component.

The photoresist is applied by a spin-on or a dipping procedure.
The layer thickness depends on the processing conditions; the
relationship between spin speed, viscosity and thickness is
shown in Figure (2).

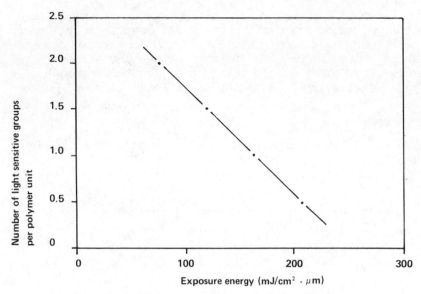

Figure 1. Dependence of sensitivity on number of light sensitive groups per polymer unit.

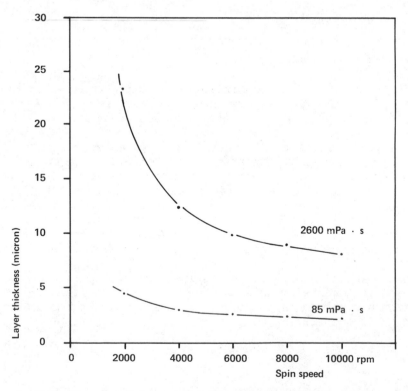

Figure 2. Dependence of layer thickness on viscosity and spin speed.

The right softbake conditions are important in achieving good adhesion and consistent sensitivity. We have found that light sensitivity is not affected between 40 $^\circ$C and 80 $^\circ$C, but adhesion problems appear below 50 $^\circ$C. Therefore, 60 $^\circ$C to 80 $^\circ$C is the optimum softbake range.

Resolution is dependent on layer thickness. Using hard or vacuum contact printing, an aspect ratio of better than one can be obtained. Figure (3) shows an example of a well resolved grid of 2.5 micron wide lines and spaces in a 2.6 micron thick resist layer on aluminum.

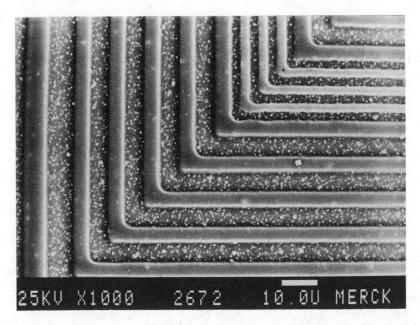

Figure 3. Example of resolution of 2.5 micron wide lines and spaces in a 2.6 micron thick resist layer on aluminum.

The exposure time required to obtain the best resolution depends on resist thickness. The values shown in Figure (4) were obtained with a contact printer (MJB 55 from K.Süss).

Thickness loss between coating and development steps occurs with all negative photoresists, it is very process-dependent and can be influenced by exposure time, exposure under nitrogen and development. A comparison of various photoresists representing

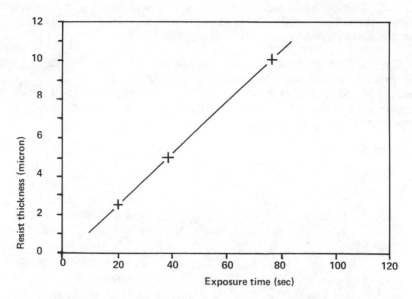

Figure 4. Dependence of exposure time on resist thickness.

the same chemical basis has been made for polyisoprene negative photoresists showing that thickness loss is reduced when the sensitivity of the photoresist is enhanced[8]. The same trend has been found when we compare the low sensitivity system described by Rubner and coworkers[6] with the photoresist discussed here. Regarding the entire overall decrease in thickness due to the photolithographic and curing step[5], the greater decrease in the photoresist layer of high sensitivity (during curing) is nearly compensated by a lower decrease during the photolithographic process compared to the low sensitivity system $\underline{/}$ Figure (6)$\underline{\,/}$.

To form the final polyimide pattern, the crosslinked photoresist layer is cured at temperatures between 275 °C and 400 °C. Due to these high temperatures the polyimide ring closure occurs via cleaving the ester groups. The crosslinked light sensitive groups are depolymerized and volatilized leaving the polyimide.

926

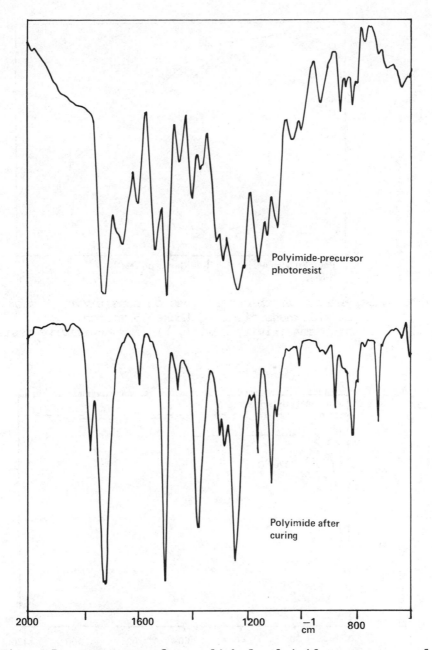

Figure 5. IR spectra of crosslinked polyimide-precursor and final polyimide.

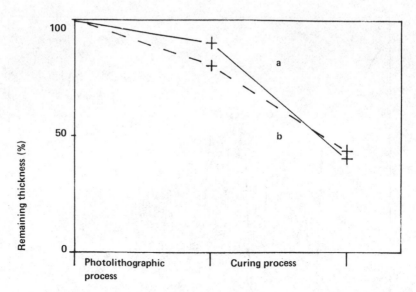

Figure 6. Dependence of thickness loss on sensitivity.
Layer thickness after softbake: 5 micron;
a) high sensitivity resist, b) low sensitivity resist.

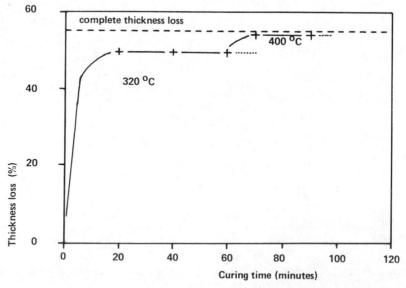

Figure 7. Thickness loss curve of crosslinked polyimide precursor.

Taking the amide IR-absorption peaks at 1670 cm^{-1} and 1540 cm^{-1} as criteria for the degree of imidization, complete cyclization is achieved at 320 $°C$ $/$ Figure (5) $/$.

The IR spectra of the fully-cured product shows no difference from the IR spectra of non-light-sensitive polyimides made from pyromellitic acid dianhydride and p,p´-diamino diphenylether[9].

The weight loss curve for a layer thickness of 2.9 micron (before curing) at 320 $°C$ under nitrogen is shown in Figure (7). After 20 minutes 95 % of the former crosslinked polymer bridges are decomposed and volatilized.

The curing process does not affect contour image profile or the evenness of the surface of the polyimide pattern. The weight loss results in a thickness loss in the vertical direction while shrinkage of the lateral pattern dimension is negligible. Figure (8) shows an example of a 6 micron thick polyimide pattern after curing which was sputtered with gold before curing. A small 0.2 to 0.3 micron wide border, which is no longer covered with gold after curing, indicates the difference of the position of the polyimide-precursor pattern before and the polyimide pattern after curing. The width of the border does not depend on the pattern size and, therefore, the border appears only due to the thickness loss of the resist edge in the horizontal direction and is not a result of lateral shrinkage of the polyimide-precursor.

Very high resolution can not only be obtained by contact printing but also by projection printing. Figure 9 shows as example 4 by 4 micron wide contact holes in a layer thickness of 2.6 micron. The resist was exposed with a Micralign 130 (Perkin-Elmer).

CONCLUSION

The chemistry and application of a high-sensitivity polyimide-precursor photoresist are presented. It is shown that a polyamic acid esterified with a polyfunctional allylalcohol exhibits a higher sensitivity compared to a polyamic acid esterified with a monofunctional acrylic group. The main properties of the system described are:
(i) high sensitivity up to 70-80 mJ/cm^2 . μm, (ii) excellent resolution, and (iii) negligible pattern distortion during the curing process.

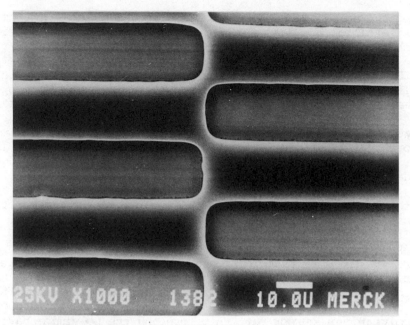

Figure 8. Example of imidized pattern, sputtered with gold before curing.

Figure 9. 4 by 4 micron wide contact holes in 2.6 micron thick resist (exposure tool: Micralign 130).

ACKNOWLEDGEMENT

The authors wish to thank Dr. Neisius for valuable discussions and Mr. Matlok for the extended investigation of the infrared spectra.

REFERENCES

1. G. Samuelson, ACS Symposium Series, 184, 93 (1982).
2. J. I. Jones, J. Polymer. Sci., C22, 773 (1969).
3. P. VanPelt, D. J. Belton and S. P. DiIorio, paper presented at the Electrochemical Society Meeting, Denver, Co., October 1981.
4. F. Brunner, P. Frasch and F. Schwerdt, DE 2,541,624 (IBM Deutschland GmbH) (1975).
5. R. K. Agnihotri, Polymer Eng. Sci., 17 (6), 366 (1977).
6. R. Rubner, H. Ahne, E. Kühn and G. Kolodziej, Photograph. Sci. Eng., 23(5), 303 (1979).
7. R. Rubner, Siemens Forsch. Entw. Ber., 5(2), 92 (1976).
8. K. Neisius and H. J. Merrem, J. Electronic Materials, 11 (4), 761 (1982).
9. D. Hummel, Farbe + Lack, 74, 11 (1968).
10. a. A. M. Wilson, D. Laks and S. M. Davis, ACS Symposium Series, 184, 137 (1982).
    b. Y. K. Lee, paper presented at the Electrochemical Society meeting, Minneapolis, MN, May 1981.
    c. A. Saiki and S. Harada, J. Electrochem. Soc., 129, 2278 (1982).
    d. D. Suryanarayana and K. L. Mittal, J. Appl. Polym. Sci., 29, 2039 (1984).

CHARACTERISTICS OF A HIGHLY PHOTOREACTIVE POLYIMIDE PRECURSOR

Fumio Kataoka, Fusaji Shoji, Issei Takemoto, Isao Obara
Mitsumasa Kojima*, Hitoshi Yokono and Tokio Isogai

Production Engineering Research Laboratory, Hitachi, Ltd.
292 Yoshida-machi, Totsuka-ku, Yokohama, Japan
*Yamazaki Works, Hitachi Chemical Co. Ltd., Hitachi, Japan

Some representative characteristics of a newly
developed photoreactive polyimide precursor, PAL, and
the polyimide produced directly by heating PAL are
described. The most striking characteristics of this
new material are the high photosensitivity, which is
comparable to that of the commercially available
negative working photoresist, and the high pattern
resolution. It should also be pointed out that the
polyimide patterns resulting from PAL patterns
retain their initial size satisfactorily, and the
properties of the polyimide such as thermal stability,
electrical and mechanical properties are good enough to
use it in the microelectronic devices as an insulating layer. Accordingly, the use of PAL instead of
ordinary polyimide precursors or photoresists should
open a promising process for the fabrication of
microelectronic devices.

INTRODUCTION

It has been well known that polyimides are useful to the electronics industry, e.g., as an insulator which retains its electrical and mechanical characteristics at relatively high temperature.[1] Furthermore, polyimides become more important with increasing packing density requirments for the microelectronic devices. In large scale integrated (LSI) circuits, for example, multilayered interconnections become unavoidable to satisfy high density requirments. In order to construct multilayered structures, it is desirable to provide a flat surface for insulating inter-layers. One of the most striking advantages of using polyimide instead of the more commonly used chemical vapor deposited silicon dioxide (SiO_2) or silicon nitride (Si_3N_4) is their ex-cellent ability to planarize stepped geometries. This property is attributed to the fluidity of polyimide precursor solution which is spun on a substrate, and then converted to polyimide by thermal curing. This method has been known as Planar Metalliza-tion with Polymer technology developed by Harada and coworkers of Hitachi.[2]

In recent years,much attention has concentrated on the photo-reactive polyimide precursor because of its direct patterning ability.[3-7] The most striking advantage, when this sort of photopolymer is applied as an insulating layer to microelectronic devices instead of an ordinary polyimide, is to shorten the polyimide layer patterning process.

In the fabrication process of an ordinary polyimide layer, two patterning steps are needed. One of which is a photoresist patterning step and the other is a polyimide patterning step with alkaline etchants. In addition, a step to remove the photoresist is also necessary. When the above process is replaced by the new process utilizing photoreactive polyimide precursor, the polyimide etching step and photoresist removing step can be eliminated and the total number of steps are reduced to about one half of that necessary if ordinary polyimide is used.

It is also noteworthy that this new process does not need to use alkaline etchant such as alkali hydroxides or hydrazine hydrate. The detrimental effects of alkali metal ions in insulating layers on device characteristics are well documented.[8] As to hydrazine hydrate, it strongly irritates the skin. So it has been highly desirable to avoid such etchants.

Figure 1 shows a schematic diagram for the new pattern formation process of polyimides utilizing its photoreactive precursors. As indicated, pattern formation is achieved at the stage of the polyimide precursor, which is then converted to the polyimide pattern by thermal curing.

934

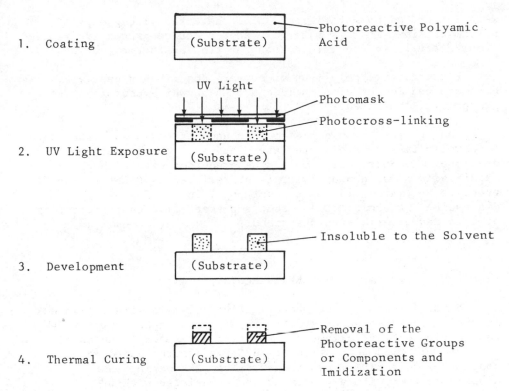

1. Coating — Photoreactive Polyamic Acid

2. UV Light Exposure — UV Light, Photomask, Photocross-linking

3. Development — Insoluble to the Solvent

4. Thermal Curing — Removal of the Photoreactive Groups or Components and Imidization

Figure 1. Pattern formation process in polyimide utilizing its photoreactive precursor.

The polyimide precursors are usually rendered photoreactive by either addition of the photoreactive additives or introduction of the photoreactive groups in the polymer side chains. Photocrosslinking reactions resulting from the UV exposure give rise to insolubility in the developers in the exposed areas of the film. Thus the most photoreactive polyimide precursors possess negative working properties.

Kerwin and Goldrick were the first to report about this kind of photopolymer which contained sodium dichromate as a photoreactive additive .[3] Use of this photopolymer for production of electronic devices is inadequate because of its instability and contamination by residual chromic ions.

Rubner and coworkers described photoreactive polyimide precursors, in which photoreactive acrylic or methacrylic groups

935

were introduced by esterification of the carboxylic groups of polyamic acid. In this material, photoreactive groups are removed simultaneously in the imidization reaction by thermal curing.[4-6]

Photoreactive polyimide precursors described recently by Hiramoto and Eguchi also include a step in which photoreactive groups are eliminated.[7]

An investigation had also been undertaken in our laboratory to develop this sort of materials. And as a result, we also developed a photoreactive polyimide precursor with high photosensitivity. Our photopolymer is abbreviated as PAL (Photoreactive Polyamic Acid for Lithography) in the following discussion. In this paper, some representative characteristics of this new material are described.

EXPERIMENTAL

Sensitivities of the photoreactive polymers were estimated by their exposure characteristic curves. An experiment for PAL was carried out as follows. A solution of PAL was spun on a silicon wafer and then dried at 70°C for 30 minutes to obtain a coating of 3 to 6 μm in thickness. The PAL coating was closely covered with a photomask and irradiated with UV light using a 500 W high pressure mercury lamp. Intensity on the exposed surface was 1-4 mW/cm^2 at a wavelength of 365 nm. After the exposure, the PAL coating was treated with a developer, the main component of which was N-methyl-2-pyrrolidone, for 3 minutes followed by rinsing with ethanol to give a relief pattern. The residual coating thickness was measured with a Talystep film thickness gauge(Rank Taylor Hobson Co. Ltd) as a function of time. The sensitivity was determined in terms of an exposure dose at which the thickness was reduced by 50% of the initial film coating thickness ($D_{0.5}$).

Resolution was evaluated from the scanning electron micrographs of the resulting patterns made with contact printing. A resolution test mask with lines and spaces and through-hole patterns of different sizes was employed for the evaluation.

Spectral sensitivity was determined using a Narumi RM-23 monochromator equipped with grating and 450 W xenon arc lamp. The irradiated film, 0.5 μm in thickness formed on a silicon wafer, was treated with developer to obtain the spectral sensitivity curve.

UV absorption of PAL coatings was measured at wavelengths of 365, 405, and 436 nm using Hitachi 200-10 spectrophotometer.

Thermal stability of the polyimide was estimated by the thermal gravimetric analysis using a TGD-500 Differential Thermal Micro-Balance (Shinku-Riko Co. Ltd). A coating of PAL was stripped off from the substrate and irradiated with UV radiation (exposure dose was 100 mJ/cm^2 in the present experiment). The film was heated at 200°C for 30 minutes, then at 450°C for 1 hour in N_2 to give the polyimide film which was subjected to thermal analysis.

Infrared transmission spectrum of the polyimide film produced as above was taken on a Nicholet 170SX FT-IR Spectrophotometer.

Electrical and mechanical properties of the polyimide were evaluated using 4-6μm thick films which were prepared by heating the PAL films at 350°C for 1 hour in N_2 flow. Instruments used in the present experiments were:GR 1615-A Capacitance Bridge;YHP 4329A High Resistance meter; Instron TTC-5000 Universal Testing Instruments.

RESULTS AND DISCUSSION

Exposure Characteristics of PAL

PAL is a negative working photopolymer like the previouly reported photoreactive polyimide precursors. One of the most striking characteristics of PAL is its high sensitivity. Figure 2 shows the exposure characteristic curve for PAL, Results for the commercially available negative working photoresist, consisting of polyisoprene and a small amount of bisazide, are also given for the sake of comparison. As can be seen in the Figure, PAL has a sensitivity($D_{0.5}$) of about 10 mJ/cm^2 at 365 nm. The value is as high as that of the commercially available negative photoresist.

The high sensitivity of PAL is closely related to its spectral sensitivity region which is in the 250-470 nm range. Since high pressure mercury lamps have strong emission peaks at 365, 405, 436 nm and PAL has a high sensitivity at each of these wavelengths as shown in Figure 3, an efficient energy transfer from the UV radiation to PAL can be achieved. This, in turn, should give rise to the high sensitivity of PAL.

High contrast, $\gamma = [\log_{10} D_1/D_0]^{-1} = 1.8$, and the non-swelling properties are other striking characteristics of PAL. The resolution of negative working resists is considered to be severely limited by the developer solvent-induced swelling[9] and poor contrast[10] resulting in non-vertical wall profile or bridging between patterns. But a high contrast, non-swelling negative working photopolymer should have inherently high pattern

937

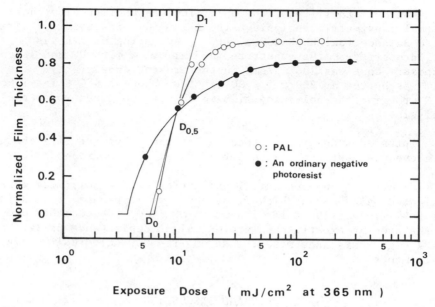

Figure 2. Exposure characteristic of PAL to UV irradiation.

resolution. This is the case for PAL. Figure 4 shows the scanning electron micrograph of 1.2 μm thick PAL patterns formed with 16 mJ/cm² UV radiation using contact printing. As can be seen, 0.5 μm lines and 1 μm spaces patterns indicate a high resolution profile. High aspect ratio, above 2, also demonstrates the high resolution characteristic of PAL.

The PAL patterns obtained show a vertical or somewhat over-hanged wall profile with no evidence of swelling. The degree of overhanging depends largely on soak time. The longer soak time leads to greater overhanged profile; this is attributed to the

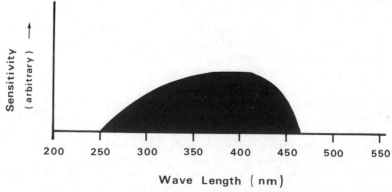

Figure 3. Spectral sensitivity for PAL.

938

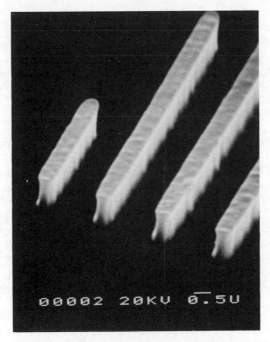

Figure 4.   Patterns in 1.2 μm thick PAL made by contact printing.

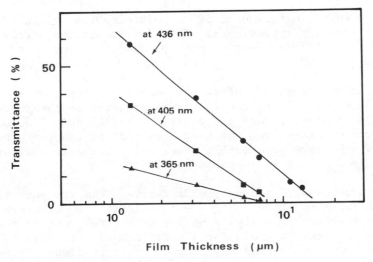

Figure 5.   Dependence of UV absorption on thickness of PAL.

lower photocrosslinking at the bottom of the film. As shown in
Figure 5, UV absorption depends strongly on the film thickness.
In the case of 2 μm thick film, for example, UV absorptions at the
bottom at 365, 405, 436 nm are about 10, 30, 50 percent of those
at the top, respectively. Accordingly, the extent of photocrosslinking which gives rise to insolubility in the solvent is lower at
the bottom. This, in turn, leads to higher solubility at the
bottom.

Besides the high contrast and non-swelling properties, the
above mentioned UV absorption feature is considered to be one of the
factors controlling the high resolution of PAL. Photolithography
ultimately suffer from the diffraction of light and the reflection
of light from interfaces. Since PAL absorbs UV light strongly at
the top of the film, undesirable influences on resolution by the
diffraction and reflection can be reduced significantly. Therefore, PAL reveals a high resolution even for relatively thick
films (3-10 μm) or at relatively high dosage of UV radiation . For
example, 3 μm lines and spaces and a 5 μm through-hole in 4.1 μm
thick patterns could be formed with 50 mJ/cm^2 UV radiation.

Characteristics of the Polyimide Produced from PAL

Photocrosslinked polyamic acid resulting from UV radiation
is then converted to polyimide by thermal curing which eliminates
the photoreactive group(s) or component(s) and water which is
produced by the dehydration reaction. Thus the thickness of
PAL film is reduced when the heat treatment is applied as shown in
Figure 6. The reduction in film thickness increases with an
increase in temperature and about 60 percent of initial thickness
is left at temperature in excess of 300°C.

Since the film formed at 300°C maintains its initial
thickness even after heating at 450°C for 30 minutes, transformation to polyimide must be accomplished at least at a
temperature of 300°C.

A Fourier transform infrared transmission spectrum for the
polyimide is shown in Figure 7. Strong absorption peaks due to
imide ring carbonyl groups are observed at 1720 and 1775 cm^{-1}.
The spectra obtained from the heat treatment at temperatures of
300°C and 450°C are practically the same. This also indicates
that the imidization is completed by 300°C.

One problem with the polyimide derived from its photoreactive
precursor is to maintain a high heat resistivity. Figure 8 shows
a thermal gravimetric analysis of the polyimide produced from
PAL after heating at 450°C for 1 hour in N_2 flow. Results for
an ordinary polyimide prepared from 4,4'-diaminodiphenylether and

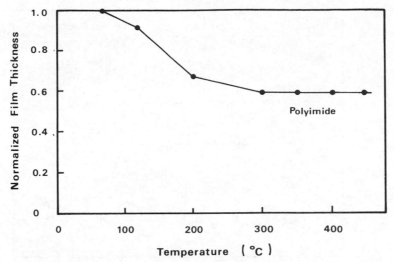

Figure 6. Remaining film thickness after heating at each temperature for 30 minutes.

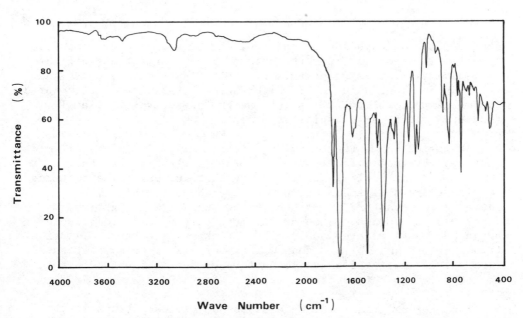

Figure 7. FT-IR transmission spectrum for the polyimide produced from PAL.

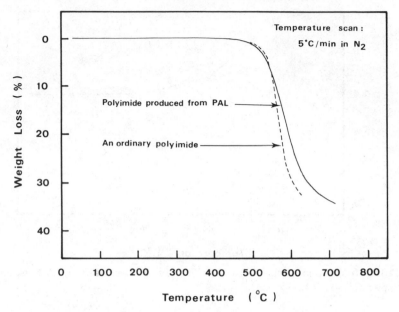

Figure 8. Thermal gravimetric analysis of the polyimide produced from PAL after heating at 450°C for 1 hour in N_2 flow.

pyromellitic dianhydride are also given for the sake of comparison. A polyimide produced from the present precursor reveals high heat resistivity which is comparable to that of an ordinary polyimide as shown in the Figure. This high thermal stability is attributed to the almost complete removal of the photoreactive functional groups by the thermal curing treatment.

Figure 9 shows the scanning electron micrographs of polyimide patterns obtained directly from the PAL patterns after heating at 400°C for 1 hour. It is worth noting that the polyimide patterns retain their initial dimensions, namely those of the PAL patterns, satisfactorily. Although the shrinkage in pattern thickness in the conversion of PAL into polyimide is relatively large as described previously, the shrinkage in pattern dimension is negligibly small (less than 0.1 μm for 1.2 μm thick patterns). This rather small shrinkage in pattern dimension might be due to the adhesion to the substrates.

Another remarkable feature of the polyimide pattern is its wall profile. In contrast to the PAL pattern, the profile of the polyimide pattern is considerably tapered. The side wall

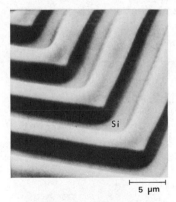

5 μm

a) Lines and Spaces

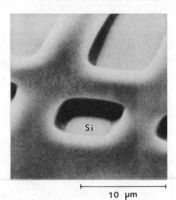

10 μm

b) Through-hole

Figure 9. Patterns of polyimide obtained directly from the PAL
patterns with heat treatment.

angles measured from the horizontal were evaluated roughly from the
scanning electron micrograph, and they are in the range of 50-70°
depending on the PAL profiles. This observation is attributable
to the shrinkage in volume. When polyimide is applied as an
insulating interlayer, a tapered through-hole is needed to avoid
breaking of wires. Accordingly, the finding above is very desirable
for its application as an insulating interlayer.

Since semiconductor device characteristics are very sensitive
to alkali metal ions[8], the alkali metal ions contents in PAL
solution were analyzed by means of atomic absorption spectroscopy.
Sodium and potassium ion were found to be 0.5 and less than 0.1
ppm, respectively. These values are acceptable for its application
in semiconductor devices.

It should also be noted that the polyimide produced from PAL

Table I. Electrical and Mechanical Properties of the Polyimide Produced from PAL.

Properties	
Electrical	
Dielectric strength, kV/mm	275
Dielectric constant at 1 kHz	3.0
Dielectric loss tangent at 1 kHz, percent	0.22
Volume resistivity, 10^{16} $\Omega \cdot$cm	0.6
Mechanical	
Tensile modulus, kg/mm^2	377
Tensile strength, kg/mm^2	12.8
Elongation, percent	10

possesses excellent electrical and mechanical properties, some representative data are listed in Table I. These values are comparable to those of an ordinary polyimide and are adequate for its application as an insulator.

CONCLUSION

A new class of photoreactive polyimide precursor, PAL, has been developed to shorten the polyimide layer patterning process in microelectronic devices. The most striking characteristics of this interesting material are the high photosensitivity and the high resolution. It should also be pointed out that the properties of the polyimide resulting from PAL are adequate for its application as an insulator. Namely, the thermal stability and the electrical and mechanical properties are comparable to those of an ordinary polyimide. Applications, other than as an insulator, such as heat resistant photoresist or a masking material for ion implantation may also be possible because of its high thermal stability. Accordingly, the use of PAL instead of ordinary polyimide precursors or photoresists should open up the prospect of a promising process for the fabrication of micro-electronic devices.

ACKOWLEDGMENT

The authors wish to thank Mr. Masayuki Okumura, General Manager of the Production Engineering Research Laboratory of Hitachi Ltd., and Mr. Tsuyoshi Tanno, General Manager of the Yamazaki Works of Hitachi Chemical Co. Ltd., for their continuing valuable advice and helpful discussions. The authors also wish

to acknowledge the suggestion and criticism of Mr. Hiroshi Kato
and Mr. Tokio Kato of the Takasaki Works , Hitachi Ltd. Dr.
Daisuke Makino, Senior Reseacher of Hitachi Chemical Co. Ltd.,
contributed greatly in the evaluation of the purity of PAL.

REFERENCES

1. A. M. Frazer, "High Temperature Resistant Polymers", in
    Polymer Reviews, Vol. 17, Ch. 7, p. 315, John Wiley and Sons
    Inc., New York, 1968.
2. K. Sato, S. Harada, A. Saiki, T. Kimura, T. Okubo and K. Mukai,
    IEEE Trans. Parts, Hybrid and Packaging, PHP-9, 176 (1973).
3. R. W. Kerwin and M. R. Goldrick, Polymer Eng. Sci., 11, 426
    (1971).
4. R. Rubner, Siemens Forsh. u. Entwickl., 5, 92 (1976).
5. R. Rubner, B. Bartel and G. Bald, ibid., 5, 235 (1976).
6. R. Rubner, H. Ahne, E. Kuhn and G. Kolodziej, Photographic
    Sci. Eng., 23, 303 (1976).
7. H. Hiramoto and M. Eguchi, Denshizairyo, p. 47, Nov. 1981.
8. A. Saiki, S. Harada, T. Okubo, K. Mukai and T. Kimura,
    J. Electrochem. Soc., 124, 1619 (1977).
9. M. J. Bowden and T. F. Thompson, Solid State Technol.,
    p. 62, May, 1979.
10. G. N. Taylor, G. A. Coquin and S. Someekh, Polymer Eng.
    Science, 17, 420 (1977).

DEEP-ULTRAVIOLET CHARACTERISTIC OF A PHOTOSENSITIVE POLYIMIDE

W.-H. Shen, A.-J. Yo and B.-M. Gong

Department of Applied Chemistry
Shanghai Jiao Tong University
Shanghai, People's Republic of China

We have prepared a photosensitive polyimide precursor, which is a negative photoresist. Heating it at about 300°C results in its conversion to pyromellitic polyimide (PI). Its decomposition temperature is above 500°C in air. Therefore, it has the same excellent properties as pyromellitic PI, such as dielectric property, heat-resistance, plasma etching resistance and others. It is not only a photoresist but also an excellent heat resistant insulation material. This is an ideal dielectric film for LSI.

Deep ultraviolet absorption is another valuable property of this photosensitive PI. Its ultraviolet absorption peak is at 275 nm, which is between deep UV and common UV regions. So it can be exposed by either deep UV source or common UV source.

INTRODUCTION

Polyimide (PI) is used in microelectronics as a material for multilevel interconnection and surface passivation. However, as the PI has no sensitivity for pattern-generation, the patterns are made with the aid of a photoresist and multilayer etching. This is inconvenient. So a photosensitive PI will greatly enhance the development in microelectronics. A comparison of the following processes using non-photosensitive and photosensitive PI shows that the second process (B) is greatly simplified. As the photosensitive PI is not only a photoresist but also an excellent heat-resistant insulation material, so it should be of great interest in microelectronics.

A: Non-photosensitive PI

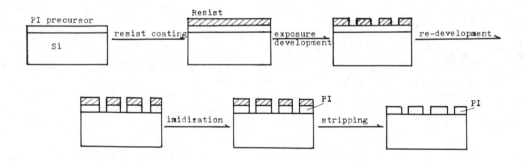

B: Photosensitive PI

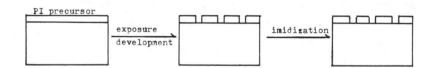

Compared with common photoresists, the heat-resistant photoresists are comparatively new on the electronics scene. Rubner et al. have succeeded in synthesizing a photosensitive PI precursor containing hydroxyethyl methacrylic groups[1]. In this paper we report some characteristics of a photosensitive PI containing acrylic groups. It structure is as follows:

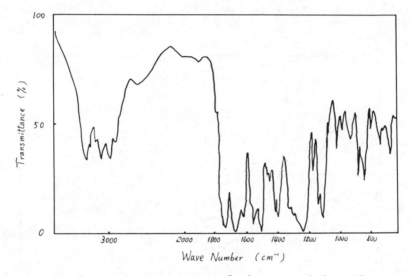

R*: photosensitive acrylic group.

Its infrared absorption spectrum is shown in Figure 1.

Figure 1.   IR absorption spectrum of photosensitive PI precursor.

RESULTS AND DISCUSSION

1.   Sensitivity:

This photosensitive PI precursor, which is dissolved in di-methylformamide (DMF), is applied to a silicon substrate by spin coating. After prebaking, exposure, and development, the pattern can be clearly formed. Ultraviolet absorption spectrum of the PI precursor is illustrated in Figure 2 (curve 1). Its UV absorption peak is at 275 nm, which is different from common photoresists. As the peak at 275 nm is between deep UV and common UV regions, so it can be exposed either by deep UV or common UV sources. When it is exposed by common UV source, proper crosslinking agents and sensitizers are required in order that the absorption peak can be shifted to longer wavelength. After carrying out a great number of tests, the authors selected acrylamide, Michler's ketone (MK), and benzoin ethylether (BEE) as additives. There is a synergistic effect between MK and BEE[2], so we decided to use them both at the same time.

As the concentration of sensitizers reached 5% by weight of polymer, a new absorption peak appeared at 350–360 nm.

Figure 2 indicates that the polymer can be exposed by common UV source. Proper additives help to increase the polymer sensitivity. If deep UV source is used, the sensitivity can be improved.

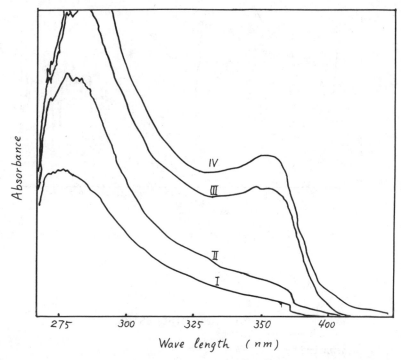

Figure 2. The UV absorption spectrum of the PI precursor without and with additives.

      Curve I: Photosensitive PI precursor solution in DMF.
      Curve II: Precursor: acrylamide: (MK+BEE) = 100:5:1.5.
      Curve III: Precursor: acrylamide: (MK+BEE) = 100:10:3.
      Curve IV: Precursor: acrylamide: (MK+BEE) = 100:10:5.

2.   Heat-resistance

This photosensitive PI has excellent heat resistance. Its decomposition temperature is above 500°C in air. This can be seen in the differential scanning calorimetry data in Figure 3. There are three peaks: solvent volatile peak, imidization peak and decomposition peak. Figure 4 is an enlarged view of the decomposition peak. The decomposition starts at 545°C and ends at 633°C.

These figures illustrate that this photosensitive PI can withstand temperatures up to 500°C in air.

3.   Imidization

Imidization peak in Figure 3 is magnified as shown in Figure 5. The peak temperature is 306°C. This result provides an important basis for our selection of optimum imidization temperature of 300°C. After heating the polymer for 1 hour at 300°C, imidization reaction can be completed. The IR spectrum of the final product is shown in Figure 6.

This spectrum is exactly the same as that for typical pyromellitic PI[3]. Some of the electrical properties of the final product are given below:

Volume resistivity: $1.62 \times 10^{16} \Omega cm(20°C)$, $1.28 \times 10^{15} \Omega cm$ (200°C).
Dielectric constant: 2.7 (50Hz), 2.78 (1MHz).
Dielectric loss factor: 0.003 (50 Hz).
Breakdown Voltage: 147 KV/mm.

These results show that the photosensitive PI precursor can thermally convert to pyromellitic PI, and, therefore, the former has the same excellent properties as the latter, such as dielectrical property, heat-resistance, plasma etching resistance and so on.

CONCLUSION

Because of its special structure, this polymer is sensitive to deep UV. Unlike commercial photoresists, this polymer can withstand up to 500°C in air. In addition, the storage stability of this precursor is much better than that of polyamic acid. This precursor does not degrade at room temperature. The solution is quite stable in the dark. This advantage results from the presence of photosensitive group R*, instead of active proton, in this PI precursor.

In summary, this functional polymer which is both a dielectrical material as well as a photoresist should play an important role in developments in microelectronics.

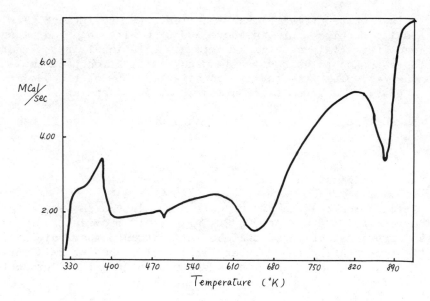

Figure 3.   DSC spectrum of photosensitive PI precursor.

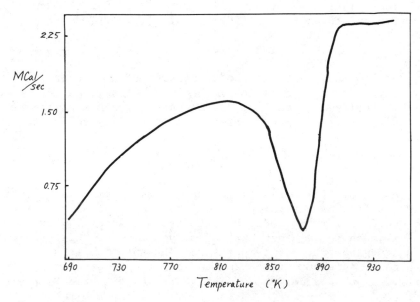

Figure 4.   Enlarged view of the decomposition peak (see Figure 3).

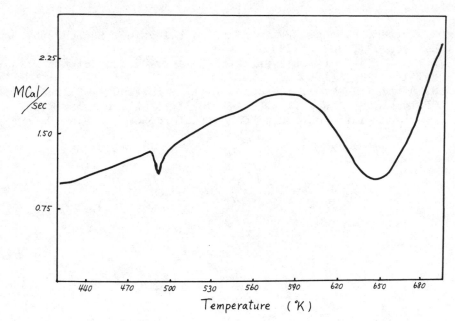

Figure 5.   Enlarged view of imidization peak (see Figure 3).

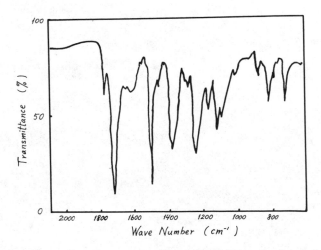

Figure 6.   IR spectrum of the imidization product.

REFERENCES

1. R. Rubner, H. Ahne, E. Kuhn and G. Kolodziej, Photograph. Sci. Eng., 23(5), 303 (1979).
2. X.-D. Fong, "Polymer Synthetic Chemistry", The Science Press, Beijing, PRC, (1980).
3. N. A. Adrova, M. I. Bessonov, L. A. Laius and A. P. Rudakov, "Polyimides, A New Class of Thermally Stable Polymers", p. 194, Technomic Publishing Co., Stamford, CT, 1970.

PART V. AEROSPACE AND OTHER APPLICATIONS

PMR POLYIMIDE COMPOSITES FOR AEROSPACE APPLICATIONS

Tito T. Serafini

National Aeronautics and Space Administration
Lewis Research Center
Cleveland, Ohio 44135

Fiber reinforced PMR polyimides are finding increased acceptance as engineering materials for high
performance structural applications. Prepreg materials
based on this novel class of highly processable, high
temperature resistant polyimides, which were first developed at the NASA Lewis Research Center, are commercially available and the PMR concept has been
incorporated in several industrial applications. This
paper reviews the current status of PMR polyimides.
Emphasis is given to the chemistry, processing, and
applications of the first generation PMR polyimide
known as PMR-15.

INTRODUCTION

Advanced fiber reinforced polymer matrix composites are achieving acceptance as engineering materials for the design and fabrication of high performance aerospace structural components. Epoxy resins are the most widely used polymer matrix materials because of their excellent mechanical properties and processing characteristics. In fact, the processing characteristics of epoxies have been adopted as "practical standards" for measuring the processability of other potential matrix resins. A disadvantage of epoxy resins is that their upper-use temperature is limited to approximately 175° C. Until recently, the intractable nature of higher temperature resistant polymers, which in theory could provide nearly a two-fold increase in use-temperature, has prevented the realization of the full potential of these polymers as matrix resins for high temperature polymer matrix composites.

In response to the need for high temperature polymers with improved processability, investigators at the NASA Lewis Research Center developed the novel class of addition-type polyimides known as PMR (for in situ polymerization of monomer reactants) polyimides.[1-3] The PMR concept consists of impregnating the reinforcing fibers with a monomer mixture dissolved in a low boiling point alkyl alcohol. The monomers are essentially unreactive at room temperature, but undergo sequential in situ condensation and ring-opening addition crosslinking reactions at elevated temperatures to form a thermo-oxidatively stable polyimide matrix resin. Because the final in situ reaction occurs without the release of volatile materials, high quality void-free composites can be fabricated by either compression or autoclave molding techniques. Thus, the highly processable PMR polyimides have made it possible to realize much of the potential of high temperature resistant polymers.

Our research has identified monomer reactant combinations for several PMR polyimides differing in chemical composition. The earliest, or "first generation", PMR matrix is designated PMR-15. Prepreg materials employing PMR-15 are commercially available from the major suppliers of composite materials.

The purpose of this paper is to review the current status of PMR polyimides. Emphasis is given to reviewing the aerospace applications of PMR-15 polyimide composites.

DISCUSSION

PMR Polyimide Chemistry

Condensation-type aryl polyimides are generally prepared by reacting aromatic diamines with aromatic dianhydrides, with aromatic tetracarboxylic acids, or with dialkyl esters of aromatic tetracarboxylic acids. The diamine-dianhydride reaction is preferred for preparing films and coatings, whereas the latter two combinations of reactants are generally employed for the preparation of matrix resins. Prepeg solutions are prepared by dissolving the appropriate reactants in high boiling point aprotic solvents (e.g., N-methyl-2-pyrrolidone, NMP) or in solvent mixtures containing an aprotic solvent. During composites fabrication, volatilization of the solvent and the condensation reaction by-products results in high void content composites having inferior mechanical properties and thermo-oxidative stability.

Figure 1 outlines some salient features of the PMR polyimide approach for the fabrication of composites. In the PMR approach, the reinforcing fibers are hot-melt or solution impregnated with a monomer reactant mixture dissolved in a low-boiling-point alkyl alcohol, such as methanol or ethanol. The monomer reactant mixture contains a dialkyl ester of an aromatic tetracarboxylic acid, an aromatic diamine, and a monoalkyl ester of 5-norbornene-2,3-dicarboxylic acid (NE). The number of moles of each reactant is governed by the following ratio:

$$n:(n + 1):2$$

Where n, (n + 1), and 2 are the number of moles of the dialkyl ester of the aromatic tetracarboxylic acid, the aromatic diamine, and NE, respectively. In situ polymerization of the monomer reactants (PMR) occurs when the impregnated fibers are heated. A condensation reaction forming low molecular weight norbornenyl-endcapped oligomers occurs in the temperature range of 121°-232° C. At temperature in the range of 275°-350° C, the norbornenyl groups undergo a ring-opening reaction generating maleic end-groups and cyclopentadiene. The maleic functions immediately copolymerize with the cyclopentadiene and unreacted norbornenyl groups to form a crosslinked polyimide without the evolution of volatile materials.

In the initial study[1] which established the feasilibity of the PMR concept, it was noted that a PMR matrix prepared from a monomer mixture consisting of the dimethyl ester of 3,3',4,4'-benzophenonetetracarboxylic acid (BTDE), 4,4'-methylene-dianiline (MDA), and NE exhibited a higher level of thermo-oxidative stability at 316° C than a PMR matrix made from the dimethyl ester of pyromellitic acid, MDA and NE. This unexpected finding was confirmed in a subsequent study[4] and the number of moles of BTDE which provided the best overall balance of processability and 316° C thermo-oxidative stability was found to be 2.087, corresponding to a PMR polyimide having a formulated molecular weight (FMW) of 1500. PMR matrices employing BTDE are referred to as "first generation" materials. Also, the stoichiometry of a PMR resin is generally denoted by dividing the FMW by 100. Thus, the first generation PMR matrix prepared from BTDE,

MDA, and NE having an FMW of 1500 is widely known as PMR-15. Prepreg materials based on PMR-15 are commercially available from the major prepreg suppliers. The structures of monomers used in PMR-15 are shown in Table I.

The early investigations[1,4] also clearly demonstrated the efficacy and versatility of the PMR approach. By simply varying the chemical nature of the either the dialkyl ester acid or aromatic diamine, or both, and the monomer reactant stoichiometry, PMR matrices having a broad range of processing characterisitics and properties were prepared. A modified PMR-15, called LARC-160, has been prepared by substituting an aromatic polyamine for MDA.[5] Other studies[6] have shown that the PMR approach has excellent potential for "tailor making" polyimide matrix resins with specific properties. For example, as shown in Figure 2, the resin flow characteristics (amount of resins exuded during processing) can be varied, or "tailored", over a broad range by simple adjustment of FMW (monomer stoichiometry). However, as shown in Figure 3, the higher-flow formulations exhibited decreased thermo-oxidative stability at 288° C. The reduction in resin flow and increase in thermo-oxidative stability at 288° C with increasing FMW clearly show the sensitivity of these properties to imide ring or alicyclic contents. The reduction in resin flow with increased FMW (increased imide ring content) also serves to qualitatively account for the intractable nature of linear condensation polyimides at an early stage in their process history.

Partial or complete replacement of BTDE with the dimethyl ester of 4,4'-(hexafluoroisopropylidene)-bis(phthalic acid) (HFDE) significantly improved the thermo-oxidative stability of "first generation" PMR resins.[7,8] The HFDE-PMR compositions are referred to as "second generation" materials to distinquish them from the "first generation" BTDE-PMR materials. The "second generation" HFDE-PMR resins have not been accepted as matrix resins because of the lack of a commercial source for the anhydride of HFDE.

As mentioned earlier, current technology PMR-15 polyimide prepreg solutions are generally prepared by dissolving the monomer mixture in a low-boiling-point alkyl alcohol. Although the volatility of these solvents is highly desirable for obtaining void free composites, it does limit the tack and drape retention characteristics of unprotected prepreg exposed to the ambient. PMR-15 monomer reactants and a mixed solvent have been identified which provide prepreg materials with improved tack and drape retention characteristics without changing the basic cure chemistry, processability, or thermo-oxidative stability.[9] The modifications consist of substituting higher alkyl esters for the methyl esters in NE and BTDE and using a solvent mixture (3:1 methanol/1-propanol) in lieu of pure methanol.

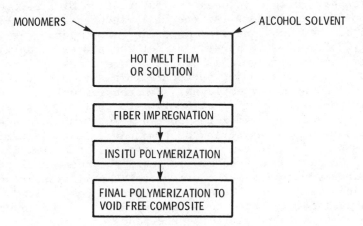

Figure 1. - Lewis PMR polyimide technology.

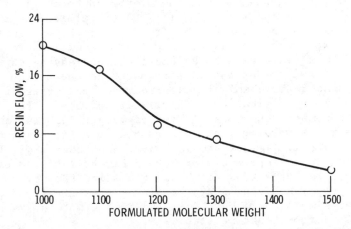

Figure 2. - Resin flow of HTS graphite/PMR composites (Resin flow is based on weight of resin flash formed during molding).

High pressure (compression) and low pressure (autoclave) molding cycles have been developed for fabrication of fiber reinforced PMR composites. Although the thermally-induced addition-cure reaction of the norbornenyl group occurs at temperatures in the range of 275° to 350° C, nearly all of the processes developed use a maximum cure temperature of 316° C. Cure times of 1 to 2 hours followed by a free standing post-cure in air at 316° C for 4 to 16 hours are also normally employed. Compression molding cycles generally employ high rates of heating (5 to 10° C/min) and pressure in the range of 3.45×10^6 to 6.0×10^6 N/m^2. Vacuum bag autoclave processes at low heating rates (2 to 4° C/min) and pressures of $1.38 \times 10^6 N/m^2$ or less have been successfully used to fabricate void-free composites. The successful application of autoclave processing methodology to PMR polyimides results from the presence of a thermal transition, termed "melt-flow", which occurs over a fairly broad temperature range.[10] The lower limit of the melt-flow temperature range depends on a number of factors including the chemical nature and stoichiometry of the monomer reactant mixture, and the prior thermal history of the PMR prepreg. Differential scanning calorimetry studies have shown the presence of four thermal transitions which occur during the overall cure of a PMR polyimides.[11] The first, second and third transitions are endothermic and are related to the following: (1) melting of the monomer reactant mixture below 100° C, (2) in situ reaction of the monomers at 140° C, and (3) melting of the norbornenyl terminated prepolymers at the range of 175° to 250° C, referred to as the melt-flow temperature range. The fourth transition, centered near 340° C, is exothermic and is related to the addition crosslinking reaction. To a large extent the excellent processing characteristics of PMR polyimides can be attributed to the presence of these widely separated and chemically distinct thermal transitions.

The recommended final cure temperature of 316° C for PMR-15 exceeds the temperature capabilities of many industrial autoclave facilities which were originally acquired for processing of epoxy matrix composites. Recent studies have shown that a significant reduction in cure temperature of PMR-15 can be achieved by replacing 50 mole percent of the NE with p-aminostyrene.[12] Cure studies of the mixed endcap system showed that the final cure temperature of PMR-15 could be reduced to 260° C without sacrificing its 316° C thermo-oxidative stability.

Many studies have been conducted to determine the effects of various hostile environments on the physical and mechanical properties of PMR-15 composites. Glass,[13] graphite,[14-16] and Kevlar[17] PMR-15 composites have been investigated. Figure 4 shows the excellent 316° C interlaminar shear strength retention characteristics of PMR-15 composites made with four different

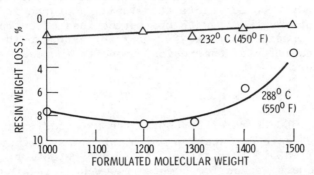

Figure 3. - Percent resin weight loss for PMR PI/HTS graphite fiber composites after 600-hr exposure in air at 232° C (450° F) and 228° C (550° F).

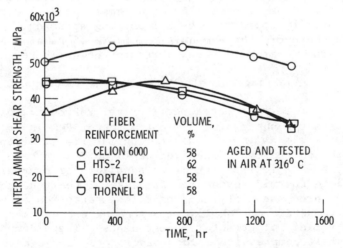

Figure 4. - Interlaminar shear strength of graphite fiber/PMR-15 composites.

graphite fibers.[15] The difference in strength levels between
the Celanese Celion 6000 composites and the three other composites
was attributed to differences in fiber surface morphologies.
    In the study[18] to correlate the stability of the individual
resin components in PMR-15 with the weight loss of Celion
6000/PMR-15 composites during 2200 hours of exposure in air at
316° C, the stability of the resin components increased in the
following order:

                        MDA < NE << BTDE

After 1200 to 1500 hours of exposure, the NE crosslinker exhibited
a marked increase in degradation which correlates with the 316° C
performance of PMR-15 composites. Excellent retention of composite
mechanical properties and low composite weight loss are generally
observed for 316° C exposure in air up to about 1500 hours. Thus,
it would appear that the 316° C thermo-oxidative stability of the
nobornenyl crosslinker determines the useful lifetime of PMR-15
composites at 316° C. However, in a recent study it was found
that the addition of low levels (4 and 9 mole percent) of a monofunctional norbornenyl crosslinker to PMR-15 increased the composite thermo-oxidative stability and resin flow during compostie
processing.[19] Although the modified PMR-15 composites exhibited
lower initial properties at 316° C than unmodified PMR-15 composites, the addition of small quantities of a monofunctional
norbornenyl crosslinker to PMR-15 to increase the 316° C thermo-
oxidative stability appears to be a promising approach. At the
present time, however, the composition of PMR-15 remains unchanged
from the original composition developed in the early seventies.

            Aerospace Applications of PMR-15 Polyimide Composites

    One of the most rewarding aspects of the PMR polyimide devel-
opment has been the successful demonstration of PMR-15 polyimide
composite materials as viable engineering materials. Prepregs,
molding compounds and even adhesives based on PMR-15 have been
commercially available from the major suppliers of composite
materials since the mid seventies. Because of their commercial
availability, processability, and excellent retention of properties
at elevated temperature, PMR-15 composites have been used to
fabricate a variety of structural components. These components
range from small compression molded bearings to large autoclave
molded aircraft engine cowls and ducts. Processing technology and
baseline materials data are being developed for the application of
PMR-15 composites in aircraft engines and nacelles, space structures, and weapon systems. Some representative applications of
PMR-15 composites are listed in Table II. None of the components
listed in the table, except the ion engine beam shield, are applications in the sense that the components are currently being

Table I. Monomers Used for PMR-15 Polyimide.

STRUCTURE	NAME	ABBREVIATION
	MONOMETHYL ESTER OF 5-NORBORNENE-2,3-DICARBOXYLIC ACID	NE
	DIMETHYL ESTER OF 3,3',4,4'-BENZOPHENONETETRACARBOXYLIC ACID	BTDE
	4,4'-METHYLENEDIANILINE	MDA

Table II. Applications of PMR-15 Composites.

Component	Agency	Contractor
Ultra-high tip speed fan blades	NASA-LeRC	PWA/TRW
QCSEE inner cowl	NASA-LeRC	GE
F404 outer duct	Navy/NASA-LeRC	GE
F101 DFE inner duct	Air Force	GE
T700 swirl frame	Army	GE
JT8D reverser stang fairing	NASA-LeRC	McDonnel-Douglas
External nozzle flaps		
PW1120		PWA[a]
PW1130	Air Force	PWA
Shuttle orbiter aft body flap	NASA-LeRC	Boeing
Ion thruster beam shield	NASA-LeRC	Hughes

[a] Company funded

produced. However, several of the components listed in Table II are scheduled for production introduction in the near future. A brief discussion of each of the components listed in Table II follows.

The blade illustrated in Figure 5 was the first structural component fabricated with a PMR-15 composite material. The reinforcement is HTS graphite fiber. The blade design was conceived by Pratt and Whitney Aircraft (PWA) for an ultra-high-speed fan stage.[20] Blade tooling and fabrication were performed by TRW Equipment.[21] The blade span is 28 cm and the chord is 20 cm. The blade thickness ranges from about 1.3 cm just above the mid-point of the wedge shaped root to 0.06 cm at the leading edge. At its thickest section the composites structure consists of 77 plies of material arranged in varying fiber orientation. The "line of demarkation" visible in Figure 5 at approximately one-third the blade-span from the blade-tip resulted from a required change in fiber orientation from 40 degrees in the lower region to 75 degrees in the upper region to meet torsional stiffness requirements. Ultrasonic and radiographic examination of the compression molded blades indicated that they were defect free. Although some minor internal defects were induced in the blade during low cycle and high cycle fatigue testing, the successful fabrication of these highly complex blades established the credibility of PMR-15 as a processable matrix resin.

Figure 6 shows the inner cowl installed on an experimental engine, called QCSEE for a Quiet Clean Short-Haul Experimental Engine, developed by General Electric (GE) under contract with NASA Lewis.[22] The cowl defines the inner boundary of the fan air flowpath from the fan frame to the engine core nozzle. The cowl was autoclave fabricated by GE from PMR-15 and Union Carbide's T300 graphite fabric. The cowl has a maximum diameter of about 91 cm and is primarily of honeycomb sandwich construction. Hexcel's HRH327 fiberglass polyimide honeycomb was used as the core material. The honeycomb core was bonded to the inner surface of the premolded outer skin with duPont's NR150B2G adhesive. The inner skin was then co-cured and bonded with NR150B2G to the core surface of the honeycomb core/outer skin assembly. Complete details about the cowl fabrication process are given in reference 23. The cowl was installed on the QCSEE engine and did not exhibit any degradation after more than 300 hours of ground engine testing. The maximum temperature experience by the cowl during testing was 260° C.[24] The successful autoclave fabrication and ground engine test results of the QCSEE inner cowl established the feasibility of using PMR-15 composite materials for large engine static structures.

Under a jointly sponsored Navy/NASA Lewis program (NAS3-21854), GE is developing a T300 Graphite fabric/PMR-15 composite outer duct to replace the titanium duct presently used on the F404 engine for the Navy's F18 strike fighter. The titanium duct is a sophis-

Figure 5. – Graphite fiber/PMR-15 polyimide fan blade (span, chord, and maximum thickness = 28, 20, and 1.3 cm, respectively).

Figure 6. – Graphite fiber/PMR-15 polyimide inner cowl installed on QCSEE engine.

ticated part made by forming and machining titanium plates followed by chem milling to reduce weight. A preliminary cost-benefit study indicated that significant cost and weight saving[25] could be achieved by replacing the titanium duct with a composite duct. The F404 composite outer duct differs from the QCSEE inner cowl in several important respects. The F404 duct is a monolithic composite structure, needs to withstand fairly high loads and, perhaps most importantly, the F404 duct is to be a production component and not a "one-of-a-kind" demonstration component. A full-scale composite duct (97 cm diameter by 165 cm length by 0.2 cm wall thickness) has been autoclave fabricated. The overall duct fabrication process consists of many operations. The sequence of the major operations is as follows: (1) autoclave fabrication of the composite shell, (2) ultrasonic inspecton, (3) adhesive bonding of T300 fabric/PMR-15 ply buildups to the shell using PMR-15 as the adhesive, (4) drilling of the buildups and cutting of the shell into two halves and (5) attachment of the split line stiffeners and titanium end-flanges. Figure 7 shows a photograph of the completed outer duct. The duct was installed on an F404 engine (Figure 8) and has successfully undergone nearly 900 hours of engine testing. The T300/PMR-15 composite outer duct is tentatively scheduled for production introduction in 1985.

Figure 9 shows a full-scale composite forward inner duct fabricated by GE for their F101 DFE (Derivative Fighter Engine). The approximate dimensions of the duct which was autoclave fabricated from T300 graphite fabric/PMR-15 are 102 cm diameter by 38 cm length by 0.15 cm wall thickness. Engine testing demonstrated that the composite duct was fully functional and met engine performance requirements. The T300/PMR-15 composite inner duct is currently scheduled for production introduction in 1985.

The current bill-of-materials inlet particle separator swirl frame on GE's T700 engine is an all metal part that involves machining, shape-forming, welding, and brazing operations. Design studies conducted under U. S. Army contract number DDAK51-79-C0018 indicated that the fabrication of a metal/composite swirl frame could result in a cost and weight savings of about 30 percent. Figure 10 shows a schematic diagram of a section of the metal/composite swirl frame that was fabricated from 410 stainless steel and various kinds of PMR-15 composite materials. The outer casing uses stainless steel in the flow path area to meet anti-icing temperature requirements and T300 and glass fabric/PMR-15 hybrid composite to meet structural requirements. The T300/glass hybrid composite was selected on the basis of both cost and structural considerations. An aluminum-coated glass fabric/PMR-15 composite material is utilized in the inner hub flowpath to meet heat transfer requirements for anti-icing. The glass fabric/PMR-15 composite utilized for the front-edge and front-inner surfaces was selected because of cost as well as temperature considerations. A full-scale (O.D. ä 51 cm) metal/composite swirl frame has been subjected

Figure 7. - Graphite fiber/PMR-15 outer duct after final machining.

Figure 8. - Graphite fiber/PMR-15 outer duct installed on an F404 engine (composite duct located directly above carriage).

Figure 9. - Graphite fiber/PMR-15 F101 DFE inner duct (Diameter = 102 cm, length = 38 cm, and thickness = 0.15 cm.).

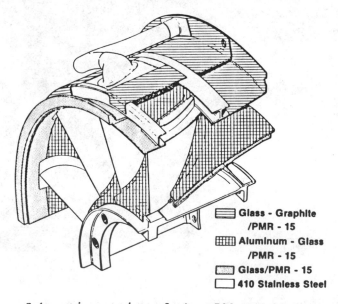

- ▭ Glass - Graphite /PMR - 15
- ▦ Aluminum - Glass /PMR - 15
- ▢ Glass/PMR - 15
- ▢ 410 Stainless Steel

Figure 10. - Schematic section of the T700 PMR-15 composite swirl frame (O. D. ≅ 51 cm).

to sand erosion and ice ball impact tests. The metal/composite
swirl frame provided improved particle separation and successfully
met the impact test requirements. Fabrication feasibility has
been demonstrated and if the metal/composite swirl frame successfully meets all of the performance rquirements, the metal/com-
posites T700 swirl frame is scheduled for engine testing in 1985.

Figure 11 shows a photograph of a DC-9 aircraft. The inserts
schematically depict the design of the presently used metal
reverser stang fairing and a composite redesigned fairing developed
by Douglas Aircraft Company under the NASA Lewis Engine Component
Improvement Program.[26] Studies had shown that a redesigned
fairing provided an opportunity to reduce baseline drag and would
result in reduced fuel consumption. The fairing serves as the aft
enclosure for the thrust reverser actuator system on the nacelle
of the JT8D engine and is subjected to a maximum exhaust temperature of 260° C during thrust reversal. A Kevlar fabric/PMR-15
composite fairing has been autoclave fabricated and flight-tested.
Compared to the metal components, the composite fairing resulted
in a one percent airplane drag reduction (1/2 percent had been
anticipated) and a 40 percent reduction in component weight.

Figure 12 is a schematic showing "committed" and "possible"
applications of graphite/PMR-15 composite materials on the PW1120
turbojet currently being developed by Pratt and Whitney Air-
craft/Government Products Division (PWA/GPD). A committed applications is an application for which a metal back-up component is
not being developed. The only committed applications for
graphite/PMR-15 composites at this time are the external nozzle
flaps and the airframe interface ring. PWA/GPD is in the process
of completing its assessment of the various "possible" applications
and anticipates that many of these will also become "committed",
if engine test schedules can be met. The PW1120 engine is currently scheduled for production deliveries in 1986. Graphite/PMR-15
external nozzle flaps have been committed for production by PWA/GPD
for its PW1130 turbofan engine. Production deliveries of the
PW1130 are scheduled for 1984. Prepregs made from T300 or Celion
3000 uniweave fabrics and PMR-15 are being evaluated for fabrication of the nozzle flaps used on both the PW1120 and PW1130
engines.

The CASTS (Composites for Advanced Space Transportation
Systems) program was undertaken by NASA Langley in 1975 to
develop graphite/polyimide composites for application to future
aerospace vehicles. As part of the CASTS Program, Boeing Aerospace Company developed manufacturing processes for fabrication of
graphite/PMR-15 structural elements and demonstrated their structural integrity at temperature up to 316° C.[27] The structural
elements consisted of flat panels, corrugated stiffeners, I-beams,
hat stiffeners, honeycomb panels, and chopped fiber moldings. A
section (76 cm wide x 203 cm long x 56 cm tapered to 15 cm thick)
of a simulated shuttle orbiter aft body flap (full scale, 213 x

CURRENT   MODIFIED

Figure 11. – Douglas DC-9 airplane. Inserts schematically show the current production metal fairing and the redesigned lower drag composite fairing made with Kelvar fabric/PMR-15.

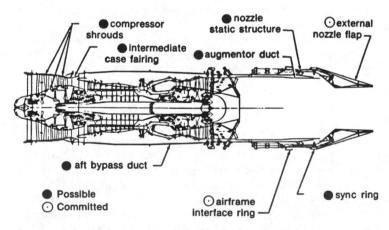

Figure 12. – Schematic of PW1120 showing applications of graphite/PMR-15 materials.

Figure 13. – Mercury ion thruster equipped with glass fabric/PMR-15
beam shield (shield diameter = 25 cm).

640 cm) was successfully fabricated. Thus, demonstrating that the manufacturing processes which were developed for structural elements could be adapted to large scale airframe hardware.

Figure 13 shows a mercury ion thruster for an auxiliary propulsion system being built by Hughes Space and Communications Group under contract to NASA Lewis. The ion propulsion system is scheduled for launch and testing on a future space shuttle flight. The thruster is equipped with a glass fabric/PMR-15 composite beam shield to protect the solar arrays and sensitive instrumentation on the spacecraft from ion-beam damage. The composite shield (approximate dimensions: 25 cm diameter x 20 cm length x 0.1 cm thickness) was selected over tantalum and titanium because of weight and structural considerations. The feasibility of using a glass fabric/PMR-15 composite shield was demonstrated by in-house fabrication and testing of full-scale beam shields.

CONCLUDING REMARKS

The in situ polymerization of monomer reactants (PMR) approach has been demonstrated to be a powerful approach for solving many of the processing difficulties associated with the use of high temperature resistant polymers as matrix resins in high performance composites. PMR-15, the PMR polyimide discovered in the early seventies, provides the best overall balance of processing characteristics and elevated temperature properties. The excellent properties and commercial availability of composites materials based on PMR-15 have led to their acceptance as viable engineering materials. PMR-15 composites are currently being used to produce a variety of high quality structural components. Increased use of these materials is anticipated in the future.

REFERENCES

1. T. T. Serafini, P. Delvigs, and G. R. Lightsey, J. Appl. Polym. Sci., 16, 905 (1972).
2. T. T. Serafini, P. Delvigs, and G. R. Lightsey, U. S. Patent 3,745,149, (1973).
3. T. T. Serafini in "Proc. International Conference on Composite Materials," E. Scala, Editor, Vol. 1, p. 202, AIME, New York, 1976.
4. P. Delvigs, T. T. Serafini, and G. R. Lightsey, "Materials Review for '72," SAMPE, Azusa, CA, 1972.
5. T. L. St. Clair, and R. A. Jewell, NASA TM-74994, National Aeronautics and Space Administration, Washington, D. C., 1978.
6. T. T. Serafini, and R. D. Vannucci, in "Reinforced Plastics - Milestone 30," p. 14-E1, Society of the Plastics Industry, Inc., New York, 1975.
7. R. D. Vannucci, and W. B. Alston, NASA TMX-71816, National Aeronautics and Space Administration, Washington, D. C., 1975.

8. T. T. Serafini, R. D. Vannucci, and W. B. Alston, NASA TMX-71894, National Aeronautics and Space Administration, Washington, D. C., 1976.

9. R. D. Vannucci, NASA TM-82951, National Aeronautics and Space Administration, Washington, D. C., 1982.

10. R. D. Vannucci, in "Materials and Processes - In Service Performance," p. 171, SAMPE, Azusa, CA., 1977.

11. R. W. Lauver, J. Polym. Sci., Polym. Chem. Ed., _17_, 2529 (1979)

12. P. Delvigs, NASA TM-82958, National Aeronautics and Space Administration, Washington, D. C., 1982.

13. P. J. Cavano, and W. E. Winters, TRW-ER-7884-F, TRW Equipment Labs, Cleveland, OH, 1978. (NASA CR-135377.)

14. P. Delvigs, W. B. Alston, and R. D. Vannucci, in "The Enigma of the Eighties: Environment, Economics, Energy," Book 2, p. 1053, SAMPE Azusa, CA., 1979.

15. R. D. Vannucci, in "Materials 1980," p. 15, SAMPE, Azusa, CA., 1980.

16. T. T. Serafini, and M. P. Hanson, in "Composites for Extreme Environments," ASTM-STP-768, p. 5, ASTM, Philadelphia, PA., 1982.

17. M. P. Hanson, in "Materials 1980," p. 1, SAMPE, Azusa, CA., 1980.

18. W. B. Alston, in "Materials 1980," p. 121, SAMPE, Azusa, CA., 1980.

19. R. H. Pater, NASA TM-82733, National Aeronautics and Space Administration, Washington, D. C., 1981.

20. J. E. Halle, E. D. Burger, and R. E. Dundas, PWA-5487, Pratt and Whitney Aircraft, East Hartford, Conn., 1977. (NASA CR-135122.)

21. P. J. Cavano, TRW-ER-7677-F, TRW Equipment Labs, Cleveland, OH., 1974. (NASA CR-134727.)

22. A. P. Adamson, in "Quiet Powered-Lift Propulsion," NASA CP-2077, p. 17, National Aeronautics and Space Administration, Washington, D. C., 1979.

23. C. L. Ruggles, R78AEG206, General Electric Co., Cincinnati, OH., 1978. (NASA CR-135279.)

24. C. L. Stotler, in "Quiet Powered-Lift Propulsion," NASA CP-2077, p. 83, National Aeronautics and Space Administration, Washington, D. C., 1979.

25. C. L. Stotler, in "The 1980's - Payoff Decade for Advanced Materials," p. 176, SAMPE, Azusa, SA., 1980.

26. R. T. Kawai, and F. J. Hrach, AIAA Paper 80-1194, 1980.

27. C. H. Sheppard, J. T. Hoggatt, and W. A. Symonds, D180-20545-2, Boeing Aerospace Co., Seattle, WA, 1979. (NASA CR-159129.)

THE DEVELOPMENT OF AEROSPACE POLYIMIDE ADHESIVES

A. K. St. Clair and T. L. St. Clair

Materials Division
NASA-Langley Research Center
Hampton, Virginia 23665

Few materials are available which can be used as aerospace adhesives at temperatures in the range of 300°C. The Materials Division at NASA-Langley Research Center has developed several novel high-temperature polyimide adhesives to fulfill the stringent needs of current aerospace programs. These adhesives are the result of a decade of basic research studies on the structure-property relationships of both linear and addition aromatic polyimides. The development of both in-house and commercially available polyimides will be reviewed with regards to their potential for use as aerospace adhesives.

INTRODUCTION

The U.S. Government and aerospace industries have escalated materials research over the past fifteen years in order to achieve ever higher levels of thermal and structural performance. However, the development of appropriate adhesives is lagging behind the production of structural aerospace materials. In an attempt to correct this situation, the government/aerospace community has begun to recognize specific adhesive needs generated by programs requiring advanced structural materials for elevated temperature applications on aircraft and spacecraft. The service life for such materials can be expected to range from the single flight of a missile lasting only a few minutes at temperatures approaching 500°C to the supersonic transport which will require at least 50,000 hours of cumulative service at 177°-232°C (Figure 1).

977

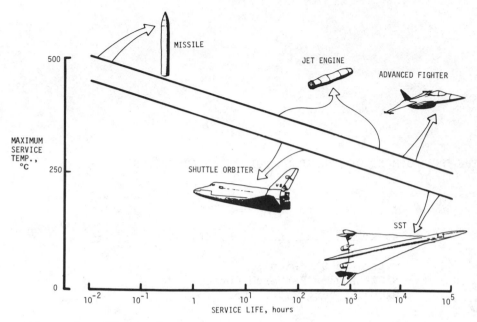

Figure 1. Time and temperature needs of advanced materials.

During the past decade, the need for high-temperature adhesives has been addressed by the National Aeronautics and Space Administration (NASA)-Langley Research Center in the Composites for Advanced Space Transportation Systems (CASTS), Supersonic Cruise Research (SCR), and the Solar Sail programs. CASTS is directed toward the development of technology to reduce the structural weight of vehicles such as the Space Shuttle through the use of high-temperature composite materials. The SCR program is designed to select and develop materials for structural applications on future supersonic transports. The Solar Sail is an outgrowth of the concept of solar wind propulsion of vehicles in interplanetary space requiring the long-term service of materials at temperatures as high as 300°C.

At present, commercially available adhesives do not meet all of the requirements for these programs. The Materials Division at NASA-Langley is therefore conducting research to develop high-temperature adhesives for joining metals, fiber-reinforced/polymer-matrix composites and films. The materials under investigation include both linear-condensation and addition-type aromatic polyimides. This paper will review the development of both NASA in-house and commercial high-temperature polyimides. Discussions will include synthetic methods for the preparation of both linear and addition polyimides and their applicability as aerospace adhesives.

978

LINEAR POLYIMIDE ADHESIVES

Linear aromatic condensation polyimides are attractive candidates for aerospace adhesives because of their toughness and flexibility, remarkable thermal and thermooxidative stability, radiation and solvent resistance, low density, and excellent mechanical and electrical properties. The processing of these materials, however, is tedious compared to other engineering plastics. Traditionally, linear polyimides have been processed in the polyamic acid form because, once converted to the polyimide, they become intractable. The usual preparation of a linear aromatic polyimide (Figure 2) involves the reaction of an aromatic dianhydride with an aromatic diamine (A) in a high-boiling solvent to produce a high molecular weight, soluble polyamic acid (B). The polyamic acid is then imidized to an insoluble and thermally stable polyimide (C) by chemical conversion at ambient temperature or by heating at a high temperature (300°C) to drive off both water of condensation and solvent. Because of the volatiles generated during the cure, the preparation of large void-free adhesive joints with such a material can become exceedingly difficult.

Solvent Studies on Linear Polyimide Adhesives

Early linear polyimide adhesives[1] were prepared as amic acids in highly polar solvents such as dimethylacetamide (DMAc) or dimethylformamide (DMF). These amide solvents are retained by the polymer during the bonding process which causes the formation of detrimental voids and promotes degradation of the adhesive through transamidization (chemical interaction between the amic acid and amide solvent). In an effort to circumvent this problem with conventional amide solvents, an investigation was begun at NASA-

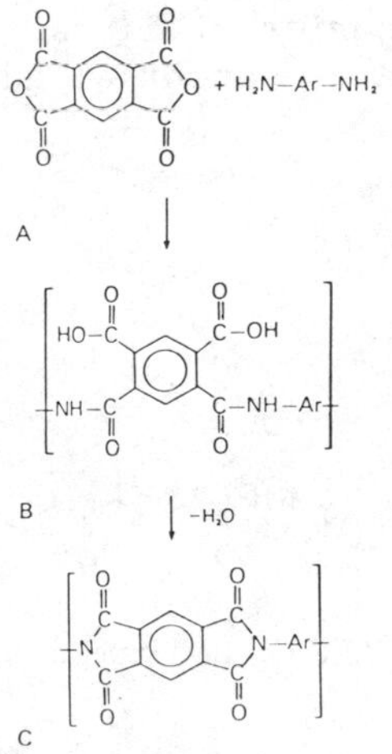

Figure 2. Method for preparing a linear aromatic polyimide.

Langley in the early 1970s to find a more suitable solvent for polyimide adhesives.

Many solvents were screened using the polymer LARC-2[2] which had been found earlier to have excellent adhesion to glass. LARC-2 (Figure 3) when prepared in DMAc or DMF is a high molecular weight polyamic acid. Polyimide films prepared by casting a 15% solids solution of this polymer onto plate glass remained flexible after a thermal cure at 300°C. However, a darkening in the yellow color of these films occurred upon curing in air which was an indication of polymer degradation. Many other solvents used in making LARC-2 also yielded high molecular weight polymer, but darkened the films during thermal imidization.

One particular series of solvents, however, did permit the polyamic acid to build to a high molecular weight and caused very little discoloration during imidization. These advantageous solvents are aliphatic ethers. Simple ethers such as diethyl ether would not properly dissolve the diamine or dianhydride monomers, but solvents such as tetrahydrofuran (THF), dioxane, and bis-(2-methoxyethyl)ether (diglyme) proved to be good solvents for both the monomers and for the resulting polyamic acid. These aliphatic ether solvents seemed to interact much less with the polyamic acid during cure than the amide solvents previously used. In addition, a remarkable increase in the adhesive lap shear strength of LARC-2 was obtained when the polymer was prepared in diglyme as shown in Table I.[3] As a result of this finding, diglyme is currently in use as a solvent for several linear polyimide adhesive systems.[4,5]

Structure-Property Relationship Studies on
Linear Polyimide Adhesives

In order to observe the effect of molecular structure changes on adhesive strength, a series of linear polyimides prepared in diglyme has been screened as potential adhesives.[3,6] The effect of

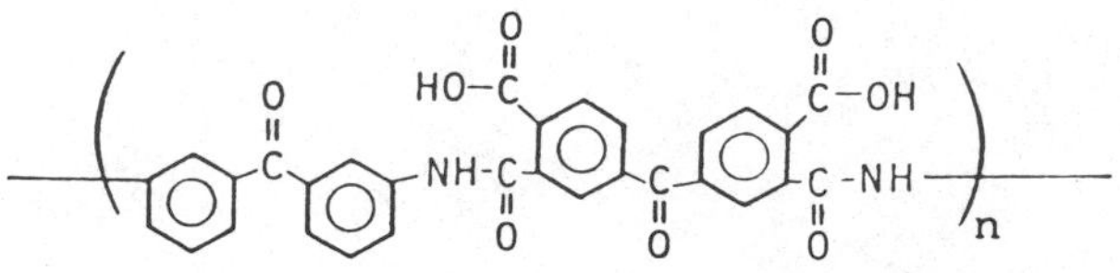

Figure 3. LARC-2 polyamic acid adhesive.

Table I. Effect of Solvent on LARC-2 Adhesive Strength.

Solvent	Lap Shear Strength, psi (MPa)*
DMF	500 (3.5)
DMAc	2500 (17)
DMAc/Dioxane	2900 (20)
Diglyme	6000 (41)

*RT test, titanium adherends

the anhydride monomer on adhesive strength is shown in Table II.
In this study, an added flexibilizing group between the benzene
rings of the anhydride portion of the polymer increased the adhesive lap shear strength. The effect of varying the structure of
the diamine portion of the polymer is shown in Table III. A major
finding in this study was the dependence of adhesive strength upon
the position of attachment of the amine groups to the benzene
rings. Adhesive strengths were always higher when the amine ($-NH_2$)
groups were situated in the meta (3,3') position as opposed to the
para (4,4') position. This amine isomer effect far outweighed any
changes in flexibilizing groups between the benzene rings.

Polyimide Film Bonding Applications

During the mid 1970's, NASA had an urgent need for a flexible
high-temperature adhesive that could bond ultrathin polyimide
Kapton® film* as part of the proposed NASA Solar Sail program. The
heliogyro sail (artist's impression shown in Figure 4) was intended
for a rendezvous mission in space with Halley's Comet in 1986. An
adhesive was needed to join strips of polyimide film at intervals
across each of the 12 blades measuring 8 m in width by 7350 m in
length. A series of linear polyimide film adhesives having both
the necessary high-temperature stability and flexibility were
developed for this application.[7] The series was based on LARC-2
polyamic acid (Figure 3) prepared in diglyme with the incorporation
of varying amounts of pyromellitic dianhydride and the para(4,4')-

---

*Use of trade names or manufacturers does not constitute an offi-
cial endorsement, either expressed or implied, by the National
Aeronautics and Space Administration.

Table II.   Effect of Anhydride on Polyimide Adhesive Strength.

ANHYDRIDE	AMINE	LAP SHEAR STRENGTH, psi, (MPa)*
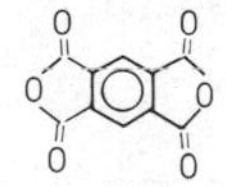	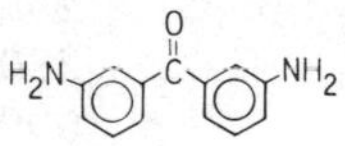	0 (0)
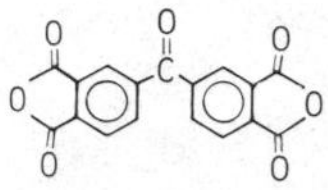	"	6000 (41)
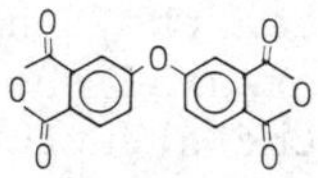	"	4700 (32)

*Titanium Adherends

Table III.   Effect of Diamine Isomer on Polyimide Adhesive Strength.

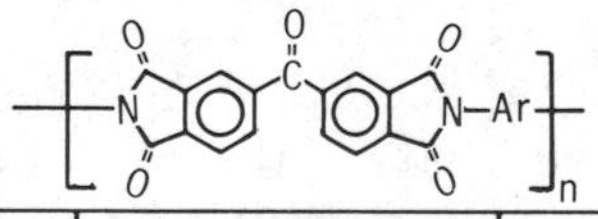

AMINE STRUCTURE (Ar)	AMINE ISOMER	LAP SHEAR STR., psi(MPa)*
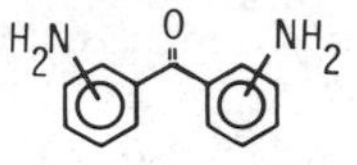	3, 3'	6000 (41)
	3, 4'	2500 (17)
	4, 4'	2600 (18)
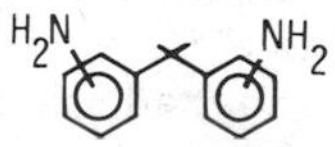	3, 3'	4200 (29)
	4, 4'	1900 (13)
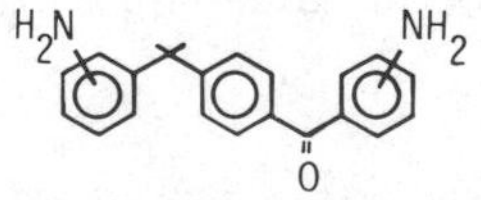	3, 3''	4200 (29)
	4, 4''	2600 (18)

*TITANIUM ADHERENDS

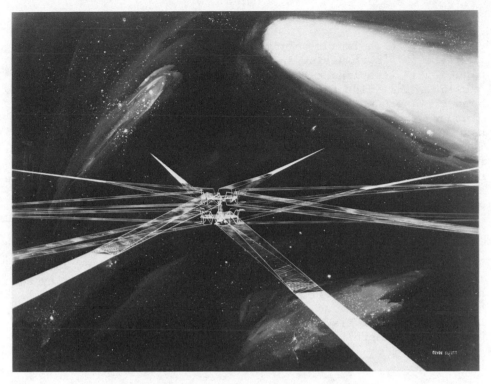

Figure 4. NASA heliogyro solar sail.

linked diamine to increase the ultimate use temperature of the
adhesive. The sail joints required only a 0.64 cm overlap of film
(Figure 5) which was a small enough area to allow for the escape
of volatiles during the bonding process. The lap shear strengths
of film specimens bonded at 316°C and aged for 6000 hrs at 275°C
were 70–90 psi (480–620 kPa). This was approximately 8 times the
strength required for the application. Although the LARC–2 based
series of adhesives met the immediate needs of the Solar Sail program, a need still remained for an adhesive which could bond larger
areas of film, metal or composites without the evolution of volatiles and entrapment of voids during cure.

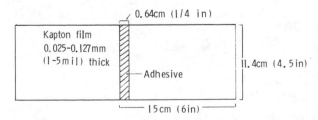

Figure 5. Schematic of bonded polyimide film specimens.

LARC-TPI Thermoplastic Polyimide Adhesive

Research on linear polyimide film adhesives progressed during the late 1970's resulting in the development of LARC-TPI in 1980. LARC-TPI[8,9] is a linear thermoplastic polyimide which can be processed in the imide form to produce large-area, void-free adhesive bonds. Like LARC-2, it is based on the BTDA and 3,3'(m,m')-DABP monomers shown in Figure 6. Unlike conventional polyimides, LARC-TPI is imidized and freed of water and solvent prior to bonding. Its thermoplastic nature is undoubtedly due to the structural flexibility introduced by bridging groups in the monomers and by the meta-linked diamine.[10] This material shows considerable potential as an adhesive because it allows the formation of large-area, voidless bonds due to thermoplastic flow.

LARC-TPI is currently being investigated as an adhesive for laminating large areas of polyimide film for both aerospace and industrial applications in the production of flexible electrical circuitry. A process[8] was recently developed at NASA-Langley for laminating polyimide films to each other or to conductive metal foils using LARC-TPI as an adhesive (Figure 7). Multi-ply laminates including metal foils may also be prepared in one step. Kapton film laminates prepared by this method do not fail in the adhesive

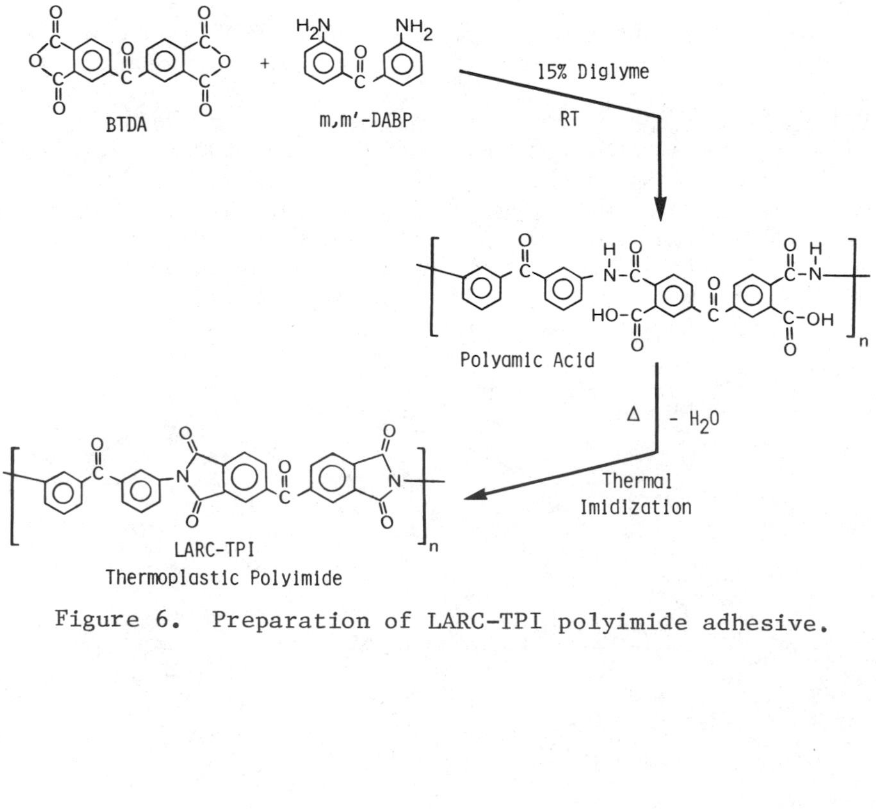

Figure 6. Preparation of LARC-TPI polyimide adhesive.

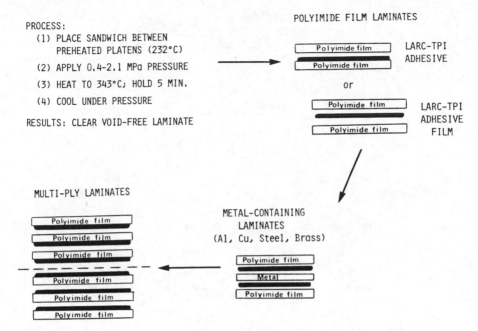

Figure 7. Process for preparing polyimide film laminates.

when subjected to standard peel tests; failure occurs through tearing of the Kapton film. Circuits (Kapton/copper/Kapton) laminated with LARC-TPI have been evaluated at the Rogers Corporation, Rogers, CT and found to withstand a ten-second immersion in a molten solder bath without blistering or delaminating.[11]

Simultaneously, LARC-TPI is being evaluated for the large-area bonding of an experimental graphite composite wing panel as part of the NASA SCR program. Preliminary bonding results on the adhesive have been generated both in-house at NASA-Langley and in studies at Boeing Aerospace Company[12] (Figure 8).

This ability to form large-area, void-free bondlines is unique for fully aromatic linear polyimide adhesives, and it makes LARC-TPI a leading candidate for future structural bonding in NASA programs. LARC-TPI is presently in the early stages of commercialization and may soon be marketed through a licensing agreement with NASA.

Linear Polyimide Adhesives Containing Metal Ions

Concurrent with the development of film adhesives was an investigation to study the effects of adding metal ions to a linear polyimide matrix.[13] Previously, the upper use temperature of linear polyimide adhesives prepared in diglyme had been extended by the addition of 30-70% by weight of aluminum powder prior to

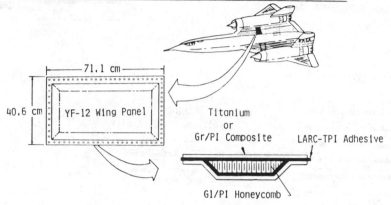

DATA FROM	T1/T1 LAP SHEAR STRENGTH, MPa (psi)		T1/T1 LAP SHEAR STRENGTH AFTER 3000 hrs at 232°C MPa (psi) 232°C
	RT	232°C	
BOEING AEROSPACE	36.5 (5300)	13.1 (1900)	20.7 (3000)
NASA-LANGLEY	41.4 (6000)	17.9 (2600)	—

Figure 8.   Structural adhesive application of LARC-TPI.

bonding.[3,6] The use of heavy metal fillers causes a large increase in bond weight and a loss in flexibility of the adhesive joint. It was discovered in the late 1970's that the high-temperature strength of a polyimide adhesive can be greatly improved by adding a small amount of aluminum ions (2% by weight) without embrittling the adhesive.[14] The structure of this metal ion-containing polymer is represented in Figure 9. Al ions are incorporated into the polyamic acid solution in the form of tri(acetylacetonato) aluminum, Al(acac)$_3$. As shown in Table IV, the addition of this organometallic complex raises the glass transition temperature (T_g) of the adhesive by 20°C thus increasing its upper use temperature. This adhesive shows much promise for bonding high-temperature films for applications in space where flexibility of the bondline is a necessity.

Experimental Linear Polyimide Adhesives

A novel linear aromatic polyimide (Figure 10) having both the characteristics of linear polyphenylene oxides and sulfides has been synthesized and characterized.[15] Having a T_g of 161°C, this polyimide prepared from the flexible BDSDA and APB monomers is easily processed as a hot-melt thermoplastic. As of the latter part of 1982, this material is still in the preliminary experimental stages of development as an adhesive. Room-temperature lap shear

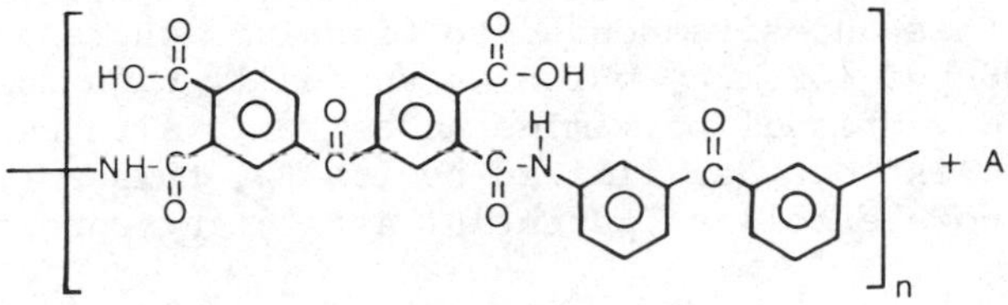

Figure 9. Aluminum ion-containing polyamic acid.

Table IV. Adhesive Strength of Al Ion-Containing Polyimide

Polyimide Adhesive	T_g, $^{\circ}C$	Lap Shear Strength,[*] psi (MPa)			
		RT	250°C	275°C	300°C
Polymer Alone	251	3000 (20.7)	1600 (11.0)	0	0
With Al(acac)$_3$	271	2400 (16.5)	1900 (13.1)	1600 (11.0)	700 (4,8)

[*]Titanium adherends

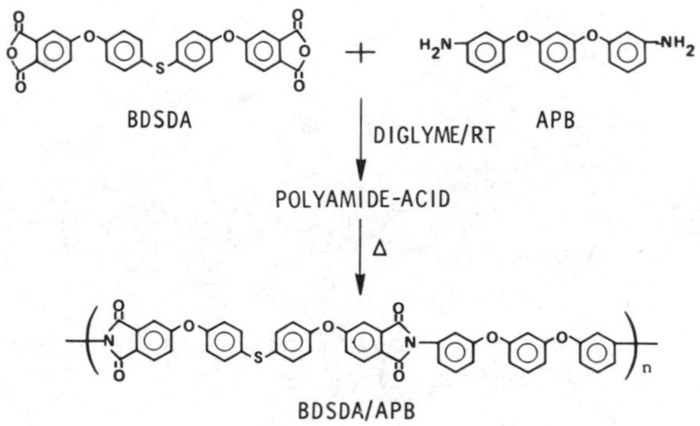

Figure 10. Polymer synthesis of a hot-melt adhesive.

strengths for this adhesive bonded to titanium adherends are averaging 6,000 psi (42 MPa) which is unusually high compared to strengths of conventional polyimide adhesives. Although its upper use temperature is somewhat limited by its T_g, this polyether-sulfideimide shows excellent potential as a highly processable hot-melt adhesive.

An experimental polyimidesulfone[16] is also under current investigation at NASA-Langley for use as an engineering thermo-plastic. This novel polymer system (Figure 11) has the advan-tageous thermoplastic properties generally associated with poly-sulfones, and the added solvent resistance and thermal stability of aromatic polyimides. Because of its improved processability over the base polyimide, the polyimidesulfone can be processed in the range of 260°–325°C to yield strong, high-temperature resistant adhesive bonds. Initial studies indicate high lap shear strengths of 4500 psi (31.0 MPa) for polyimidesulfone adhesive bonds pre-pared using titanium adherends, and good retention of strength (2600 psi, 17.9 MPa) at 232°C. The combination of properties offered by this material make it attractive for structural adhe-sives applications for aircraft and spacecraft.

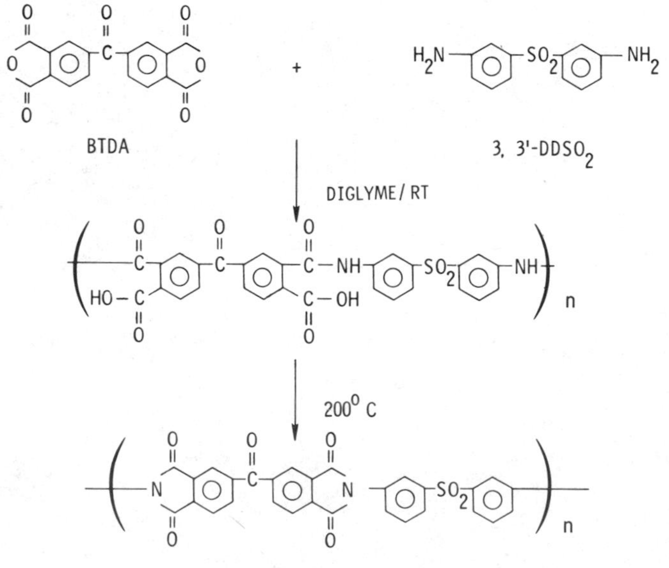

Figure 11.   Synthetic route for preparing thermoplastic polyimidesulfone.

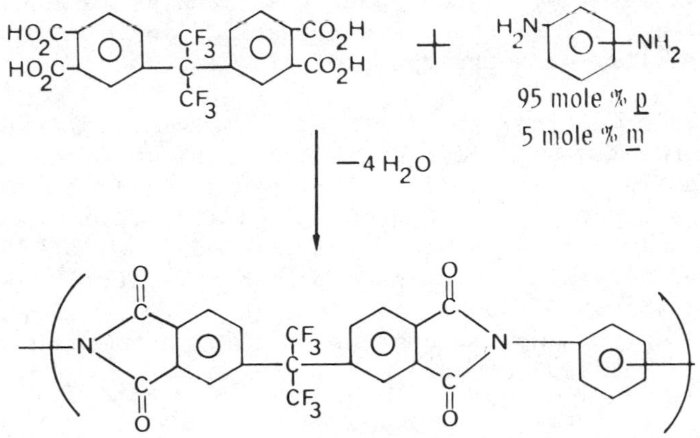

Figure 12.   Preparation of DuPont's NR-150B2.

Commercial Linear Polyimide Adhesives

Linear polyimides have rarely been commercially available for use as adhesives. In 1972, DuPont began marketing a series of linear aromatic polyimides called NR-150.[17] This series was based on the highly thermooxidatively stable monomer 2,2-bis(3',4'-dicarboxyphenyl)hexafluoropropane (6F). NR-150B2 shown in Figure 12 displayed excellent high-temperature adhesive properties, but was difficult to process because of the presence of condensation volatiles and a high-boiling solvent. In spite of these processing difficulties, this material was widely evaluated as a structural adhesive until its removal from the market in 1980.

Several other linear aromatic polyimide adhesives have been available over the past decade from companies such as Monsanto (Skybond series), American Cyanamid (FM-34), Amicon, Hexcel and U.S. Polymeric. However, the chemical structures of the polyimide products from these companies have been withheld as proprietary. Due to a general lack of information, these polyimide products will not be discussed here in any detail. Of the products mentioned, American Cyanamid's FM-34 has probably been the most widely used as an adhesive. This adhesive was recently evaluated in the NASA CASTS program for bonding titanium and graphite/polyimide composites.[18] It is capable of bonding small areas for applications at 316°C and is suited for large-area bonding if the joint is well vented to allow for escape of volatiles.

A new linear polyimide called Ultem was commercialized in early 1982 by the General Electric Company.[19] This new poly-etherimide is reported to be unique in high-performance plastics because of its processability. Ultem exhibits good mechanical strength, flame retardance, and chemical and radiation resistance. Although this material has not been evaluated as an adhesive, it may have potential for adhesive applications where extended use temperatures will be in the range of 170°C.

As of this writing, the most recent polyimide products to become commercially available are FM-36 adhesive film and FM-30 adhesive foam introduced in late 1982 by American Cyanamid.[5] These materials are currently being promoted as aerospace adhesives, serviceable over the temperature range of -55°C to 287°C. FM-36 is a modified polyimide which is supplied as a film supported by 112 glass cloth. A follow-up to FM-34, these materials are thought to be linear polyimides although no chemical structures have been disclosed by the manufacturer.

ADDITION POLYIMIDE ADHESIVES

Attempts have been made over the past decade to alleviate the processing difficulties associated with high-temperature linear polyimides through use of addition-type polymers. Addition polymers are easily processed in the form of short-chained oligomers which thermally chain extend by an addition polymerization involving reactive unsaturated end groups. Although much has been gained toward alleviating the evolution of volatiles associated with linear systems using this method of polymerization, the addition-type polyimides which have been developed are thermosets which cure to form a highly crosslinked network. In comparison to a linear system, these crosslinked networks are very brittle.

Development of Nadic-Terminated Addition Polyimides

In the late 1960's TRW developed a route for preparing short-chained aromatic imides endcapped with 5-norbornene-2,3-dicarboxylic (nadic) anhydride (Figure 13).[20] The final cure of this polymer called P-13N was effected by the thermal chain extension of olefinic end groups on the nadimide prepolymer to produce an essentially void-free, highly crosslinked polymer.

Unfortunately, a problem exists with the nadic-capped addition polyimides which is two-fold. First, the nadic portion of this material contains aliphatic groups which somewhat reduce the long-term thermal stability of the resulting polymer. Second, nadic-capped imides must be cured under pressure in order to prevent a

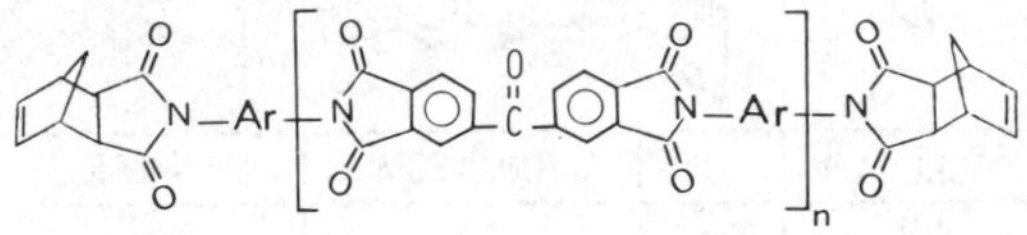

Figure 13. Nadic-terminated imide oligomer.

retrograde Diels–Alder reaction from occurring which would dis-
sociate the molecule into two components[21] (Figure 14). One of
the possible components is a volatile chemical, cyclopentadiene,
which can be detrimental in preparing void–free components or
adhesive bonds.

Structure–Property Relationship Studies of Nadic–Terminated Addition Polyimide Adhesives

Although the first nadic-terminated imides were more readily
processable than linear polyimides, they did not perform well as
adhesives. For the purpose of developing a nadic–capped system
with good adhesive properties for aerospace applications, NASA–
Langley began a structure-property relationship study of addition
polyimides in the mid 1970's to better understand how molecular
structures affect adhesive properties.[22]

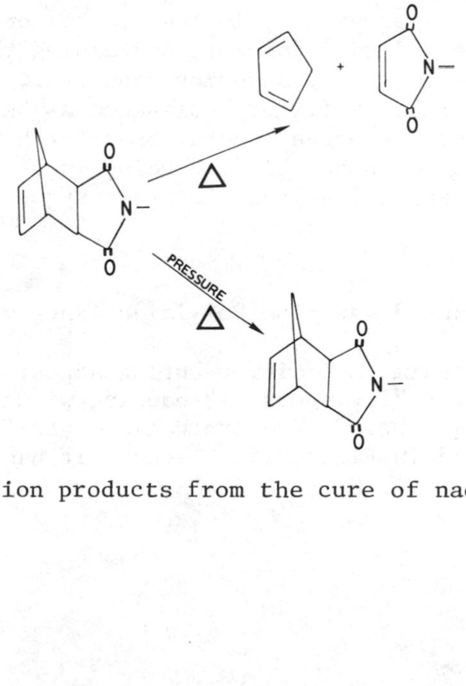

Figure 14. Reaction products from the cure of nadic–capped imides.

Table V. Adhesive Properties of Nadic-Terminated Imides.

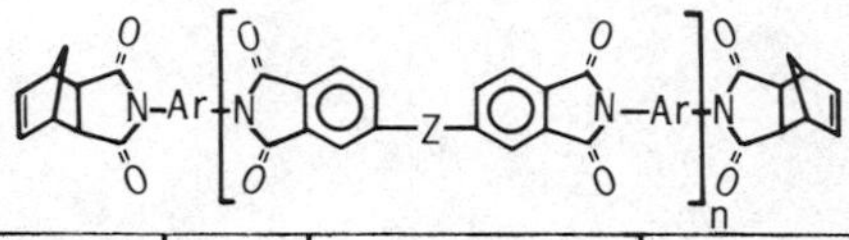

AMINE STRUCTURE (Ar)	Z	AMINE ISOMER	LAP SHEAR STR., psi(MPa)*
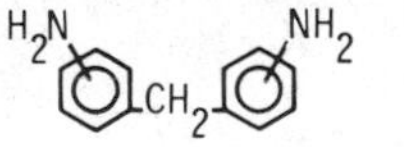 H$_2$N—⬡—CH$_2$—⬡—NH$_2$	-C- ‖ O	3,3'	2800 (19)
	-C- ‖ O	4,4'	600 (4)
	-O-	3,3'	2500 (17)
	-O-	4,4'	1300 (9)
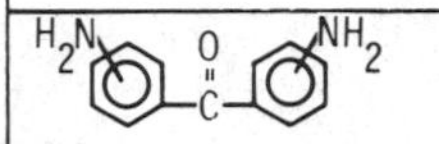 H$_2$N—⬡—O‖C—⬡—NH$_2$	-C- ‖ O	3,3'	2100 (14)
	-C- ‖ O	4,4'	1300 (9)
	-O-	3,3'	3000 (21)
	-O-	4,4'	1300 (9)

*TITANIUM ADHERENDS

The addition polyimides which were screened for their potential as structural adhesives for bonding titanium adherends are listed in Table V. The material which is listed second in Table V prepared from the para (4,4')-methylenedianiline was the only commercially available material at the time this study was conducted. It also had the poorest adhesive strength of any of the systems tested. As was found previously in the evaluation of linear polyimide adhesives, the addition polymers containing the meta (3,3')-linked aromatic diamines were superior adhesives. The first nadimide listed in Table V (later designated as LARC-13) exhibited the highest lap shear strength and has been further developed as a structural adhesive for specialized bonding in the NASA CASTS[18] and SCR[23] programs as well as for other aerospace applications.[24,25,26]

LARC-13 Addition Polyimide Adhesive

LARC-13 is a nadimide addition-curing adhesive which has been developed by NASA for "low pressure" bonding without the generation of volatiles during cure.[23] The synthetic route for preparing LARC-13 is displayed in Figure 15. Because it has a high degree of

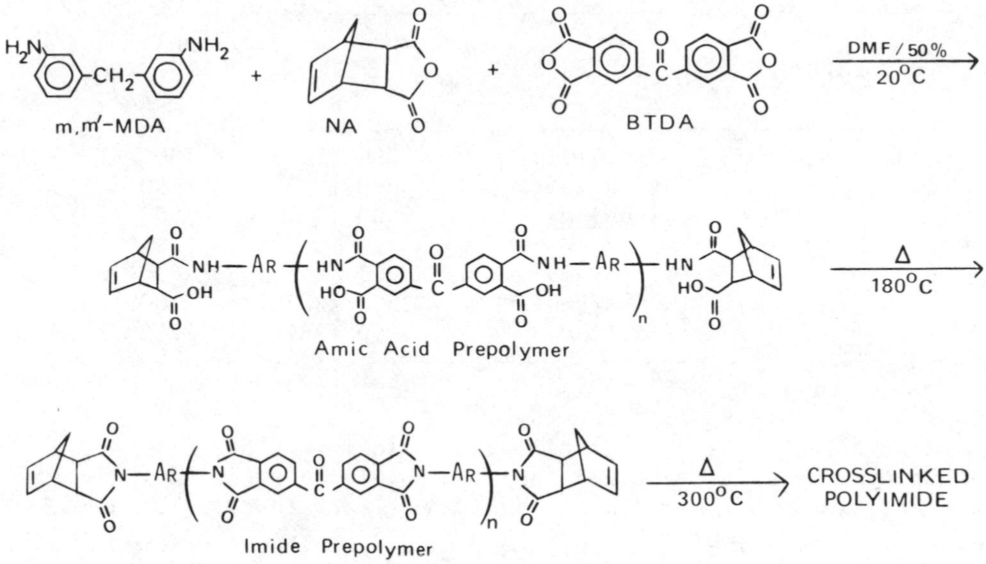

Figure 15.   Synthetic route for preparing LARC-13 adhesive.

flow during cure, it is easily autoclave processable. This ease of
flow is attributed to the presence of the _meta_-linked aromatic
diamine.

Because of its high crosslink density, LARC-13 can be used
well in excess of its glass transition temperature of 270°C. It
has been used successfully to bond a high-temperature composite to
a ceramic for missile applications requiring several seconds performance at 595°C. LARC-13 has also met criteria for NASA CASTS
applications exhibiting adequate adhesive strength after 125 hrs.
aging at 316°C. A major usage of this adhesive has been in the
bonding of honeycomb sandwich structures where the following are
required: (1) bond at low pressure of 50 psi (345 kPa) in order
not to collapse the honeycomb structure and (2) fillet around the
cell structure. LARC-13 has met both of these requirements. As
part of the NASA SCR program, LARC-13 is being evaluated by both
NASA-Langley and Boeing Aerospace Company for bonding titanium and
polyimide/graphite composites. Bonding data from both centers are
shown in Table VI.

Elastomer-Toughened Addition Polyimide Adhesives

In early 1979 an investigation was begun at NASA-Langley to
determine the effects of incorporating elastomers into the brittle

Table VI. LARC-13 Lap Shear Strengths.

Laboratory	Adherend	Lap Shear Str., psi (MPA)	
		RT	316°C
NASA–Langley	Gr/PMR–15	2300 (16)	1900 (13)
	Gr/NR–150B2	5000 (35)	1700 (12)
	Titanium	3000 (21)	1500 (10)
Boeing	Gr/PMR–15	2000 (14)	2000 (14)
	Glass/PMR–15	3300 (23)	2200 (15)
	Titanium	4000 (28)	2200 (15)

LARC–13 adhesive matrix. Butadiene/acrylonitrile and silicone elastomers were incorporated into the polyimide resin to study the effects on polymer glass transition temperature (T_g), thermal stability, adhesive strength, and fracture toughness.[27] Two approaches were used to add elastomers to LARC–13 (Figure 16). Fluorosilicone (Silastic) and vinyl-terminated silicone (Sylgard) elastomers were physically blended into the LARC–13 amic acid

ELASTOMER-TOUGHENED POLYIMIDE ADHESIVES

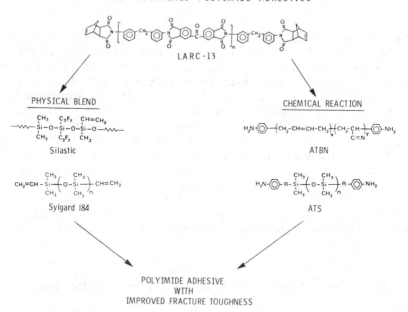

Figure 16. Preparation of elastomer–modified LARC–13 addition polyimides.

Table VII. Lap Shear Strengths (LSS) of Elastomer-Toughened
LARC-13/Titanium Bonds.

Adhesive	LSS, psi (MPa) Samples Unaged			LSS, psi (MPa) Aged 500 hrs @ 232°C		
	RT	232°C	288°C	RT	232°C	288°C
LARC-13	2900 (20)	2800 (19)	1300 (9)	2900 (20)	2900 (20)	2200 (15)
LARC-13/ Silastic	2200 (15)	1800 (12)	500 (3)	1600 (11)	1800 (12)	1300 (9)
LARC-13/ Sylgard	3300 (23)	2400 (17)	-	2500 (17)	2300 (16)	-
LARC-13/ ATBN	3700 (25)	2800 (19)	-	2000 (14)	1900 (13)	-
LARC-13/ ATS	2600 (18)	1800 (12)	900 (6)	1700 (12)	1600 (11)	1400 (10)

adhesive; and aromatic amine-terminated butadiene/acrylonitrile (ATBN) and silicone (ATS) elastomers were chemically reacted into the prepolymer backbone.

Each of the four modified resins displayed a separate phase for the elastomer and LARC-13 moieties as indicated by a T_g at low temperature (-115° to -65°C) and at high temperature (277° to 300°C). Incorporation of elastomer particles at a concentration of 15% solids by weight resulted in a six- to seven-fold increase in peel strength and a three- to five-fold increase in the fracture toughness of LARC-13. This improvement in toughness was accomplished at a sacrifice in the elevated temperature adhesive strengths of the elastomer-modified materials as shown in Table VII. It was noted, however, that the 288°C strengths of the elastomer-modified LARC-13 bonds improved after aging at elevated temperature. This behavior is probably due to latent crosslinking of the elastomer with the polyimide.

An additional study was conducted in mid 1980 to observe the effect of the elastomer chain length on the properties of LARC-13 addition polyimide adhesive.[28] A new series of polymers was synthesized using amine-terminated silicones of varying chain lengths. Four ATS elastomers (Figure 16) prepared with chain lengths varying

Table VIII. Lap Shear Strengths of LARC-13 Modified With ATS Elastomers of Varying Chain Lengths.

Adhesive	LSS, psi (MPa) Unaged Samples			LSS, psi (MPa) [*] Aged 500 hrs @ 232°C		
	RT	232°C	288°C	RT	232°C	288°C
LARC-13	2900 (20)	2800 (19)	1300 (9)	2900 (20)	2900 (20)	2200 (15)
LARC-13/ ATS$_{105}$	2600 (18)	1800 (12)	900 (6)	1700 (12)	1600 (11)	1400 (10)
LARC-13/ ATS$_{63}$	2600 (18)	1400 (10)	700 (5)	1600 (11)	1600 (11)	1300 (9)
LARC-13/ ATS$_{41}$	2500 (17)	1800 (12)	800 (6)	1700 (12)	1600 (11)	1400 (10)
LARC-13/ ATS$_{10}$	2900 (20)	1800 (12)	700 (5)	1800 (12)	1700 (12)	1400 (10)
LARC-13/ ATS$_{105}$ + ATS$_{10}$	3000 (21)	2100 (14)	1300 (9)	2300 (16)	1900 (13)	1700 (12)

* Titanium Adherends

from n = 10 to n = 105 were reacted into the LARC-13 backbone at a concentration of 15% solids by weight. An additional LARC-13/ATS formulation was prepared for this study using a 50:50 combination of ATS elastomers having repeat units of 105 and 10. The adhesive lap shear strengths of these five resins are compared to that of LARC-13 in Table VIII. Results from the lap shear tests showed that variation of the elastomer chain length had very little effect on the adhesive properties of ATS-modified LARC-13. An important finding, however, was the significant increase in strength displayed by the LARC-13 formulation containing a 50:50 bimodal distribution of elastomers ATS$_{105}$ and ATS$_{10}$. This adhesive containing elastomers of both long and short chain lengths shows potential as an aerospace adhesive for selected high-temperature applications.

Variation of End Caps on Addition Polyimide Adhesives

Although nadic-terminated imides offer considerable processing advantages, they cannot withstand long-term elevated temperatures in air without gradually degrading. An investigation was conducted to find a crosslinking end group which would improve the thermo-oxidative stability of LARC-13 adhesive. A second objective was to find a latent crosslinking end cap which would allow a cure without the evolution of cyclopentadiene (Figure 14). The end groups evaluated in this study are shown in Figure 17. Six novel addition polyimides were synthesized using these end caps and were charac-terized for solubility, melt-flow and cure properties, T_g, thermal stability and adhesive strength.[29] Of the polymers tested, only the hexachloro-nadic and nadic-capped LARC-13 outgassed a cyclopentadiene on curing. The LARC-13 nadic, acetylene, and N-propargyl-containing polymers exhibited excellent adhesive bond strengths on titanium at both ambient and elevated temperature compared to the other four polymers. As shown in Table IX, the acetylene-capped addition polyimide had 40% higher adhesive lap-shear strength than nadic-capped LARC-13 after aging 1000 hrs at elevated temperature.

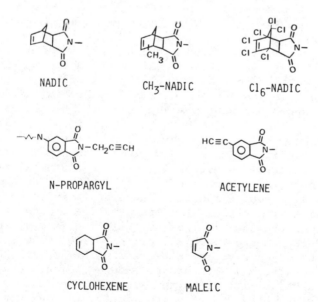

NADIC   CH₃-NADIC   Cl₆-NADIC

N-PROPARGYL   ACETYLENE

CYCLOHEXENE   MALEIC

Figure 17. Crosslinking end groups for LARC-13 adhesive.

Table IX. Lap Shear Strengths of Titanium/ Addition Polyimide Bonds Aged for 1000 Hours at 232°C

Oligomer End Group	LSS, psi (MPa) Unaged Sample		LSS, psi (MPa) after 1000 hrs @ 232°C	
	RT	232°C	RT	232°C
Nadic	3200 (22)	2600 (18)	2600 (18)	2000 (14)
Acetylene	2900 (20)	2500 (17)	2500 (17)	2800 (19)
N-Propargyl	3100 (21)	2800 (19)	800 (6)	1000 (7)

Commercial Addition Polyimide Adhesives

Of the few aromatic addition polyimide adhesives that have been
available commercially, Thermid 600 is the most attractive for use
at elevated temperature and therefore shows the most potential as
an aerospace adhesive. In 1969, Hughes Aircraft Company prepared
a series of novel acetylene-terminated imide oligomers which could
be thermally polymerized through the ethynyl end group. Initiated
under the sponsorship of the U.S. Air Force Materials Laboratory
(AFML)[30] the work was not made public until 1974.[31,32] The oligomer
shown in Figure 18 was designated HR600 by Hughes and is presently
marketed by Hughes Aircraft Company under the tradename of
Thermid 600.[33] Thermid MC-600 is the fully imidized molding
powder; and Thermid LR-600 is the amic acid form of the oligomer
in N-methylpyrrolidone. Both show promise as high-temperature
structural adhesives and have been effective in bonding titanium,
aluminum, copper, and composites. The LR-600 has been most widely
evaluated as an adhesive resin. Though it exhibits excellent
thermomechanical properties at temperatures approaching 316°C, it
has an inherent drawback in that its gel time is very short on the
order of 100-180 seconds. This short gel time (flow life) is the
cause of reduced adhesive lap shear strengths at both ambient and
elevated temperatures. Work has been done to improve upon this
system at the University of Dayton Research Institute and AFML.[34]
Researchers have found that the addition of various cure inhibitors
greatly increase the gel time of Thermid 600 while improving flow
and titanium alloy wetting. Room temperature lap shear strengths
of 3800 psi (26 MPa) have been obtained on titanium adherends by

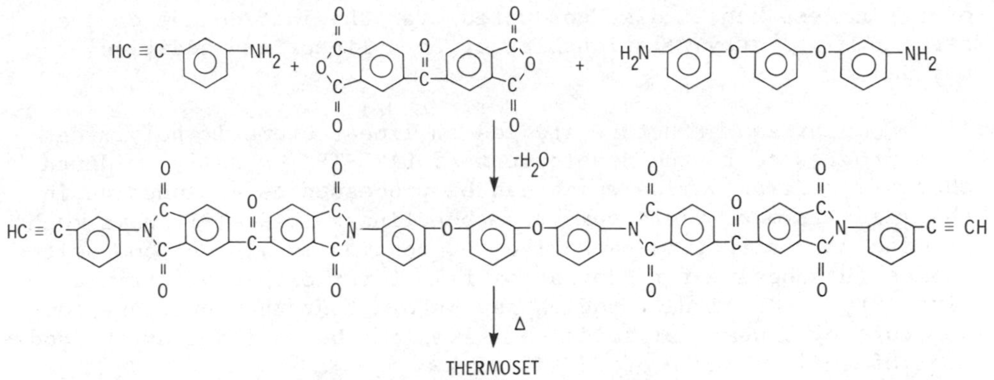

Figure 18. Thermid 600 resin chemistry.

the addition of 5% hydroquinone to the LR-600 resin. The addition of hydroquinone also enhances the high-temperature (316°C) strength of this adhesive (1650 psi, 11.4 MPa), but at the same time somewhat reduces the long-term thermooxidative stability of the polymer. Current studies at Hughes involve the examination of modified HR600 resins geared at improving heat resistance and resistance to fatigue.[35]

Another class of addition-type polyimides which have been offered commercially are the bismaleimides. In 1973, the development of maleimide-capped oligomers was reported by Rhone-Poulenc.[36] These materials were then marketed in the U.S. by Rhodia as a product line called Kerimid. Although bismaleimides are being used for a variety of applications, they have not shown strong potential as high-temperature adhesives due to their limited thermal stability.

SUMMARY

High-temperature adhesives are becoming increasingly important for aerospace applications. At the present time there are very few high-temperature adhesives available commercially, and none of these are able to satisfy all of the stringent needs of current aerospace programs. The Materials Division at NASA-Langley Research Center is screening both commercial and in-house experimental aromatic polyimides for use as structural adhesives on NASA programs. The polyimide adhesives being developed at NASA-Langley are the result of basic research studies conducted over the last decade on the structure-property relationships of both linear and addition polyimides.

Solvent and structure studies on linear aromatic polyimides have progressed to the development of LARC-TPI, a fully imidized thermoplastic polyimide which can be processed as an adhesive in the imide form to yield void-free bondlines. LARC-TPI shows much promise as a structural adhesive for bonding metals and composites and as an adhesive for laminating film for flexible electronic circuitry. Additional studies have shown that the upper use temperature of linear polyimide adhesives can be extended by the addition of small amounts of Al(III) ions.

Structure-property studies on addition polyimides have led to the development of LARC-13, a nadic-capped imide oligomer which is currently being used to bond both titanium and composites for several aerospace applications. Research on LARC-13 has shown that this type of adhesive is more readily processable than a linear system, but is less thermooxidatively stable and more brittle. However, because of its high crosslink density, the addition adhesive is able to perform at least for short terms at elevated temperatures up to 600°C where linear systems fail thermoplastically.

Active NASA research goals on both linear and addition aromatic polyimide adhesives geared to meet future needs include: (1) Improved processability of thermoplastic polyimides for use as hot-melt adhesives; (2) Use of a polyimidesulfone adhesive to obtain both improved processability and solvent resistance; (3) Improved fracture toughness of addition polyimide adhesives by incorporating heat-resistant elastomers; and (4) Improved thermooxidative stability of addition polyimide adhesives through chemical modifications.

REFERENCES

1. H. A. Burgman, J. H. Freeman, L. W. Frost, G. M. Bower, E. J. Traynor, and C. R. Ruffing, J. Appl. Polym. Sci., 12, 805 (1968).
2. D. J. Progar, V. L. Bell, and T. L. St. Clair, U.S. Patent 4,065,345 to NASA-Langley Research Center, 1977.
3. T. L. St. Clair and D. J. Progar, in "Adhesion Science and Technology," L-H. Lee, Editor, Vol. 9A, pp. 187-198, Plenum Press, New York, 1975.
4. P. S. Blatz, Adhesives Age, 21 (9), 39 (1978).
5. "Aerospace Adhesives," Product Bulletin, American Cyanamid Co., Havre de Grace, Maryland, 1982.

6. D. J. Progar and T. L. St. Clair, Proc. of the 7th National SAMPE Technical Conference, 7, 53 (1975).

7. A. K. St. Clair, W. S. Slemp, and T. L. St. Clair, Adhesives Age, 22 (1), 35 (1979).

8. A. K. St. Clair and T. L. St. Clair, SAMPE Quarterly, 13 (1), 20 (1981).

9. A. K. St. Clair, T. L. St. Clair, W. P. Barie, and K. R. Bakshi, Proc. of the Printed Circuit World Exposition, pp. 3-5 (1981).

10. V. L. Bell, U.S. Patent 4,094,862 to NASA-Langley Research Center, 1978.

11. R. T. Traskos, Lurie R&D Center, Rogers Corporation, Rogers, Connecticut, private communication, 1980.

12. S. G. Hill, P. D. Peters, and C. L. Hendricks, "Evaluation of High Temperature Structural Adhesives for Extended Service," NASA Contractor Report 165944, 1982.

13. L. T. Taylor, V. C. Carver, T. A. Furtsch, and A. K. St. Clair, in "Modification of Polymers," C. E. Carraher and M. Tsuda, Editors, ACS Symposium Series No. 121, pp. 71-82, American Chemical Society, Washington, D.C., 1979.

14. A. K. St. Clair, T. L. St. Clair, and L. T. Taylor, U.S. Patent 4,284,461 to NASA-Langley Research Center, 1981.

15. H. D. Burks and T. L. St. Clair, these proceedings, Vol. 1, pp. 117-136.

16. T. L. St. Clair and D. A. Yamaki, these proceedings, Vol. 1, pp. 99-116.

17. H. H. Gibbs, Proc. of the 17th National SAMPE Symposium, 17, IIIB-6, pp. 1-9 (1972).

18. D. J. Progar, in "Graphite/Polyimide Composites," H. B. Dexter and J. G. Davis, Editors, NASA Conference Publication 2079, pp. 123-138, NASA-Langley Research Center, Hampton, VA, 1979.

19. J. P. Bright and P. N. Foss, Industrial Research and Development, 24 (7), 82 (1982).

20. H. R. Lubowitz, U.S. Patent 3,528,950 to TRW, Inc., 1970.

21. M. A. Ogliaruso, M. G. Romanelli, and E. I. Becker, Chem. Rev., 65 (3), 261 (1965).

22. A. K. St. Clair and T. L. St. Clair, Polym. Eng. Sci. 16 (5), 314 (1976).

23. T. L. St. Clair and D. J. Progar, Proc. of the 24th National SAMPE Symposium, 24 (2), 1081 (1979).

24. A. A. Stenersen and D. H. Wykes, Proc. of the 12th National SAMPE Technical Conference, 12, 746 (1980).

25. S. G. Hill and C. H. Sheppard, Proc. of the 12th National SAMPE Technical Conference, 12, 1040 (1980).

26. V. Y. Steger, Proc. of the 12th National SAMPE Technical Conference, 12, 1054 (1980).

27. A. K. St. Clair and T. L. St. Clair, Internatl. J. Adhesion and Adhesives, 5, 249 (1981).

28. A. K. St. Clair, T. L. St. Clair, and S. A. Ezzell, NASA Technical Memorandum 83172, 1981.

29. A. K. St. Clair and T. L. St. Clair, Polym. Eng. Sci. 22 (1), 9 (1982).

30. R. H. Boschan and A. L. Landis, "Development of High Temperature Addition Cured Adhesives", Air Force Contract Report AFML-TR-75-90, August 1975; N. Bilow and A. L. Landis, "High Temperature Resins Having Improved Processability", Air Force Contract Report AFML-TR-74-89, July 1974.

31. A. L. Landis and N. Bilow, Am. Chem. Soc. Polymer Preprints, 15, (2), 533 (1974).

32. N. Bilow, A. L. Landis, and L. J. Miller, U.S. Patent 3,845,018 to Hughes Aircraft Company, 1974.

33. "Formulating Adhesives With Thermid 600," Gulf Oil Chemicals Company, Overland Park, Kansas, 1981.

34. R. J. Kuhbander and T. J. Aponyi, Proc. of the 11th National SAMPE Technical Conference, 11, 295 (1979).

35. N. Bilow, A. L. Landis, R. H. Boschan, and J. G. Fasold, SAMPE Journal, 18 (1), 8 (1982).

36. M. A. J. Mallet and F. P. Darmory, Am. Chem. Soc. Org. Coatings and Plastics Preprints, 34, (1), 173 (1974).

POLYIMIDE COMPOSITES - GRUMMAN APPLICATION/DEVELOPMENT CASE

HISTORIES

Leonard M. Poveromo
Grumman Aerospace Corporation
Bethpage, New York 11714

Advanced composite hardware exposed to
thermal environments above 127°C (260°F) must be
fabricated from materials having resin matrices
whose thermal/moisture resistance is superior to
that of conventional epoxy-matrix systems. A
family of polyimide (PI) resins has evolved in the
last 10 years that exhibits the thermal-oxidative
stability required for high-temperature technology
applications. The weight and structural benefits
for organic-matrix composites can now be extended
by designers and materials engineers to include
structures exposed to 316°C (600°F). Polyimide
composite materials are now commercially available
that can replace metallic or epoxy composite
structures in a wide range of aerospace applications. Thermal envelopes between 127°C and 232°C
(260°F and 450°F) require greater thermal-
oxidative stability and resistance to moisture
degradation than currently exhibited by state-
of-the-art epoxy systems. The bismaleimide (BMI)
polyimides provide the optimum combination of
production-compatible, epoxy-like cure cycles and
structural efficiency. Graphite/BMI polyimide prepreg
is commercially available from several sources, and
is being used in aircraft production for the first
time by Grumman on the DC-8 Nacelle Anti-Icer
Plenum. This paper will follow the progression of
polyimide composite applications at Grumman and in
the aerospace industry with particular emphasis on
the above program and the growing industry development efforts centered on BMI composites.

INTRODUCTION

Advanced polyimide composite materials have reached a level of industrial maturity that make them viable construction materials offering the potential for weight reduction and improved structural performance. Polyimide composite materials are now commercially available that can replace metallic or epoxy composites for a wide range of aerospace structures. Table 1 describes the resin systems available versus aircraft hardware service temperature requirements.

Epoxy-matrix advanced composite materials (177°C (350°F) cure temperature) were originally formulated to provide structurally efficient aerospace components for prolonged service at 177°C (350°F). However, as a result of long-term exposure to the environment, advanced composites with epoxy matrices will absorb moisture in excess of 1% by weight. This combination of absorbed moisture with corresponding reduction in glass-transition-temperature (T_g) (Figure 1) and service

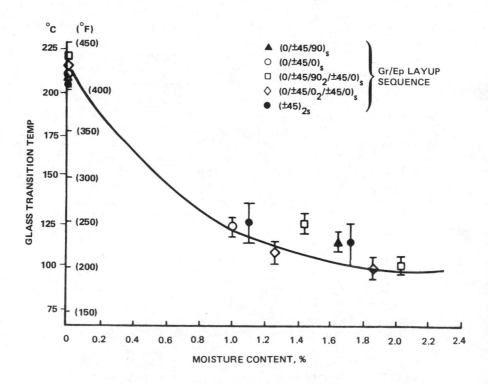

Figure 1. Effect of moisture content on glass transition temperature of AS/3501-5 Gr/Ep as measured by TMA.

Table I. Organic-Matrix Characteristics vs Aircraft Service Temperature.

23° TO 127°C (73° TO 260°F)	127° TO 232°C (260 TO 450°F)	232° TO 316°C (450 TO 600°F)	> 316°C (600°F)
EPOXY	BMI	ADDITION POLYIMIDE	CONDENSATION POLYIMIDE
• 3501 • 5208 • 976	• F-178 • V378A • CPI2272 • 4001	• LARC-160 • PMR-15 • PMR-11	• NR-150B2 • THERMID 600
• ESTABLISHED DESIGN DATA • WIDESPREAD AEROSPACE EXPERIENCE • EASILY PROCESSED • DURABILITY CUT-OFF 127°C (260°F)	• PROCESSIBILITY SIMILAR TO EPOXIES • PRELIMINARY STRUCTURAL DATA AVAILABLE • HOT-WET STATIC AND FATIGUE PROPERTIES • DESIGN EXPERIENCE LIMITED	• PROCESSING PARAMETERS INCLUDE 316°C (600°F) / 1.379 MPA (200 PSI) • BLEEDER/BREATHER SYSTEMS REQUIRE OPTIMIZATION • EXCELLENT THERMAL OXIDATIVE STABILITY • DURABILITY TBD • STRUCTURAL DATA MUST BE GENERATED	• PROCESSING PARAMETERS INCLUDE 371°C (700°F)/ 1.379 MPA (200 PSI) • BLEEDER/BREATHER SYSTEMS REQUIRE OPTIMIZATION • EXCELLENT THERMAL OXIDATIVE STABILITY • DURABILITY TBD • DESIGN EXPERIENCE LIMITED • RESINS COSTLY

temperatures in excess of 121°C (250°F) greatly reduces the
compressive strength of conventional epoxy-matrix composites
and minimizes their potential for weight savings (Figure 2).
The evolution of a family of BMI polyimide resins with epoxy-
like processing and excellent thermal resistance in the last
few years is allowing the aerospace designer to design composite structure with 121°C – 232°C (250°F – 450°F) thermal
envelopes. All state-of-the-art BMI resins have processing or
structural limitations; the optimal system is still to be
developed. However, existing BMI composites can and are being
utilized in applications tailored for their individual properties. Specific details of the Grumman programs are included
in this paper.

Grumman has been actively engaged in applying polyimide
composite materials to advanced aerospace structures since the
early 1970s. The pertinent on-going production programs and
development efforts are listed in Figure 3. The early requirements were for resins with long-term thermal stability
(260°C (500°F)) on high-speed aircraft with advanced electrical systems. Of the limited high-temperature resin systems

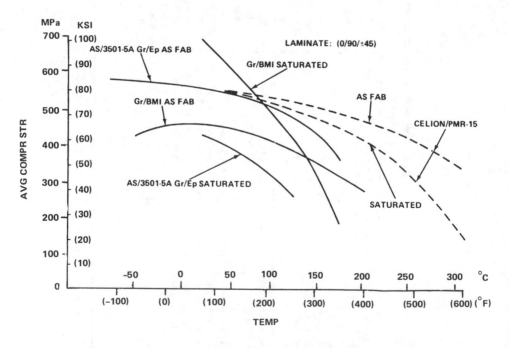

Figure 2. Comparison of compression strength of AS-1/3501-5A
Gr/Ep, AS-4/4001 Gr/BMI and Celion/PMR-15 Gr/Pi laminates (as fabricated and saturated at 95% RH).

1006

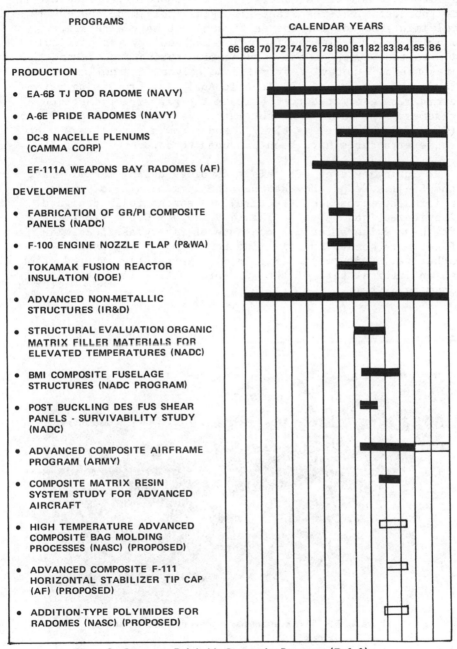

Figure 3. Grumman Polyimide Composite Programs (Ref. 1).

available at that time, Grumman selected a condensation polyimide based on Monsanto's Skybond 709 resin. This material is derived from the reaction of a diacidester mixture and an aromatic amine. A manufacturing methodology was developed to fabricate heat-resistant fiberglass/polyimide and quartz/polyimide radomes for the EA-6B and A-6E Navy aircraft. Developing a viable manufacturing methodology was difficult due to the nature of the condensation curing process and the presence of N-methyl-2-pyrrolidone solvent in the laminating resin. Using Dynamic Dielectric Analysis, processing parameters were generated to fabricate the glass/PI production radomes. The largest of the latter is the 4.5 x 0.5 x 0.6 m (15 x 1.5 x 2.0 ft) EA-6B T.J. Pod radome; more than 200 of these structures have been produced to date.

Springboarding off this technology during the mid-70s, the EF-111A Weapon Bay Radome (the largest all-polyimide-reinforced aerospace structure) was successfully designed, fabricated, and tested. This 4.9 x 0.6 x 0.6 m (16 x 2 x 2 ft) A-sandwich structure consists of fiberglass/polyimide facesheets bonded to polyimide honeycomb with polyimide film adhesive. This radome (Figure 4) is presently in production as an integral part of the Air Force's EF-111A Tactical Jamming System.

Figure 4. EF-111A weapons bay radome shell in post-curing fixture.

DC-8/CFM 56 NACELLE ANTI-ICER PLENUM

Grumman designed and is fabricating the nacelles for Cammacorp's DC-8 re-engining program. The severe thermal/acoustic environment of the nacelle de-icer plenum (Figure 5) presented a design problem. A study of the maintenance records of the baseline titanium anti-ice skins in service showed high failure rates typified by multiple fatigue cracks with a tendency towards fragmentation. Since the cause was determined to be acoustic/thermal fatigue, a graphite/polyimide design with excellent fatigue resistance was developed. The processing requirements and 232°C (450°F) maximum service temperature of the part drove the matrix selection toward the bismaleimide (BMI) addition-curing polyimides. Initially, these resins were developed for substitution in fiberglass/polyimide radomes fabricated with condensation polyimides that had exhibited extensive processing problems. The two commercially available principal matrix systems were Hexcel's F-178 and USP's V-378A. The early Grumman BMI work involved development of woven graphite cloth/F-178 resin for advanced composite aircraft structures subjected to thermal environments between 127°C and 232°C (260°F and 450°F). In

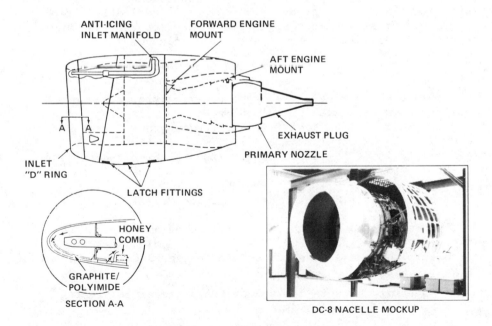

Figure 5. Grumman-fabricated graphite/polyimide de-icer plenum for DC-8 engine nacelle mockup.

addition, Grumman has been working with the V-378A resin since its introduction and has established in-house materials and processing requirements for this system. The advantages of these BMI systems are that they are fully imidized and supplied without solvent. These resins do not emit volatile byproducts during the addition/free-radical curing mechanism. Void-free laminates are produced using standard 177°C (350°F) epoxy-prepreg bagging and curing procedures and production facilities.

Since the principal plenum design condition was thermal/acoustic fatigue, a test sequence involving noise levels up to 158 dB and 232°C (450°F) was developed (Figure 6) and a series of candidate laminates (Table II) were tested. The combinations included woven graphite/F-178, woven graphite/V378A, woven fiberglass/unidirectional graphite/F-178 hybrid, and woven fiberglass/unidirectional graphite/V-378A hybrid. All the laminates successfully withstood the thermal/ acoustic fatigue test. The woven graphite/BMI laminate was selected for use in the nacelle de-icer plenum (Figure 7) because of ease of layup and thickness considerations. Subsequent fatigue and static tests (Table III) were conducted to verify the structural integrity of the graphite/BMI plenum configuration for FAA certification. The composite design resulted in a 30% weight savings and significant manufacturing/tooling cost savings over the alternative titanium plenum.

BMI POLYIMIDE DEVELOPMENT PROGRAMS

Prior to and during the above pioneering production program, Grumman was actively involved in development programs geared to generating BMI polyimide composite properties and structures. These programs are also listed in Figure 3. The specific details of the selected efforts are briefly described in the following subsections.

Fabrication of Graphite/BMI Polyimide Panels (NADC)

The overall objective of this early program was the fabrication of over 300 T-300/F-178 graphite/BMI panels for environmental exposure to provide data for an NADC composite data base. Specific objectives included:

- Material characterization of the as-received prepreg by chemical analysis

- Determination of the mechanical properties of autoclave-cured panels

- Determination of the environmental durability of autoclave-cured panels.

TEST PHASE	ELAPSED TIME, HR	TEST NOISE LEVEL IN ONE-THIRD OCTAVE BAND, dB	TEST TEMP °C (°F)
1	10.0	152	23(73)
2	19.5	152	232(450)
3	42.5	155	232(450)
4	3.0	158	232(450)
	75.0 (TOTAL)		

A. SEQUENCE

B. SETUP

Figure 6. Thermal/acoustic fatigue test setup.

Table II. Advanced BMI Composite Acoustic Fatigue Test Panels

PANEL DESCRIPTION	NUMBER OF PLIES	THEORETICAL THICKNESS, MM (IN.)	LAYUP SEQUENCE
HYBRID { 7781 GL/F-178 PI / T-300 GR/F-178 PI }	16	3.2 (0.124)	$+45_{GR}/-45_{GR}/0_{GR}/+45_{GL}/90_{GR}/-45_{GL}/0_{GL}/90_{GL}/90_{GL}/0_{GL}/-45_{GL}/90_{GR}/-45_{GR}/+45_{GR}$
WOVEN GR/PI { 24X23 8HS GR CLOTH F-178 & V-378A PI }	9	3.0 (0.117)	$+45_{GR}/0_{GR}/90_{GR}/-45_{GR}/0_{GR}/-45_{GR}/90_{GR}/0_{GR}/+45_{GR}$
HYBRID { 7781 GL/V-378A PI / T-300 GR/V-378A PI }	20	4.2 (0.164)	$+45/_{GL}90_{GR}/-45_{GL}/0_{GL}/90_{GL}/0_{GL}/-45_{GL}/-45_{GL}/0_{GL}/90_{GL}/0_{GL}/-45_{GL}/90_{GR}/+45_{GL}/0_{GR}/-45_{GR}/+45_{GR}$

LEGEND:

GR — UNDIRECTIONAL GRAPHITE/BMI POLYIMIDE
GL — WOVEN FIBERGLASS/BMI POLYIMIDE

Table III. Summary of Static Test Results[1] of GM301411
Woven Graphite/BMI Polyimide (F-178).

TEST	LAYUP[2] ORIENTATION	TEST TEMP, $^\circ$C($^\circ$F)	ULTIMATE STRENGTH MPa(KSI)	MODULUS GPa(MSI)
TENSION	MULTI	23(73)WET 232(450)	356(51.9) 340(49.3)	48.0(6.98) 42.7(6.21)
	0/90	23(73)WET 232(450)	466(67.8) 500(72.6)	62.1(9.07) 57.5(8.38)
TENSION (OPEN-HOLE)	MULTI	23(73)WET 232(450)	327(47.5) 336(48.8)	— —
	0/90	23(73)WET 232(450)	422(61.2) 381(55.3)	— —
COMPRESSION (IITRI)	MULTI	23(73)WET 232(450)	429(62.1) 284(41.2)	— —
	0/90	23(73) 232(450)	562(81.8) 375(54.2)	—
FLEXURE	MULTI	23(73)WET	430(62.4)	48.5(7.07)
RAIL SHEAR	MULTI	23(73)WET 232(450)	142(20.6) 110(16.0)	— —
	0/90	23(73)WET 232(450)	76(11.1) 55(8.0)	— —
BEARING AT 4% DEFLECTION	MULTI	23(73)WET 232(450)	436(63.2) 334(48.4)	— —
CTE	MULTI	—	1.5×10^{-6} [3]	—

NOTES:

(1) AVERAGE OF FIVE SPECIMENS

(2) PLY ORIENTATION:

MULTI: $(0, +45, 0 -45, \bar{0})_s$, TESTED IN 90°

BI: $(0/90)_9$ TESTED IN 90°

(3) MICRO-UNITS

A considerable amount of data was accumulated with respect to basic material properties and analytical procedures
that helped to minimize batch-to-batch variations of incoming
graphite/BMI prepreg.

Figure 7. Section of graphite/polyimide DC-8 nacelle de-icer
plenum in postcuring fixture.

Structural Evaluation of Organic Matrix Filamentary Materials for Use at Elevated Temperatures (NADC)

The purpose of this project was to demonstrate the suitability of graphite/BMI polyimide for high-temperature structural applications through the static and fatigue testing of quasi-isotropic coupons and hat-stiffened shear panels at room temperature-"dry" and at 177°C (350°F) "wet" conditions.

Specifically, this testing included:

- Static testing of unnotched and notched tension and compression coupons

- Static testing of horizontal shear specimens

- Constant-amplitude (R=0, R=-1) fatigue testing of unnotched and notched tension and compression coupons

- Spectrum fatigue testing of notched coupons

- Static and constant-amplitude fatigue testing of impacted coupons

- Static, constant-amplitude, and spectrum fatigue testing of hat-stiffened shear panels representative of a V/STOL aft-fuselage panel.

In addition, the temperature-dependent absorption characteristics of Gr/BMI were quantified in an early phase of the effort to permit accurate prediction of the expected in-service moisture contents and to enable adequate definition of the requisite conditioning and testing procedures for reliable simulation of these in-service conditions. The effects of a range of impact energy levels on graphite/BMI were also determined visually and ultrasonically in order to determine the level that causes barely visible impact damage for subsequent tests as a "worst case" condition of subvisual damage. Grumman investigated the F-178 resin matrix in this program while Northrop evaluated the V-378A resin matrix in a parallel effort.

Bismaleimide Composite Fuselage Structures for High-Temperature Applications (NADC)

The objective of this on-going Bismaleimide Composite Fuselage Structures (BCFS) Program is to design and develop an advanced composite fuselage structure which can satisfy the load and environmental requirements of a Fighter/Attack (F/A) V/STOL aircraft. In so doing, this program will develop design criteria and data, and demonstrate manufacturing techniques for generic high-temperature composite fuselage structure designed to operate in the 177°C to 204°C (350°F to 400°F) temperature range. The component selected is a 3.7 x 1.2 m (12 x 4 ft) section of V/STOL aft fuselage. The specific design is a hat-stiffened post-buckled skin with J-section longerons and frames (Figure 8). The potential advantages of this structure are:

- Weight savings of 28% over an equivalent metallic baseline

- Damage-tolerance for both low- and high-energy impact

- Structural integrity, reliability, and maintainability

- Affordable production costs (10% less than metallic baseline).

Grumman developed a design/manufacturing methodology for the BMI resin matrix (Hercules 4001) used in this program.

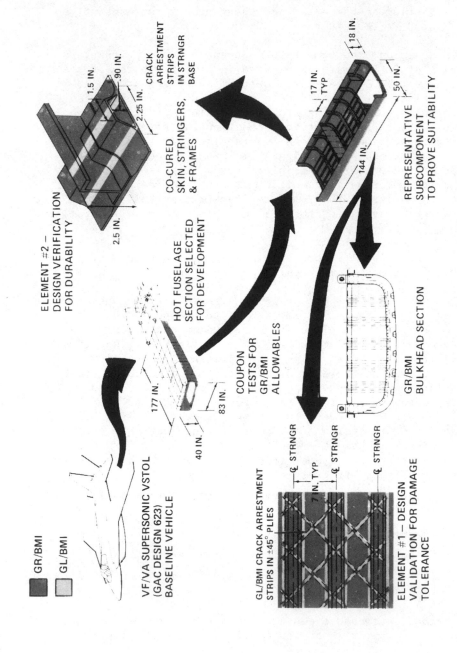

GR/BMI

GL/BMI

VF/VA SUPERSONIC VSTOL
(GAC DESIGN 623)
BASELINE VEHICLE

177 IN.

83 IN.

40 IN.

HOT FUSELAGE
SECTION SELECTED
FOR DEVELOPMENT

ELEMENT #2 –
DESIGN VERIFICATION
FOR DURABILITY

1.5 IN.
.90 IN.
2.25 IN.
2.5 IN.

CRACK
ARRESTMENT
STRIPS
IN STRNGR
BASE

CO-CURED
SKIN, STRINGERS,
& FRAMES

COUPON
TESTS FOR
GR/BMI
ALLOWABLES

18 IN.

17 IN.
TYP

50 IN.

144 IN.

REPRESENTATIVE
SUBCOMPONENT
TO PROVE SUITABILITY

GR/BMI
BULKHEAD SECTION

GL/BMI CRACK ARRESTMENT
STRIPS IN ±45° PLIES

₵ STRNGR   ₵ STRNGR   ₵ STRNGR

7 IN. TYP

ELEMENT #1 – DESIGN
VALIDATION FOR DAMAGE
TOLERANCE

Figure 8.   Schematic representation of BMI composite fuselage
structures for high-temperature applications.

Advanced Composite Airframe Program - ACAP (Army)

The objective of this program is to demonstrate how advanced composites, when applied in a helicopter airframe design, can provide weight/cost payoffs for future Army helicopters. Due to the operating service temperature (127°C (260°F)) and the fire-containment requirements of the Bell/Grumman ACAP design, a high-temperature resin system is required. Specifically, the aft roof is composed of two major components -- the engine deck and the center beam assemblies. Both of these assemblies are fabricated using graphite/BMI. The engine deck is designed to contain a 1093°C (2000°F) fire for 15 minutes without backside penetration. The construction materials are HRP honeycomb core faced with fiberglass-graphite/BMI laminates. Outside of the fire containment area, the deck design consists of an integrally stiffened graphite/BMI skin with longitudinal hat stiffeners and transverse J-frames.

Composite Matrix Resin System Study for Advanced Aircraft (NADC)

The objective of this new and innovative program is to select the optimal composite matrix system for service to 121°C (250°F) and in the 177°C to 232°C (350°F to 450°F) temperature range. The candidate graphite/resin system will be assessed for structural efficiency and environmental durability plus compatibility with cost-effective, reliable, and reproducible manufacturing processes. As a corollary to this effect, an evaluation of the compatibility of these resins with the new high-strain fibers will be investigated.

Industry-wide development efforts with BMI composites are growing, with every large aerospace company either actively evaluating these materials and/or planning their introduction into production. A partial listing of the programs that the author is aware of is shown in Table IV.

TOKAMAK FUSION REACTION - DIELECTRIC INSULATION

A unique polyimide program of interest at Grumman is the Tokamak Fusion Test Reactor (TFTR) which utilizes high-temperature polyimides. The overall TFTR is the largest project in the long-range program by the U.S. Energy Research and Development Administration (ERDA) to achieve a demonstration fusion power reactor by the late 1990s. The TRTR will be the first magnetic fusion system in the U.S. capable of producing fusion energy in significant quantities; it is being designed and developed at Princeton University's Plasma Physics Laboratory (PPPL). The prime industrial contractor

Table IV. Other Industry BMI Applications.

COMPANY	APPLICATION	MATERIAL
GENERAL DYNAMICS/ FORT WORTH	F-16XL WING COVERS	GRAPHITE/V-378A
LTV	S-3 NACELLE DOOR	GRAPHITE/V-378A
MCDONNELL DOUGLAS/ ST. LOUIS	AV-8B HARRIER STRAKES, VENTRAL ANTENNA COVER & INBOARD TRAILING EDGE	FIBERGLASS/V-378A
ROCKWELL/BRUNSWICK (SUBCONTRACT)	B-1 RADOMES	FIBERGLASS OR QUARTZ/F-178
SNECMA	• ENGINE SHROUD • CFM 56 AIR OIL SEAL	• FIBERGLASS/ KERIMID 601 • KINEL 5504
ROLLS ROYCE	RB-211 TERMINAL BLOCK & STIFFENER RB-162 STATOR VANE & ROTOR BLADE	KINEL 5504

working for PPPL is Ebasco, Inc., and their major subcontractor is Grumman Aerospace. In fabricating the TFTR torus, a doughnut-shaped container (high-temperature dielectric insulation) is used for magnetic confinement of plasma. This insulation prevents the flow of structure eddy currents that would distort the desired magnetic field of the Tokamak. The severe thermal (274°C (525°F)) and electrical (0.75 kV) environment dictated a high- performance polyimide/fiberglass insulation material. Based on previous Grumman experience and the availability of industry data, the PMR-15 (Polymerization of Monomeric Reactants) polyimide resin system was selected as the matrix construction material. This resin system, originally formulated by NASA-Lewis Research Center, is prepared by combining two ester-acids and a diamine in methyl-alcohol solvent. The 7781/PMR-15 fiberglass prepreg was available from several industry sources; Ferro's CPI-2237 material was selected.

Grumman generated material procurement and process specifications for the fiberglass/PMR-15 system. Extensive structural and electrical testing was conducted to characterize the material's performance in the unique Tokamak environment. Chemical evaluation of the prepreg consisted of establishing processing parameters and baseline control characteristics. Several structural tests were run to determine the structural performance of the material: room-temperature compression creep; 1000-hour at 295°C (520°F)

compression creep and coefficient of static friction (Table
V); compression test under electric load (Table VI); and
electrical properties–dielectric strength, insulation
resistance, dielectric constant, and dissipation factor (Table
VII). Following successful completion of the above testing,
the required Tokamak dielectric insulation including 50 0.9 x
0.9 m (3 x 3 ft.) panels were fabricated and installed (Figure
9). Ring insulations for the Tokamak torus vacuum vessel
outboard support pins were also successfully produced (Figure
10).

Table V. Coefficient of Static Friction Results: Fiberglass/
PMR–15 at 127° C (260° F) (Ref. 2).

DIRECTION OF SURFACE STRIAE, DEG.	TEST RUN NO.	ORIENTATION OF WEAVE			
		WARP		FILL	
		ANGLE, DEG	μ	ANGLE, DEG.	μ
0	1	16.0		15.0	
	2	15.7		14.7	
	3	16.0		15.3	
	AVG	15.9	0.285	15.0	0.268
90	1	25.2		21.7	
	2	23.0		21.7	
	3	25.5		23.0	
	AVG	24.2	0.449	22.1	0.406

Table VI. Electrified Compression Properties of 7781 Fiber-
glass/PMR–15 (Ref. 2).

TEST TEMP, °C (°F)	APPLIED STRESS, MPa (KSI)	DC VOLTAGE, kV	MAX LEAKAGE CURRENT, μA
127(260)	276(40)	0.75	NIL
127(260)	345(50)	0.75	NIL
271(520)	207(30)	0.75	0.72
271(520)	276(40)	0.75	0.78

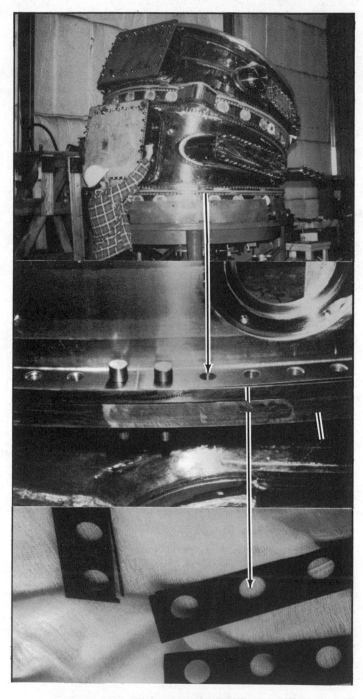

Figure 9. Location of 7781 fiberglass/PMR—15 dielectric
insulation in Tokamak vacuum vessel (Ref. 2).

Table VII. Electrical Properties of 7781 Fiberglass/PMR—15 (Ref. 2)

PROPERTY	AVERAGE VALUE
DIELECTRIC STRENGTH, V/MIL	1,023
INSULATION RESISTANCE, 10^{14} OHMS	>5.31
DIELECTRIC CONSTANT	5.17
DISSIPATION FACTOR	0.00810
TESTS PERFORMED AT ETL TESTING LABORATORIES, CORTLAND, N.Y.	

Figure 10. Ring insulator for Tokamak torus support pin (Ref. 2).

ACKNOWLEDGEMENTS

The author wishes to acknowledge the contributions of the following personnel and companies who helped make the programs described herein possible: R. Holden, P. Palter, P. Pittari, J. Roman, M. Martin, S. DeMay, J. Lundgren, and C. Parente of Grumman; W. Redden and A. Brooks of Ebasco Services; Princeton Plasma Physics Laboratory; and Cammacorp.

REFERENCES

1. L. Poveromo, "Polyimide Composites – Grumman Application Case Histories", in Proc. 27th National SAMPE Symposium/Exhibition, May 1982, pp. 729-738.
2. L. Poveromo and J. Mahon, "Fiberglass/polyimide Dielectric Insulation for Tokamak Fusion Reactor Vacuum Vessel", Final Report, Grumman Aerospace Corp., February 1982.

POLYIMIDE BLENDS

Jerry M. Adduci

Department of Chemistry
Rochester Institute of Technology
Rochester, New York 14623

Terms such as blends, polyblends and alloys have
been used to describe physical mixtures of substances.
The chemical systems to which these terms have been
applied are quite complex and varied. They range
from (1) physical mixtures of structurally different
homo- and copolymers, (2) covalently bonded copolymers
with varying ratios of comonomer units, block and
graft copolymers, (3) three-dimensional cross-linked
network systems. These systems can be considered
dispersions of one polymer in another polymer system.
These dispersions or blends can be a homogeneous single
phase or a heterogeneous multiphase system.

Among the advantages of blending polymer systems
are: potential ease of processing or fabrication,
improved properties and reduced cost. In addition,
synergistic interaction of different polymer systems
can impart superior properties to the polymer blend
over that of either polymer alone.

The purpose of this paper is to give an overview
of the various types of polyimide blends, their syntheses,
properties and applications. The following outline
will be utilized in the discussion of the polyimide
blends:
1. Functional Groups
2. Copolyimides
   (a) Heterocyclic Copolyimides
   (b) Copoly(Amide-Imides)
   (c) Block Copolyimides
   (d) Cross-Linked Polyimides
3. Blends-Alloys

INTRODUCTION

Polyimides, produced by the chemical reaction of an aromatic dianhydride and an aromatic diamine, were the first of the aromatic thermally stable polymers introduced in the mid-1950's. Polyimides did not behave as thermoplastics even though they had linear structures. Polymer backbones comprised of rigid, inherently stable, aromatic phenylene and imide rings imparted polyimides with excellent thermal oxidative properties and at the same time made them exceedingly difficult to process because of their high and sometimes indeterminate melting points. At that time, research interest was concentrated solely upon thermal stability with little regard given to cost, ease of fabrication, performance and end-use temperature. Emphasis has since shifted toward more practical, specific applications with the result that polyimide structures, their chemistry and end-uses have grown considerably. Today they are among the more common commercial materials available.

Their outstanding properties include continuous service at 500° F, wear resistance, low friction, good strength, toughness, thermal stability, high dielectric strength and radiation resistance, and low outgassing. Polyimides are available as precured films and fibers, curable enamels, adhesives and resins for composite materials. In addition, their stable backbone structure, resistance to acid hydrolysis and bacterial attack allow polyimides to be used as membranes for reverse osmosis and hyperfiltration.

Polymer modification reactions have led to novel systems where some desirable property such as thermal stability is sacrificed in order to improve other properties such as solubility, flexibility, glass transition temperature, melt viscosity, cost and/or ease of of fabrication.

I. FUNCTIONAL GROUPS

The general synthetic method for preparing polyimides involves a two-step reaction between a tetracarboxylic acid dianhydride* and a diamine yielding a soluble polyamic acid which can be converted to the polyimide by a thermal or chemical loss of water.

To date nearly every conceivable AR and R function has been investigated for this system. Wholly aromatic polyimides have been structurally modified or "blended" in the broadest sense

---

*(Not all synthetic methods utilize a tetracarboxylic acid dianhydride; some use bismaleimides and trismaleimides. See page 5.)

by incorporating various functional groups such as carbonyl, ether, sulfide, sulfone, methylene, isopropylidene, perfluoroisopropylidene, siloxane, bipyridyls, methyl phosphine oxide or various combinations of these groups into the polymer backbone in order to achieve improved properties.

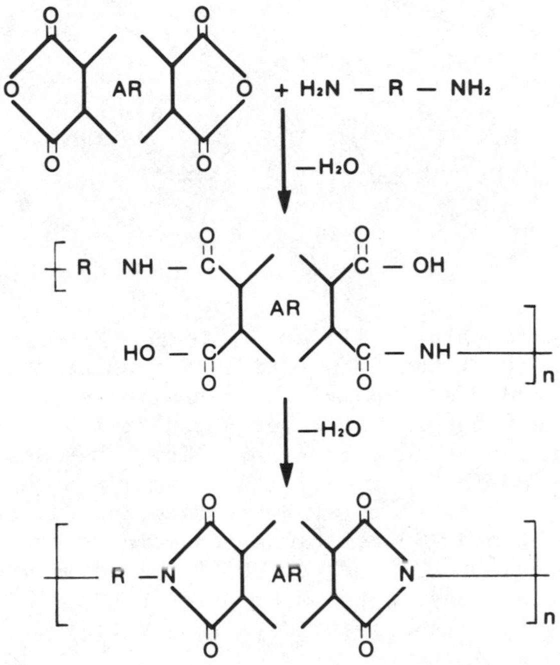

The following are examples of some AR and R modifications and the resultant polyimide properties.

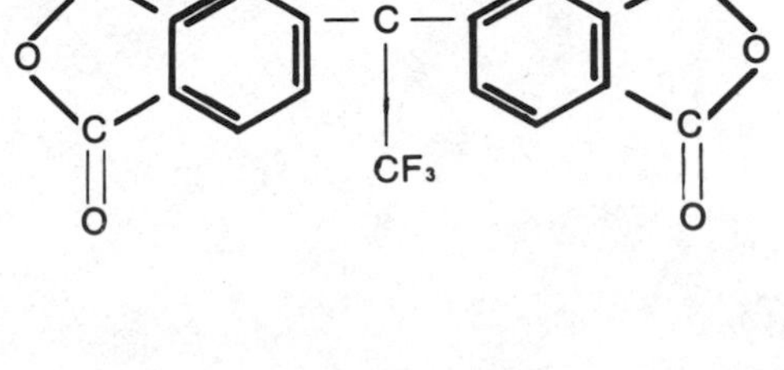

Thermoplastic polymides have been developed by reacting 2,2-bis(3,4-dicarboxyphenyl) hexafluoropropane dianhydride with a variety of diamines.[1]   The incorporation of the perfluoroisopropylidene group as a flexibilizing linkage has resulted in a new class of melt-fusible binders with less than 1% void levels ideally suited for preparing laminates. Voids were removed by applying pressure above the Tg. Depending upon the diamine structure used, Tg's varied from 229° to 385°C. These polymers were tough and had unusually good thermo-oxidative stability.

The same flexibilizing linkage has been incorporated in an aromatic diamine.[2]

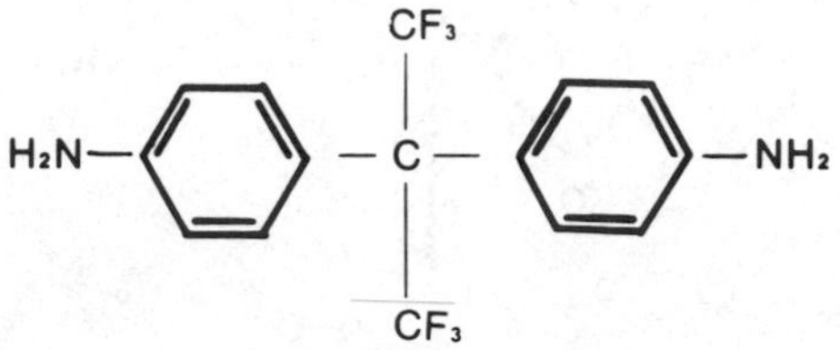

$$CF_3$$

H₂N—⬡— C —⬡—NH₂

$$CF_3$$

Polyimides containing linear perfluoroalkylene bridges in either one or both of the dianhydride or diamine monomers have also been prepared.[3-5]   A polyimide adhesive composition containing 2% arsenic thioarsenate has been prepared from 1,3-bis(3,4-dicarboxy-phenyl) hexafluoropropane anhydride.[5]   The adhesives showed improved strength retention for long periods of time at high temperatures. Stainless steel rings bonded with the adhesive and cured for 17 hours at 340°C and 200 psi had a torsional shear strength of 11,700 psi after 5000 hours at 300°C. A joint prepared with the additive-free adhesive had a shear strength of 9600 psi. These compositions were also used to bond titanium.

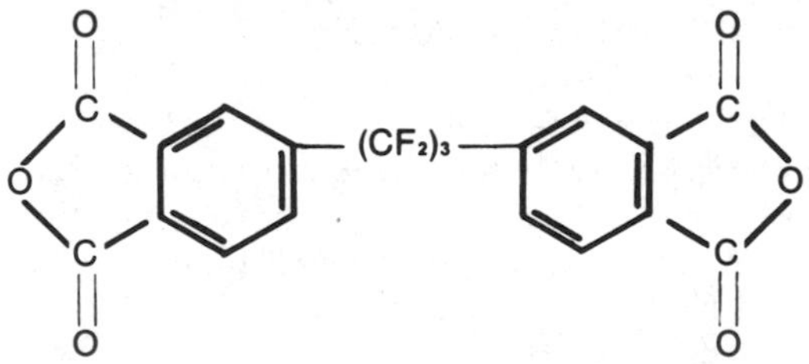

$$-(CF_2)_3-$$

Heat resistant adhesives[6] have also been prepared from perfluoroalkylene ether linked dianhydrides and diamines whose structures are shown below. They were found useful as adhesives for stainless steel and titanium.

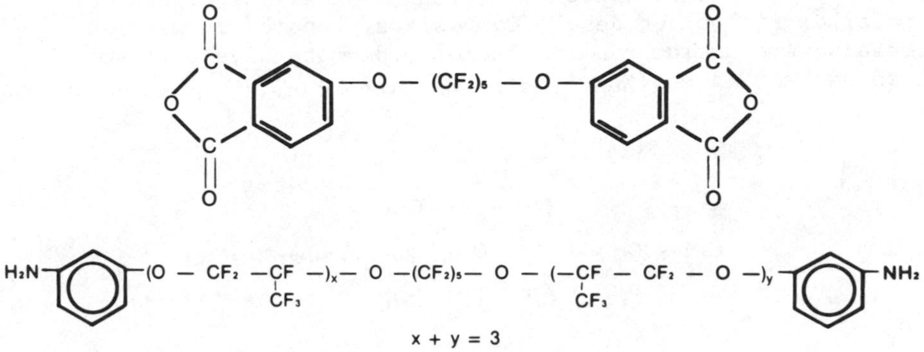

$x + y = 3$

A methyl phosphine oxide, $CH_3-P=O$, linkage introduced into the polyimide polymer backbone imparts flame retardancy as well as good mechanical properties. Bisimide resins prepared from bis(m-aminophenyl) methyl phosphine oxide and tris-(m-aminophenyl phosphine oxide) with maleic anhydride and benzophenone tetracarboxylic dianhydride have been found useful for preparing laminates with inorganic fibers.[7,8] Graphite cloth laminates fabricated from these resins had excellent flame resistance characteristics and did not burn in oxygen.

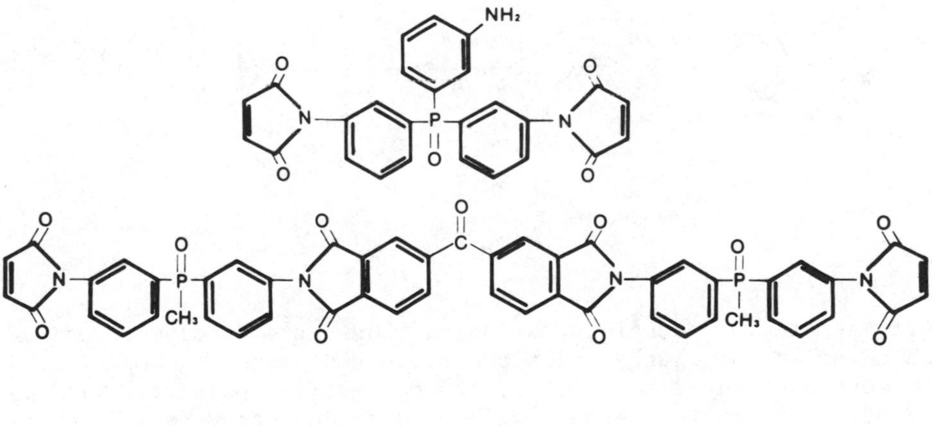

The diamines used for these preparations are examples of modified monomers, whereas the resins produced upon curing can be classified as a complex cross-linked polyimide structure. As might be expected, the resins were somewhat brittle. Siloxane containing polyimides have been prepared from the reaction of diamino siloxanes, 1,3 - bis(aminobutyl)tetramethyl siloxane, and tetracarboxylic anhydrides.[9] Composites prepared from these materials were useful for insulation and protective purposes and showed corona and heat resistant properties.

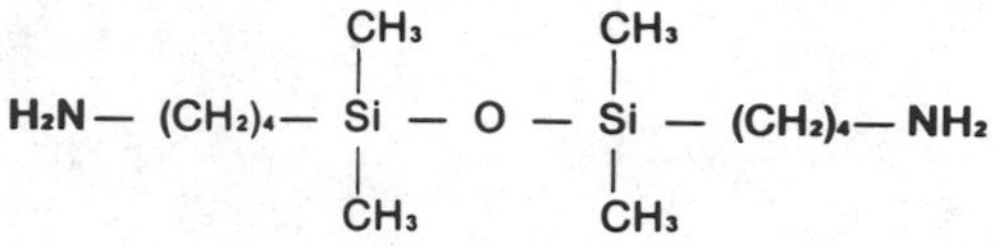

$$H_2N-(CH_2)_4-\underset{\underset{CH_3}{|}}{\overset{\overset{CH_3}{|}}{Si}}-O-\underset{\underset{CH_3}{|}}{\overset{\overset{CH_3}{|}}{Si}}-(CH_2)_4-NH_2$$

They also showed good adhesion to fibers, fabrics and various finely divided materials such as glass, boron, quartz and carbon.

Siloxane containing aromatic diamines of the general formula,

$$[H_2N-C_6H_4-R-Si(CH_3)_2]_2-O$$

can be equilibrated with a variety of cyclic siloxane materials to yield amine terminated polysiloxanes.[10]

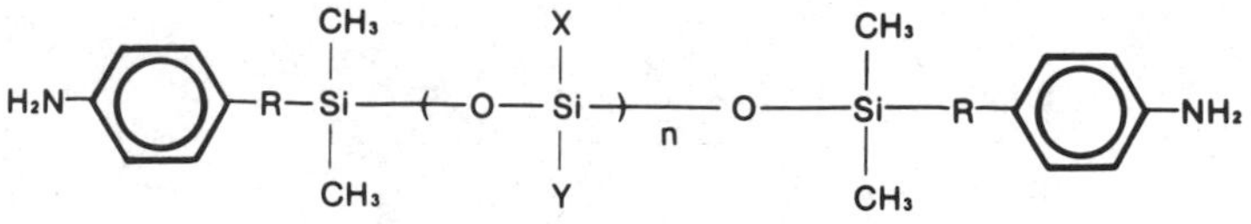

WHERE X = CH₃, C₆H₅
Y = CH₃, C₆H₅, CH = CH₂, R'F

These diamine containing siloxane blocks are copolymerized with a variety of dianhydrides and organic diamines. Complete imidization resulted in soluble, thermoplastic copolymers. The properties of these materials will be dependent upon polymer composition, the structure of the diamino siloxane and the percent of their incorporation. The copolymers show enhanced adhesion, good weatherability to UV and corona discharge and high thermal stability.

Aromatic polymers containing aryl ether and/or aryl sulfone linkages generally have greater chain flexibility, lower glass transition temperatures, greater tractability and high thermo-oxidative stability than those polymers without these groups in the repeat units. Aromatic diamines containing both ether and sulfone linkages have been synthesized and used to prepare a number of polyimides.[11,12,13]

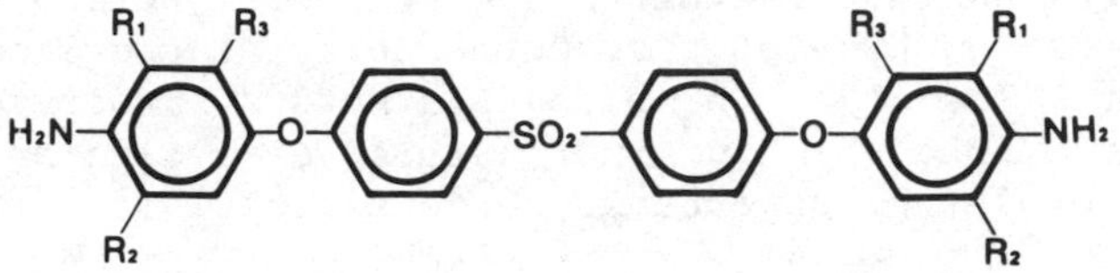

R₁ = R₂ = R₃ = H
R₁ = R₂ = H, R₃ = CH₃   R₁ = H, R₂ = R₃ = CH₃
R₂ = R₃ = H, R₁ = CH₃   R₁ = R₂ = CH₃, R₃ = H

In addition, a meta-linked diamine was prepared.

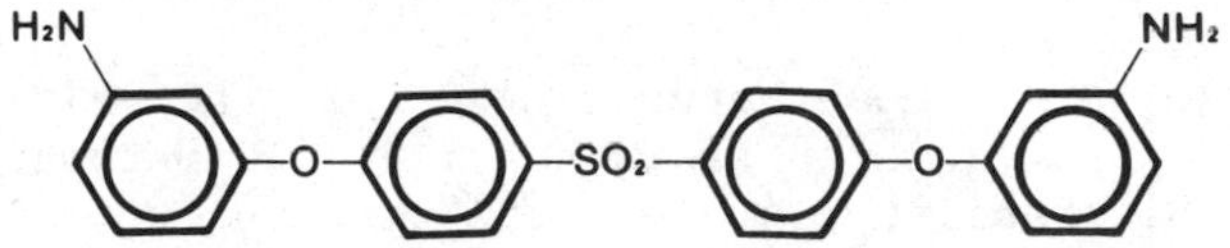

Materials containing these linkages did in fact display high heat distortion temperatures, good solvent resistance, excellent mechanical properties, high thermal and oxidative stability and in some cases thermoplastic characteristics.[12]

In general, methyl groups in these polymers tend to lower glass transition temperatures, lower the thermal stability and allow for air oxidation to occur.[13] Methyl groups ortho to the

amine functionality appear to sterically reduce the amino group's reactivity and allow cross-linking reactions to occur.
Most surprisingly, the meta-linked diamine containing polymers showed identical or superior thermal properties to the unsubstituted para-linked diamine containing polymers.

II.   COPOLYMERS

Copolymer structures are comprised of two or more different structural linking groups in the polymer repeat unit. Incorporating a second, different structural linking group into a homopolymer structure is another way of modifying its properties. The overall change in copolymer properties will be dependent upon the amount and type of structural unit incorporated into the copolymer. The order in which this second structural unit is incorporated into the copolymer structure is also an important factor in determining properties. Ordered copolymer structures generally show superior mechanical properties relative to those shown by random polymer structures.

(a)   Heterocyclic Copolyimides

The demand for thermostable polymers for special high temperature end use has led to the synthesis of many types of materials among which are the heterocyclic copolyimides. A variety of heterocyclic structures such as 1,3,4 - oxadiazole, thiazole, 1,3,4 - thiadiazole, benzothiazole, benzimidazole, 1,2,4 - triazole, quinoxaline, benzoxazole, s-triazine, phthalocyanine and bridged bipyridyls have been incorporated into polymer backbones.[14-16]

Ordered and less ordered copoly(heterocyclic imide) structures exhibited excellent thermal and thermooxidative stabilities.

Incorporation of these heterocycles into the polymer backbone on a regular, highly ordered basis has improved mechanical properties, thermal stability, processability along with increased modulus, Tg and crystallinity.[14] High molecular weight materials are formed because they are soluble and do not precipitate prematurely from solution. Ordered copoly(oxidiazole-pyromellitimide) and copoly(thiazole-pyromellitimide) neither melt nor decompose below 500°C in an inert atmosphere. The copolymers were readily soluble in organic solvents and formed films of high strength. The pyromellitimide of 2,5-bis(p-aminophenyl) 1,3,4-oxadiazole(see top of next page)wasoxidatively more stable than the pyromellitimide of 4,4'-oxydianiline as shown by isothermal aging studies.[15] Similar results were observed for a related series of benzoxazole-imide[16] and benzimidazole-imide copolymer structure.[17]

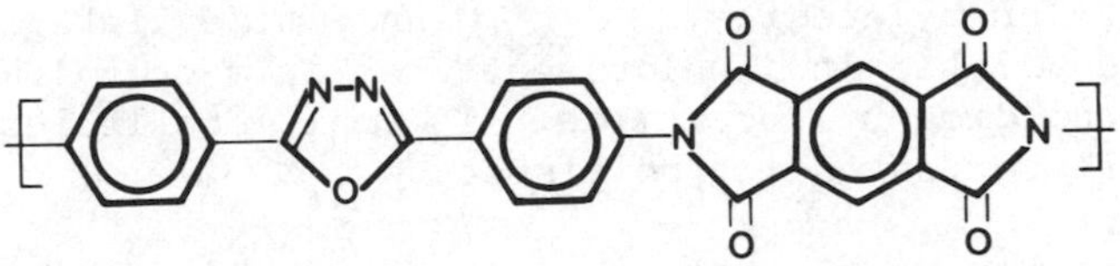

In an effort to improve thermooxidative stability, a copoly (benzheterocyclic imide) has been prepared from pyrazine tetra-carboxylic dianhydride and 2,5-diamino-1,3,4-thiadiazole.[18] The resulting polymeric structure (shown below) is notable for the absence of hydrogen in the system.

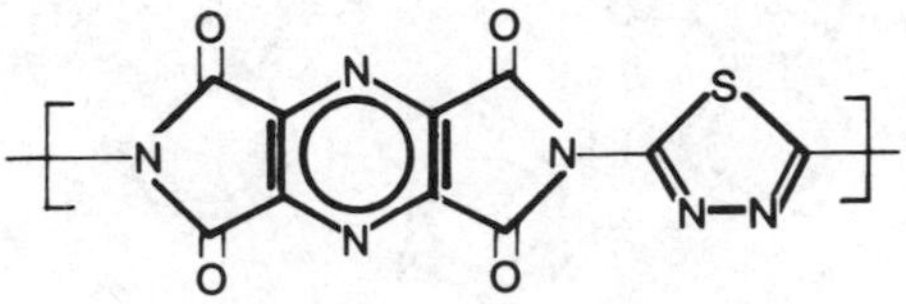

Insolubility of the polyamic acid limits the maximum molecular weight attainable. Red, clear films were obtained which were stable to 450°C in air.

Soluble polyimides containing bridged bipyridyl units have been prepared from diamine structures shown below.[19]

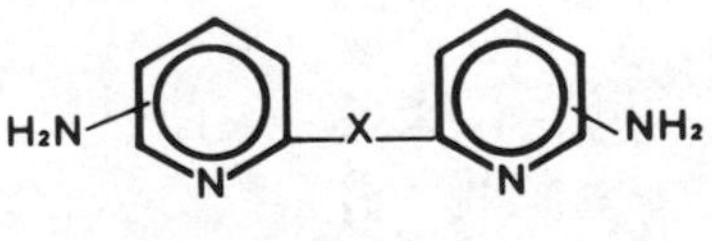

WHERE X = nil, NH, NCH₃, NCOCH₃, S, SO, SO₂, C₂H₄.

The gain in solubility resulted in a loss of thermal stability, the pyridyl systems was less stable than their corresponding carbocyclic analogs, 410° C vs. 495–505°, respectively for a 10% loss in air.

Good solubility was also exhibited by terpolymers comprised of 1,3- or 1,4-phenylenedioxy, pyridyl and imide linkages.[20] The polymers were soluble in dichloroacetic acid, m-cresol, N-methyl pyrrolidone and dimethyl acetamide. Thermal stabilities were lower than the corresponding polyimides.

Incorporation of a s-triazine ring into the polymer backbone gave soluble copolyimides.[21] These materials were prepared from 2-(diphenylamino)-4,6-bis(2,3-dicarboxylanilino)-s-triazine and 4,4'-oxydianiline. Films, coatings, laminates and moldings with low void content were prepared.

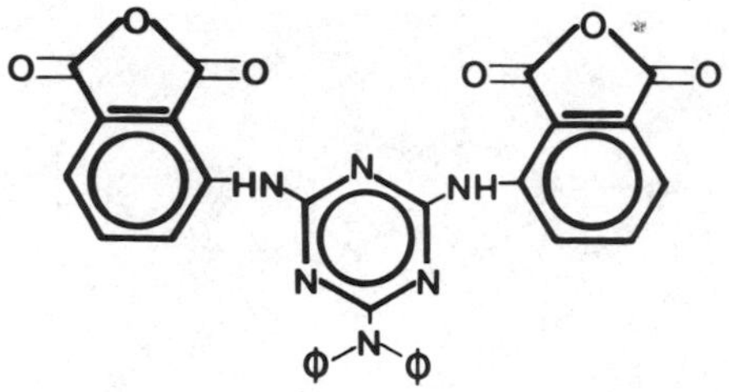

s-Triazine rings have also been incorporated with naphthalene tetracarboxydiimides.[22] A dihydrazine substituted s-triazine was used for this preparation.

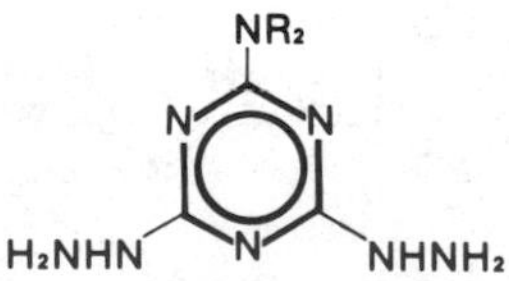

Depending upon the pendant R-groups, solubility in tetrahydrofuran or dioxane was observed. Thermal stabilities ranged from 380–410° C.

A novel type of thermally stable copolyimides containing metal phthalocyanines have been prepared by treating copper, cobalt, nickel and zinc phthalocyanino tetraamines with pyromellitic dianhydride and benzophenone tetracarboxylic di-anhydride.[23,24] These polymers showed exceptional thermal stability

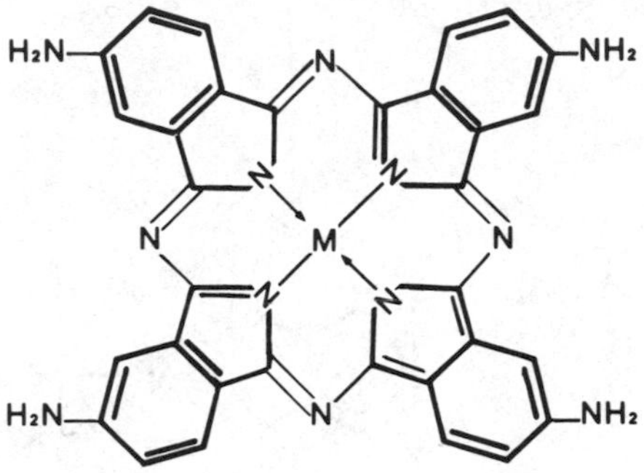

as one might expect when combining the imide and phthalocyanine structures into a polymer backbone. The polymers are self extinguish-ing and do not burn when exposed to flame. The polymers oxidize in the air at 500–600° C, leaving metal oxides. In nitrogen, they degrade at 600–800° C, leaving a 75–85% char yield and attaining a stable structure.

b)   Copoly(amide-imides)

An important class of copolyimides is one that contains the amide linkage. Incorporating the amide linkage into the polyimide backbone makes the polymer more soluble, moldable and processable. A gain in these properties understandably involves a loss in thermal stability. However, copoly(amide-imides) still have outstanding thermal stabilities, intermediate between those of polyamides and polyimides and are relatively inexpensive to synthesize. These materials are used commercially for laminating resins and wire coatings.

Trimellitic acid or its derivatives are primarily used in the synthesis of most of the copolyamide-imides. The general synthetic method involves a two step reaction between the tricarboxylic acid and a diamine yielding a soluble poly(amide-amic acid) which can be transformed to the poly(amide-imide) by thermal or chemical loss of water. The resulting resins have high strength and modulus and are stable up to 450°C.

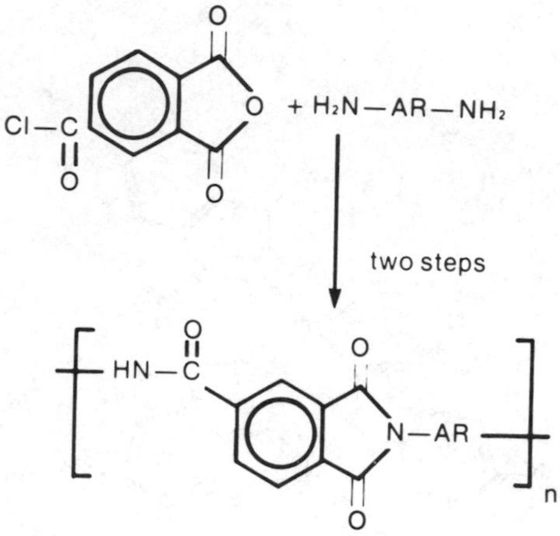

The polymer structure shown above illustrates an amide-imide linkage. It is also possible to have ordered structures which have imide-imide-amide-amide linkages.[25,26]

Thermally stable co-poly(amide-imides) have also been synthesized from tricarboxylic acid derivatives containing sulfone, terephthaloyl and biphenyl linkages (shown below) and various diamines.[27-29]

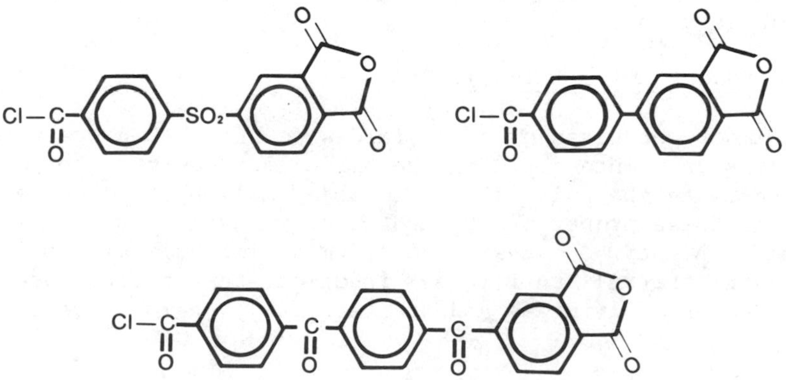

The sulfone and terephthaloyl linking groups were incorporated into the polymer backbone in order to improve tractability while still maintaining thermal stability. The polymers exhibit less than 10% weight loss at 400°C. The biphenyl containing poly(amide-imides) show better thermal stabilities than those which contain functional groups separating the aromatic rings.[29]

Aliphatic tricarboxylic acids, 1,2,3-propane tricarboxylic acid, 1,2,4-butane tricarboxylic acid and 1,3,5-pentane tricarboxylic acid, yielded essentially linear amide-imide polymers when polymerized in the melt with diamines.[30] The homopolymers were amorphous and had glass transition temperatures ranging from 50-200°C. Copolymers prepared with adipic acid and hexamethylene diamine were partially crystalline materials which could be converted to high strength fibers and high impact resistant plastics.

A series of poly(amide-imides) have been synthesized having a polyisophthalimide backbone with pendant imide groups (shown below).[31] A variety of imide-isophthaloyl chlorides, derived from 5-amino-isophthalic acid, were reacted with 4,4-oxydianiline. The presence of the pendant imide groups improved solubility characteristics. The introduction of maleimide, phthalimide, and tetrahydrophthalimide groups did not reduce thermal stability.

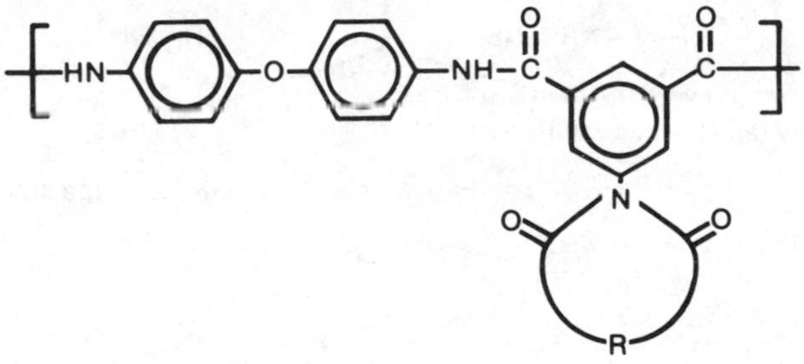

Chlorine-containing imides had decomposition temperatures of 340°C, approximately 120° lower than non-chlorine containing poly(amide-imides).

The copoly(amide-imide) backbone structures have been modified

by the incorporation of siloxane linkages.[32] The properties of
these materials are similar to those described earlier for polyimides,
i.e., corona and heat resistance and good adhesion to fibers,
fabrics, and particulate materials.

The polyimide structure has also been modified by incorporating
sulfonamide, sulfonate ester and ester linkages.[33-35] The reaction
of 4-chloroformyl-N-(4'-chlorosulfonylphenyl) phthalimide with
diamines or bisphenols yields a copoly(amide-imide-sulfonamide
or a copoly(amide-imide-sulfonate) respectively. The copoly
(ester-imidesulfonate) prepared from hydroquinone had a melting
point of 365°C. The copoly(amide-imide-sulfonamide)s prepared

copoly (amide-imide-sulfonamide)

copoly (ester-imide-sulfonate)

from aromatic diamines were thermally stable. The material prepared
from m-phenylene diamine had a decomposition temperature of 419°C.

Heat resistant copoly(ester-imides) were prepared from N-
[(chloroformyl) phenyl]-4-(chloroformyl) phthalimide (see top of
next page) isophthaloyl chloride and bisphenol A. Cast films showed
a 10% weight loss at 432°C with tensile strengths of 12.0 kg/mm^2
and 9% elongation.

Cl—C　　　N　　　C—Cl
O　　　O　　　O

c)　Block Copolyimides

The combination of blocks of two different structural types
has been used to form new materials having a "blend" of properties,
hopefully superior, of both structural types.　Block- or graft-
copolymers are structures comprised of sequences of different
structural units joined together in some set order, e.g.,
A_n-B_n diblock, A_n-B_n-A_n triblock.　With this method, it is possible
to combine incompatible polymers in the form of block-or graft-
copolymers so that each component can be distributed homogeneously
or at least firmly bonded to each other at phase boundaries.
The block- or graft-copolymer structure allows arrangements where
components of the same polymer type can form molecular aggregates.
The copolymer properties will be dependent upon the percentage
and the type of structural block or graft unit incorporated into
the material.　An infinite number of combinations are possible
with respect to the large number of different polymers with soft,
hard, elastic, brittle, water-soluble, or benzene-soluble properties.

The processing and intractability of some polyimides were
overcome by incorporating butadiene blocks into the backbone.[36,37]
Incorporation was carried out by first reacting isocyanate terminated
butadiene blocks with an aromatic diamine, the resulting amine
terminated oligomer was then reacted with a dianhydride.　The
successful reaction sequence yields a triblock copolymer having
a rigid polyimide center and two flexible, elastomeric ends.
Properties of the copolymer were modified by varying the size
of either the imide or butadiene blocks.　The poly 1,2-butadiene
blocks were used as crosslinking sites with dicumyl peroxide.
The cured, crosslinked material showed surprisingly high stability
in either air or nitrogen with a 60% and 70% imide copolymer
showing a 10% weight loss at temperatures of 450°C and 465°C
respectively.

Multiblock copolymers of polyimide and 1,2-polybutadiene have also been prepared.[37] The reaction scheme used was slightly different from that described above, however the results were similar. Chloroformyl terminated blocks of 1,2-polybutadiene were reacted with amine terminated molecules containing preformed trimellitimide amide units.

Phosphorus-containing imide resins based upon bis-(m-amino-phenyl) methyl phosphine oxide and tris-(m-aminophenylphosphine oxide) have been modified with amine terminated butadiene acrylonitrile (ATBN) and perfluoro alkylene diamine (3F) elastomers.[8] These modified resins were used for graphite cloth laminates. These laminates showed more improved properties than those prepared from unmodified resins.

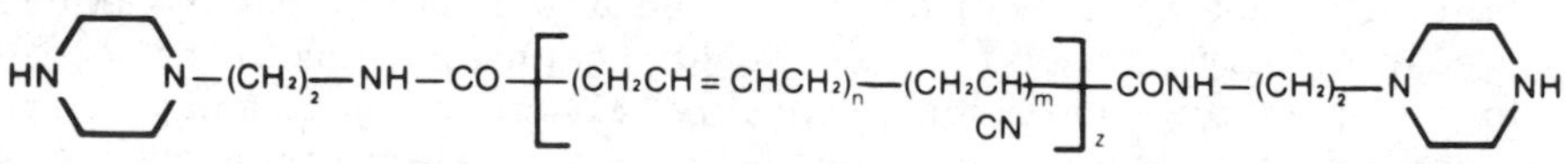

ATBN   n = 5, m = 1, z = 10

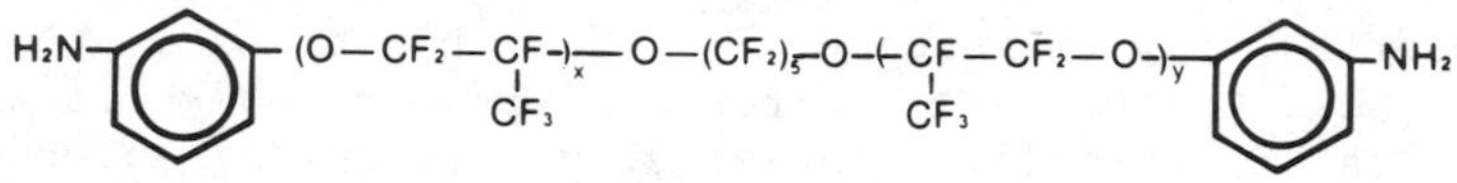

3F   x + y = 3

The fracture toughness of a high temperature addition polyimide adhesive (LARC-13) has been improved with amine terminated butadiene/acrylonitrile elastomer.[38] Fluorosilicone or silicone butadiene acrylonitrile rubber were physically blended or chemically reacted with the prepolymer backbone of LARC-13 to give elastomer toughened polyimide adhesives. All rubber modified resins showed a low Tg for the elastomeric phase and a high Tg for LARC-13 phase. Toughness was increased while thermal stability and high temperature lap shear strength were lower than that of LARC-13 alone.

Copolyimides and copoly(amide-imides) have been prepared containing blocks of variously substituted siloxanes.[10] Reaction of diamino terminated siloxane blocks with dianhydrides or trimellitic anhydride resulted in soluble thermoplastic polymers. The properties of these materials have been described earlier.

Polyimides[39] and poly(amide-imides)[40] have been prepared containing poly (ether sulfone) blocks. Incorporation of flexible ether and sulfone linkages into the backbone has improved processability and impact strength of these materials. The effect on properties can be varied by the size of the sulfone ether block incorporated. The sulfone ether diamines are prepared from the reaction of aminophenol, dichlorodiphenyl sulfone and an aryldiol. The length of the block is regulated by the amount of aryldiol used in the first step.

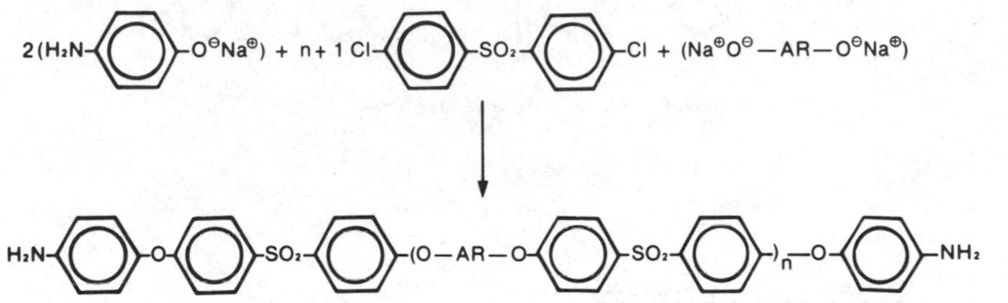

$$2(H_2N-\langle\bigcirc\rangle-O^{\ominus}Na^{\oplus}) + n+1\ Cl-\langle\bigcirc\rangle-SO_2-\langle\bigcirc\rangle-Cl + (Na^{\oplus}O^{\ominus}-AR-O^{\ominus}Na^{\oplus})$$

$$H_2N-\langle\bigcirc\rangle-O-\langle\bigcirc\rangle-SO_2-\langle\bigcirc\rangle-(O-AR-O-\langle\bigcirc\rangle-SO_2-\langle\bigcirc\rangle-)_n-O-\langle\bigcirc\rangle-NH_2$$

The sulfone ether blocks prepared as above can be reacted with PMDA, BTDA, TMA and phthaloyl chlorides to give polyimides, poly(amide-imides) and polyamides respectively. The poly(amide-imides) are the most tractable of the series. The film forming polymers are amorphous and display good thermal and mechanical properties.

Novel multiblock copolymers containing polyimidothioether blocks have been prepared.[41] Hard segments of polyimidothioether, prepared by the condensation of bismaleimides and hydrogen sulfide, are joined to soft segments of polyimidothioether blocks. The reaction sequence is shown below.

These multiblock systems show interesting elastoplastic behavior. Materials having high molecular weight and polysulfide contents of 70% or greater could be compression molded or milled.

Other block copolyimides, such as polyarylsulfone-polyimides[42], polyimide-elastomers[43] and polyimide-urethanes[44], have been prepared and characterized.

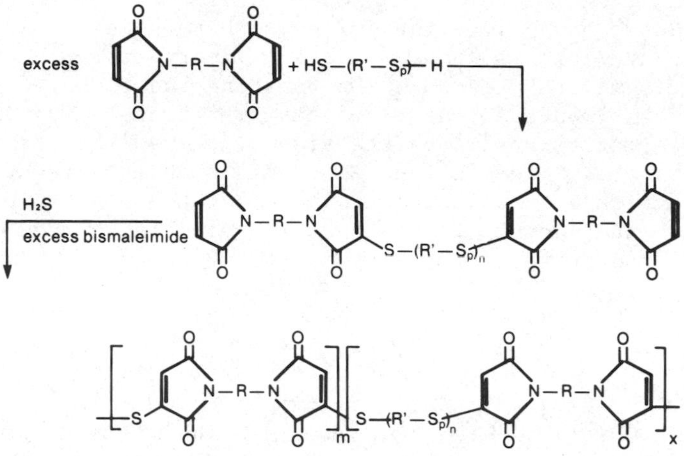

excess ... $+ HS-(R'-S_p)-H$

H_2S

excess bismaleimide

$S-(R'-S_p)_n$

where $n = 2$, $R' = -C_2H_4-O-CH_2-O-C_2H_4-$

(d) Cross-Linked Polyimides

The use of polyimides in certain composite applications has resulted in problems during the curing stage. The problems are related to the strongly retained solvents and volatile by-products that are evolved which result in a high void content in the cured product. The problems of controlling the volatiles given off during cure has been overcome by a variety of polyimide formulations. Polyimides derived from 2,2-bis (3,4-dicarboxyphenyl) hexafluoropropane dianhydride and various diamines are thermoplastic.[45] Polyimide properties have been tailored so as to allow them to flow at temperatures near their use temperatures which obviates the need for solvents. The reactions during cure are simple chain extensions, not cross-linking.

Processing related problems have also been overcome by incorporating unsaturated end groups in low molecular weight polyimide chains. Examples of the kinds of groups utilized are maleimides[8, 46, 47], nadic anhydride[48, 49], nitriles[50], and acetylenes[51,52]. In these systems, chain extension and cross-linking can occur via both amine addition to double bonds and/or polymerization through double bonds. Evidence has been accumulated which indicates that incorporation of aliphatic moieties into polyimide backbones results in reduced thermal and oxidative stability. However, polyimide formulations having acetylene groups when polymerized had a high degree of thermal stability,

almost equivalent to the conventional polyimide types. This
suggested the possibility that acetylenic groups were polymerizing
to form aromatic moieties. All in all, the disadvantage of a
loss in thermal stability is far outweighed by the advantages
of eliminating offgases, reduced void content, increased hardness,
modulus and strength in composite structures.

Bis-maleimides containing aryl sulfone ether blocks have
been polymerized thermally above 150°C.[47] The thermal addition
mechanism can be facilitated by peroxide catalysts during the
final cure. The crosslinked polymers showed weight losses of
5.5 - 12.5% at 450°C and have been proposed for adhesives, coatings
and composites.

(crosslinked)

Low temperature curable adhesives and coating compositions
have been prepared from aliphatic and aromatic bismaleimide prepolymers reacted with aromatic or cycloaliphatic epoxy resins.[53]

Novel amine terminated bisimides have been used as curing
agents for epoxy resins.[54] The 2,2-bis(3,4-dicarboxyphenyl)
hexafluoropropane dianhydride was used to prepare the bisimide
amines. The bisimide amine cured epoxy resins exhibited higher
densities and Tg's than the control even though the flexibilizing
hexafluoroisopropylidene unit had been incorporated into the
epoxy polymer network. The epoxy resins also appear to have
excellent moisture resistance.

A class of polyimides, known as PMR (for in situ polymerization of monomer reactants) has been developed which is unreactive at room temperature but at elevated temperatures react to form a high temperature resistant matrix.[49] These addition-type polyimides can be processed by either compression or autoclave molding techniques. Second generation of PMR materials contain dialkyl esters of an aromatic tetracarboxylic acid, an aromatic diamine and a mono alkyl ester of 5-norbornene-2,3-dicarboxylic acid (Nadic anhydride) dissolved in a low boiling alcohol. The solutions are used to impregnate the reinforcing fibers. The PMR approach has the potential for tailor making matrix resins for specific applications.

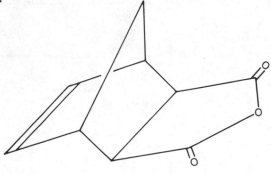

Nadic anhydride

Polyamic acid prepolymers prepared from PMDA and BTDA and various sulfone ether diamines have been blended with resole prepolymers.[55] Curing results in the formation of hard, thermally stable resins showing superior adhesion to glass. The cured resins appear to be a chain-branched polyimide-resole matrix.

An unsaturated poly(amide-imide) has been prepared from 1,5-bis(3-aminophenyl)-1,4-pentadiene-3-one and a trimellitic acid derivative, N-[(chloroformyl)phenyl]-4-(chloroformyl) phthalmide.[56] The poly(amide-imide) is soluble in polar solvents

$$H_2N \quad CH=CH-\overset{\overset{\textstyle O}{\|}}{C}-CH=CH \quad NH_2$$

and has a Tg of 225° C. It was readily cross-linked with benzoyl peroxide to give an insoluble material having a Tg of 235°C and increased thermal stability.

III. BLENDS- ALLOYS

The approach taken in the previous examples illustrates the "chemical blending" of polyimides. The most direct and simplest approach is the physical blending of a polyimide with one or more homopolymers. Again, the need for materials having a combination of properties not possible with homopolymers led to the development of blends. Many of these hybrid blends not only showed their component properties but, in addition, displayed unanticipated behavior.

Polyimides have been blended with a variety of homopolymers and copolymers including polytetrafluoroethylene (PTFE), poly-vinylidene fluoride (PVF_2), epoxy and phenolic resins, poly phenylene sulfide, polysulfone, polyesters, polycarbonates, polyamides and poly(amide-imides). Blends have been developed in which the polyimide is either a major or minor component of the mixture. In either case, the inherent properties of the major blend component (polyimide or homopolymer) are altered by the addition of the minor component until properties of the blend have been maximized.

The abrasion resistance of a poly(tetrafluoroethylene) resin was improved by the addition of 10-20 parts of polyimide (not identified in the patent).[57] Moldings of these compositions were used in dry bearing applications and showed improved abrasion resistance when in contact with soft metals such as brass.

After 150 hours and a PV (pressure X velocity rating) of 35,000, moldings having 10 and 20 parts of polyimide resin showed wear factors of 30 and 50 respectively. At the same PV rating, the polyimide resin failed in less than 5 minutes whereas PTFE moldings had a wear factor of 50,000.

Poly(vinylidene fluoride)(PVF_2) coating compositions containing minor amounts of polyimides have been developed in order to improve certain deficiencies found in PVF_2 coatings.[58] In certain applications, PVF_2 coatings have shown poor solvent resistance, a low softening point, inadequate lubricity and poor adhesion to substrates. These deficiencies have been alleviated by adding from 10 to 40 parts of a polyamide-acid to 100 parts of PVF_2 by weight. Coatings are heat cured, converting the polyamide-acid to the polyimide structure. The modified PVF_2 resins had a higher melting point, better solvent and chemical resistance, better substrate adhesion and self-lubricating properties, and better wear resistance than the PVF_2 resin itself.

In addition PTFE and various PTFE copolymers containing up to 30 mole percent of vinylidene fluoride, 1-chloro-1-fluoro-ethylene, hexafluoropropene, trifluorochloroethylene can be mixed with the above blends to increase the lubricity of coatings and films.

Coating compositions having excellent adhesion, good mechanical properties and reduced solvent crazing characteristics have been prepared from blends of aromatic polyamides and polyamide-acids or polyimides.[59] Polyamides (90-95%) prepared from either isophthalic or terephthalic acid, and toluene diisocyanate were blended with 5-10% of the polyamide-acid of pyromellitic dianhydride and 4,4'-oxydianiline. Coatings on metal showed excellent adhesion and impact resistance and substantially reduced solvent crazing characteristics. Polyamide coatings alone show a tendency to craze.

Novel polymer blends are formed when 10-95% of copolyimide A from benzophenone tetracarboxylic acid dianhydride and a mixture of 4,4'-Methylene bis(phenylisocyanate) and toluene diisocyanate, is mixed with a copolyamide imide B prepared from 4,4'-methylene bis(phenylisocyanate) and a mixture of trimellitic anhydride and isophthalic acid.[60] The above blends may be used in solution or their dry form. Films of these blends are characterized by improved elongation and tear strength values over copolyimide A alone. In addition the blend softening point is increased over the copolyamide-imide B alone.

A polyimide structure that is utilized in a large number of blends is based on regular repeating units of ether and imide linkages shown below.

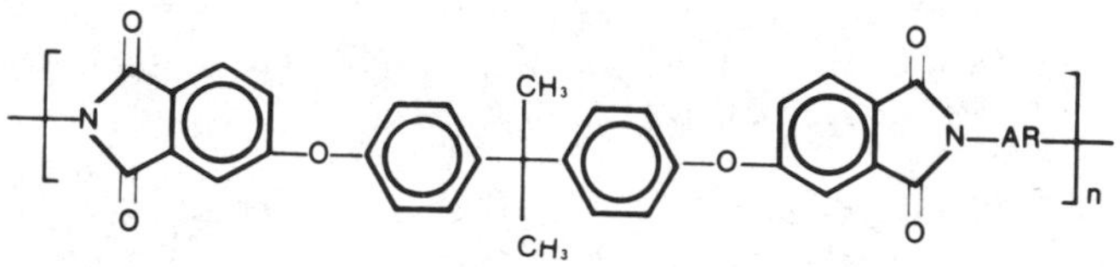

The structure of AR in the polyimide can be varied depending upon the diamine that is reacted with 2,2-bis[4-(3,4-dicarboxyphenoxy)-phenyl] propane dianhydride. The dianhydride structure is the basis for the new high performance engineering thermoplastic ULTEM[R]. This polymer exhibits high heat resistance, excellent mechanical properties, inherent flame resistance and outstanding electrical properties.[61]

ULTEM[R] is a General Electric Company Tradename.

It is therefore not surprising that this specific polyetherimide structure has been utilized in blends with other polymers.

A polyetherimide (90%) from 2,2-bis [3,4-dicarboxyphenoxy)-phenyl] propane dianhydride and 4,4'-methylene dianiline has been blended with poly(tetramethylene terephthalate) (10%).[62] The addition of the polyester produces a blend having a lower Tg, tensile strength and melt viscosity than the polyimide alone.

Blends comprised of equal parts bisphenol A-isophthaloyl chloride and terephthaloyl chloride copolymer and poly(etherimide)-m-phenylene diamine copolymer showed good mechanical properties and better environmental stress cracking resistance than the copolyester itself.[63]

Good environmental stress cracking resistance was also shown by a blend composed of 75 parts poly(etherimide)-phenylene diamine copolymer and 25 parts bisphenol A-dichlorophenyl sulfone copolymer.[64]

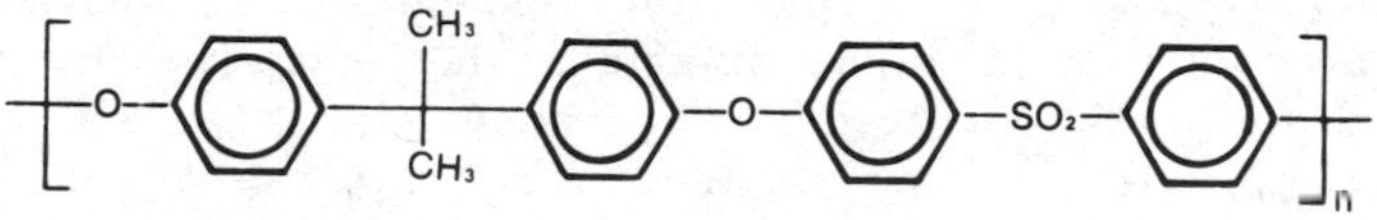

polysulfone

Poly(amide-imides) (see top of next page) containing oligomer sulfone ether diamine linkages have been blended with polysulfone copolymer (structure seen above).[65] The poly(amide-imide) structure is derived from trimellitic acid anhydride and an oligomeric diamine containing a number of arylene ether sulfone linkages. These blends showed improved environmental stress aging than the polysulfone alone, and improved processability over that of the oligomer sulfone ether diamine poly(amide-imide).

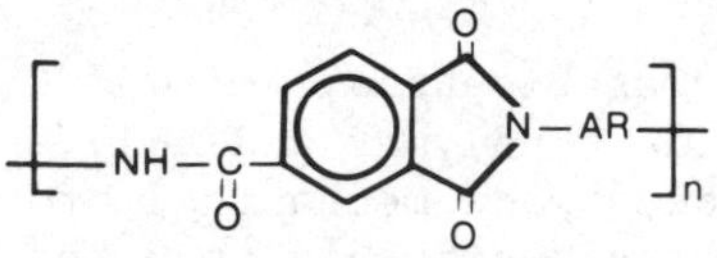

where AR =

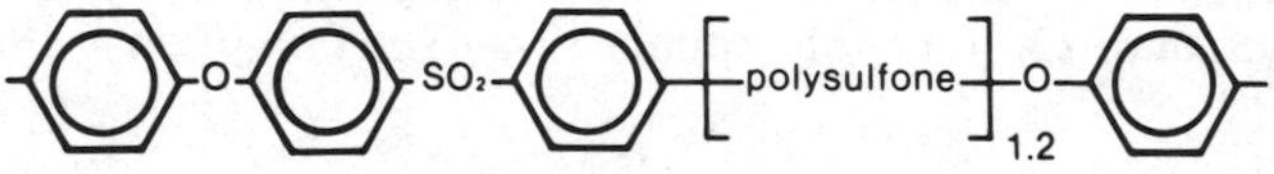

Novel halobisphenolethylene polycarbonate-poly(ether imide) blends (25:75) have been prepared.[66] These blends showed improved flame retardance and flexural strength when compared to either component alone. The components can be powder blended. The blends are suitable for the manufacture of filaments, fibers, films, sheets and laminates using conventional manufacturing techniques.

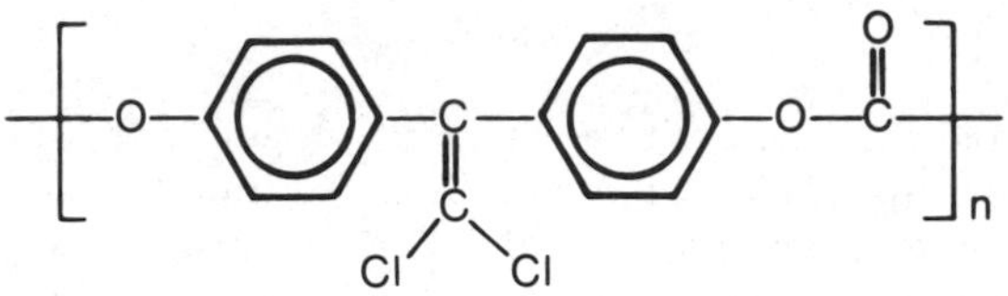

1046

Good molding properties were obtained by blending 75 parts poly(etherimide)-4,4'-methylenedianiline copolymer with 25 parts trimellitic acid-4,4'-methylenedianiline copolymer.[67] The poly(amide-imide) from trimellitic acid anhydride and 4,4'-methylenedianiline by itself is not moldable.

Poly(ether imides) have been blended with epoxy resins[68] and phenolic resins.[55, 69, 70] Blends prepared with 25% DER-661 epoxy resin and 75% poly(etherimide)-4,4'-methylene dianiline copolymer had a viscous flow temperature of 210°C, and were suitable for a solvent free application on a substrate to form an electrical insulating coating.[68] Blending the same poly(etherimide) copolymer (75%) with phenolic resins (25%) yielded a material suitable for solventless powder coating.[69, 70] The phenolic resins utilized in these blends were para substituted, oil soluble, thermoplastic and had melting points ranging from 65° to 125°C.

A blend of polyphenylene sulfide and a polyimide has been prepared which possesses surprisingly high temperature resistance characteristics, good mechanical properties at high temperatures and low friction characteristics.[71] The blend is comprised of 40-45% polyphenylene sulfide (Ryton[R])

and 60-55% Polyimide 2080[R]. The blend may be easily fabricated using conventional injection or compression molding at 300°C. A blend, based upon an aromatic polyimide such as pyromellitimide of 4,4'-oxydianiline (60%) and polyphenylene sulfide (40%), also produces a melt processable composition. Polyimide 2080[R] does not possess metal flow properties. The blending process has imparted the processing characteristics of an engineering polymer to the polyimide which is virtually impossible to mold conventionally.

Poly(amide-imides) have been mixed with prehydrolyzed kraft pulp and extracted to produce alloy fibers of the matrixfibril fiber class.[72]

RYTON[R] Trademark Phillips Petroleum
Polyimide 2080[R] Trademark Upjohn Company

Polyimide properties have been modified by adding metals to improve their high temperature adhesive properties. A NR-150 polyimide precursor adhesive solution was developed by adding aluminum powder (65%) to the copolymer of 2,2-bis(3,4-dicarboxyphenyl) hexafluoropropane dianhydride, p-phenylenediamine and 4,4'-oxydianiline in diglyme.[73] High quality, low void bonds were prepared. This is an example of a heterogenous system, where a metal is dispersed in a polymer system.

The following examples illustrate metal ions dispersed in polyimides and their unique properties.

High temperature adhesive properties of polyimides have been enhanced by incorporating a variety of organometallic dopants. The addition of aluminum (III) ions in the form of 2-3% tris (acetylacetonato) aluminum (III) to the polyamide acid of benzophenone tetracarboxylic acid dianhydride and 3,3'-diaminobenzophenone yielded a high temperature strength adhesive.[74] In the fully cured state, the polyimide-Aluminum (III) adhesive retains its flexibility and shows excellent bonding at high temperature. Other metal ions in the form of aluminum chloride hexahydrate, bis(acetylacetonato) nickel II, tris(acetylacetonato) iron (III), tris(acetylacetonato) chromium (III) and tris(acetylacetonato) cobalt (III) were found to be insoluble in the polyamide acid-diglyme solution and could not be blended.

In other studies, metals in the form of $AgNO_3$, AuI_3, $Al(AcAc)_3$, Li_2PdCl_4, $Pd[S(CH_3)_2]_2Cl_2$, $SnCl_2$, $Cu(AcAc)_3$ and $Cu(CF_3COAC)_3$ and $Cu(CF_3COAc)_3$ have been added to polyimides prepared from pyromellitic dianhydride or benzophenone tetracarboxylic acid dianhydride and various diamines.[75] Surprisingly, these metal-containing films show increased tensile strength at room temperature and elevated temperatures relative to the polyimide alone. Metal incorporation also increased tensile modulus, while percent elongation decreased at room temperature and 200°C. Curing poly(amide-acid)-metal filled cast films in air or nitrogen affected the surface characteristics of the imidized film. With certain metal dopants, the air side of the cured film possessed a metallic appearance. With a $Cu(CF_3COAc)_3$ dopant, copper in the form of $Cu(II)O$ was concentrated on the air side of the cured film, and could be etched from the surface with strong acid. Polyimide films doped with Li_2PdCl_4 and $Pd[S(CH_3)_2]_2Cl_2$ show a lower surface and volume resistivity; from an undoped value of 10^8ohm-cm to 10^6ohm-cm.[76] The metallized film surface was shown to be Pd(O) and could be removed with aqua regia. Polyimides are doped with the organometallic compound using a 4:1 polymer to metal mole ratio. Strips of these Pd(O) metallized polyimide films have been utilized as electrodes in standard electrochemical experiments.[77] These film electrodes are similar to conventional platinum electrodes in the same electrolyte system.

Tin filled poly(amide-acids) also produced metallized surfaces upon thermal curing in air.[78] A number of soluble tin complexes have been blended with poly(amide acids) from pyromellitic dianhydride or benzophenone tetracarboxylic acid dianhydride and 4,4'-oxydianiline. Most of the tin-doped films showed significantly lowered resistivity relative to the updoped polyimide.

CONCLUSION

The need for materials having a combination of properties not found in homopolymers has led to blends. The examples cited here illustrate that a large number of polyimides and copolyimides have been "blended" in order to improve properties, lower cost and improve ease of fabrication. The applications for these blends are varied. Indications are that polyimides will continue to be one of the more important and exciting materials, primarily because of their outstanding properties, their availability in a variety of forms and their utility in a variety of applications.

REFERENCES

1. H.H. Gibbs and C.V. Breder, in "Copolymers, Polyblends and Composites," N.A.J. Platzer, Editor, pp. 442-457, Adv. in Chem. Ser., No. 142, American Chemical Society, Washington, D.C. 1975.
2. F.E. Rogers, U.S. Pat. 3,356,648 (1967).
3. J.P. Critchley, P.A. Gratten, M.A. White and J.S. Pippet, J. Polym. Sci., A-1, 10, 1789 (1972).
4. J.P.Critchley and M.A. White, J. Polym. Sci., A-1, 10, 1809 (1972).
5. J.P. Critchley, Brit. Pat. 1,457,191 (1976). (CA 87, 136976B).
6. J.A. Webster, U.S. Pat., Appl. 487,852 (1974). (CA 82, 140997y).
7. I.K. Varma, G.M. Fohlen and J.A. Parker, U.S. Pat. Appl. 174,452 (1981).
8. I.K. Varma, G. M. Fohlen, J.A. Parker and D.S. Varma, These proceedings, Vol. 2, pp. 683-694.
9. J. T. Hoback and F.F. Holub, U.S. Pat. 3,740,305 (1973).
10. A. Berger. These proceeings, Vol. 1, pp. 67-76.
11. G.L. Brode, J.H. Kawakami, G.T. Kwiatkowski and A.W. Bedwin, J. Polym. Sci., Polym. Chem. Ed., 12, 575 (1974).
12. G.L. Brode, J.H. Kawakami, G.T. Kwiatkowski and A.W. Bedwin, in "Copolymers, Polyblends and Composites," N.A.J. Platzer, Editor, p. 343, Adv. in Chem. Ser., No. 142, American Chemical Society, Washington, D.C., 1975.
13. J.M. Adduci, L.L. Chapoy, E.L. Hanson, G. Jonsson, J. Kops and B.M.Shinde, in "Proceedings, IUPAC MAKRO 81, 27th Macromolecular Symposium," Strasbourg, France, Vol. 1, 61-4 (1981).
14. N. Dokoshi, S. Tohyama, S. Fujita, M. Kurihara, and N. Yoda, J. Polym. Sci., A-1, 8, 2197 (1970).

15. J. Preston and W.B. Black, J. Polymer Sci., A-1, 5, 2429 (1967).

16. J. Preston, W. DeWinter, W.B. Black, and W.L. Hofferbert, Jr., J. Polym. Sci., A-1, 7, 3027 (1969).

17. T. Kurosaki and P.R. Young, J. Polym. Sci., C23 (1), 57 (1968).

18. S.S. Hirsch, J. Polym. Sci., A-1, 7, 15 (1969).

19. K. Kurita and R.L. Williams, J. Polym. Sci., Polym. Chem. Ed., 11, 3125 (1973).

20. K. Kurita and R.L. Williams, J. Polym. Sci., Polym. Chem. Ed., 11, 3151 (1973).

21. D.A. Gordon and R. Seltzer, U.S. Pat. 3,957,726 (1976).

22. R. Takatsuka, T. Unishi, I. Honoa and T. Kakuri, J. Polym. Sci., Polym. Chem. Ed., 5, 1785 (1977).

23. B.N. Achar, G.M. Fohlen and J.A. Parker, J. Polym. Sci., Polym. Chem. Ed., 20, 773 (1982).

24. B.N. Achar, G.M. Fohlen and J.A. Parker, J. Polym. Sci., Polym. Chem. Ed., 20, 2781 (1982).

25. W. Wrasidlo and J.M. Augl, J. Polym. Sci., A-1, 7, 321 (1969).

26. F. Hayano and K. Komoto, J. Polym. Sci., A-1, 10, 1263 (1972).

27. J.M. Adduci and S.K. Sikka, Macromolecular Synthesis, 8, 53 (1982).

28. L.H. Tagle, J.F. Neira, F. R. Diaz and R.S. Ramirez, J. Polym. Sci., Polym. Chem. Ed., 13, 2827 (1975).

29. L.H. Tagle, and F.R. Diaz, Polymer, 23, 1057 (1982).

30. R.W. Campbell and H. Wayne Hill, Jr., Macromolecules, 8, 706 (1975).

31. J. DeAbajo, in "Proceedings IUPAC MAKRO 81, 27th Macromolecular Symposium," Strasbourg, France, Vol. 1, 65 (1981).

32. J.T. Holback and F.F. Holub, U.S. Pat. 3,723,385 (1973).

33. H. Kawahara and A. Noda, Japan Kokai, 78 24,393 (1978). (CA 89,111172p).

34. H. Kawahara and A. Noda, Japan Kokai, 78 24,395 (1978). (CA 89,111172q).

35. H. Kawahara and A. Noda, Japan Kokai, 78 105,596 (1978). (CA 90, 39421w).

36. W.L. Hergenrother and R.J. Ambrose, J. Polym. Sci., Polym. Lett. Ed., 12, 343 (1974).

37. R.J. Jablonski, J.M. Witzel and D. Kruh, J. Polym. Sci., Part B 8(3) 191 (1970).

38. A.K. St. Clair and T.L. St. Clair, Proceedings, 12th National SAMPE Technical Conference, 12, 729 (1980). (CA 94, 31487j).

39. G.L. Brode, G.T. Kwiatkowski and J. H. Kawakami, Polymer Preprints 15 (1), 761 (1974).

40. G.T. Kwiatkowski, G.L. Brode and L.M. Robeson, U.S. Pat., 3,658,938 (1972).

41. J.V. Crivello and P.C. Juliano, Polymer Preprints, 14 (2), 1220 (1973).

42. H.A. Vogel and H.T. Oien, French Patent, 1,559,971 (1969) (CA 71, 81912q).

43. W. F. DeWinter and J. Preston, U.S. Pat. 3,621,076 (1971)
    (CA 71, 73490p).
44. F. Kobayashi, F. Sakata, T. Mizoguchi and N. Suyama, Japan
    Kokai, 69 27,674 (1969) (CA 72, 56598e).
45. H.H. Gibbs and G.V. Breder, Polymer Preprints, 15 (1), 775
    (1974).
46. R.T. Alvarez and F.P. Darmory, Proceedings, 20th National
    SAMPE Symposium, 353 (1975).
47. G.T. Kwiatkowski, L.M. Robeson, G.L. Brode and A.W. Bedwin,
    J. Polym. Sci., Polymn, Chem. Ed. 13, 961 (1975).
48. W.R. Vaughn, R.J. Jones, M.K. O'Dell and T.V. Roszhart,
    Proceedings 20th National SAMPE Symposium, 365 (1975).
49. T.T. Serafini, in "Resins for Aerospace," C.A. May, Editor,
    p. 15, ACS Symposium Series, No. 132, American Chemical
    Society, Washington, D.C. 1980.
50. L.C. Hsu and T.T. Serafini, Organic Coatings Plastics Preprints,
    1, 193 (1974).
51. N. Bilow and A.L. Landis, Proceedings 20th National SAMPE
    Symposium, 618 (1975).
52. N. Bilow, in "Resins for Aerospace," C.A. May, Editor, p. 139,
    ACS Symposium Series, No. 132, American Chemical Society,
    Washington, D.C. 1980.
53. R.J. Jones, U.S. Pat. 4,283,521 (1981).
54. D.A. Scola and R.H. Pater, Polymer Preprints 22 (2), 228
    (1981).
55. J.M. Adduci, D.K. Dandge and J. Kops, Polymer Preprints,
    22 (2), 109 (1981).
56. S. Maiti and A. Ray, Die Makromolekulare Chemie, Rapid
    Communications, 2 (11), 649 (1981).
57. E.J. Mack and H.T. Childs, Jr., Fr. Pat. 1,580,077 (1969).
    (CA 72, 112418k).
58. J.P. King, U.S. Pat. 3,668,193 (1972). (CA 77, 76847S).
59. J.C. Fang, U.S. Pat. 3,592,952 (1971). (CA 75, 119270b).
60. K.B. Onder and F.P. Recchia, U.S. Pat. 4,225,686 (1980).
    (CA 94, 4644b).
61. I.W. Serfaty, These proceedings, Vol. 1, pp. 149–162.
62. D.M. White and R.O. Matthews, Ger. Offen. 2,622, 135 (1976).
    (CA 86, 56306x).
63. L.M. Maresca, M. Matzner and L.M. Robeson, U.S. Pat. 4,250,279
    (1981) (CA 95, 81976j).
64. L.M. Robeson, M. Matzner and L.M. Maresca, Eur. Pat. Appl.
    33,394 (1981). (CA 95, 188128m).
65. G.T. Kwiatkowski, G.L. Brode and L. Robeson, U.S. Pat. 3,658,938
    (1972).
66. P.D. deTorres, U.S. Pat. 4,225,687 (1980), (CA 94, 4637b).
67. F.F. Holub and G.A. Mellinger, U.S. Pat. 4,258,155 (1981).
    (CA 95, 81983j).
68. E.G. Banucci and E.M. Boldebuck, Ger. Offen., 2,650,019
    (1977). (CA 87, 40870m).

69. E.G. Banucci and E.M. Boldebuck, U.S. Pat. 4,199,651 (1980).
70. E.G. Banucci and E.M. Boldebuck, U.S. Pat. 4,163,030 (1979). (CA 91, 142210w).
71. R.T. Alvarez, U.S. Pat. 4,017,555 (1977). (CA 87, 24198t).
72. H. Rodier and J.P. Sacre, Proceedings, 5th International Dissolving Pulps Conference, TAPPI, 109-20, Atlanta, Ga., 1980. (CA 94, 85966h).
73. P.S. Blatz, Adhes. Age, 21, 39 (1978). (CA 89, 180932z).
74. A.K. St. Clair, T.L. St. Clair and L.T. Taylor, NASA Case No. LAR 12640-1, Pat. Appl. (1979).
75. L.T. Taylor and A.K. St. Clair, These proceedings, Vol. 2, pp. 617-646.
76. E. Khor And L.T. Taylor, These proceedings, Vol. 1, pp. 507-520.
77. T.A. Furtsch, H.O. Finklea, and L.T. Taylor, These proceedings, Vol. 2, pp. 1157-1166.
78. S.A. Ezzell and L.T. Taylor, These proceedings, Vol. 1, pp. 189-204.

POLYIMIDES - TRIBOLOGICAL PROPERTIES AND THEIR USE AS LUBRICANTS

R. L. Fusaro

National Aeronautics and Space Administration
Lewis Research Center
Cleveland, Ohio 44135

Friction, wear, and wear mechanisms of several different polyimide films, solid bodies, composites, and
bonded solid lubricant films are compared and discussed.
In addition, the effect of such parameters as temperature,
type of atmosphere, contact stress, and specimen configuration are investigated. A friction and wear transition
occurs in some polyimides at elevated temperatures and
this transition is related to molecular relaxations that
occur in polyimide. Friction and wear data from an
accelerated test (pin-on-disk) are compared to similar
data from an end use test device (plain spherical bearing),
and to other polymers investigated in a similar geometry.

INTRODUCTION

The use of polymers for tribological applications is continually increasing. In addition, technology is placing ever increasing performance demands on polymers used for tribological applications. Polymers are needed which have improved mechanical properties from cryogenic to the highest possible temperature. One class of thermally stable organic polymers which has demonstrated increased capabilities in these areas is polyimide.

The word polyimide is a generic designation and refers to a class of long-chained polymers which have repeating imide groups as an integral part of the main chain. By varying the monomeric starting materials, polyimides of different chemical composition and structure can be obtained. The polyimide chains consist of aromatic rings alternated with heterocyclic groups and due to multiple bonds between these groups, the polyimides are characterized by a high thermal stability. At the decomposition point, they crumble to a fine powder without melting. They have a high radiation stability and can withstand exposure to neutrons, electrons, ultraviolet light, and gamma radiation. They are resistant to most common chemicals and solvents, but are attached by alkalis. For a more detailed discussion of the physical properties, see references 1 to 5.

Polyimides are being considered for use in bearings, gears, seals, and prosthetic human joints[6-12]. The intended end use part can be machined or molded from the polyimide or a film of polyimide may be applied to a metallic part. In many instances, polyimide by itself will be sufficient to improve the tribological properties of the intended end use component. However, in some cases, solid lubricant additives may be needed to improve lubrication. To improve the load carrying capacity of polyimide solid bodies, they can be reinforced with fibers. If graphite fibers are used, in addition to improving the strength and stiffness of the polyimide, improved lubricating performances can be obtained, due to the graphite fibers good tribological properties[13-19].

In order to propitiously use polyimides for tribological applications, mechanisms of lubrication and wear, and the factors affecting these mechanisms must be clearly understood. This paper will review the work to date conducted at NASA Lewis Research Center on the fundamental aspects of polyimide lubrication. In addition, tribological materials formulated at Lewis for specific end use applications with polyimide as the matrix material will be discussed.

MATERIALS

Nine different polyimides were evaluated and compared. Those with known composition are given in Figure 1. Seven of the polyimides were evaluated as films (20 to 25 μm thick) applied to sandblasted AISI 440C HT (high temperature) stainless steel disks (Rockwell hardness, C-60; surface roughness, 0.9 to 1.2 μm CLA). These polyimides are designated PIC-1 to PIC-7. Three of the polyimides were evaluated as solid bodies or composites; they are designated Types "A", "C", and "V". The polyimide film PIC-7 and the polyimide solid body Type "C" are the same polyimide. Polyimide Type "V" is a common commercially available polyimide.

Polyimide-bonded solid lubricant films were formulated using the PIC-1 polyimide and 75 weight percent of molybdenum disulfide (MoS_2) or 50 weight percent of graphite fluoride (($CF_{1.1})_n$). One commercially available composite containing 15 percent graphite powder in a Type "V" polyimide matrix was also evaluated. Graphite fiber reinforced composites (GFRPI) were made from 50 weight percent low modulus graphite fibers and Type "A" polyimide. The fibers were 8.4×10^{-6} meter in diameter and chopped into lengths of 6.4×10^{-3} meter. They were randomly dispersed throughout the polyimide matrix. For more details on the films or composites see references 20 to 29.

The hemispherically tipped pins were made of AISI 440C HT stainless steel or the polyimide composite material. The 440C pins were slid against the polyimide films or composite disks and the composite pins were slid against smooth 440C disks with a surface roughness of less than 0.1 μm, CLA.

EXPERIMENTAL PROCEDURE

A pin-on-disk tribometer was used for these experiments (Figure 2). The pins were either hemispherically tipped with a radius of 0.475 cm or the same hemispherically tipped pins with flats worn on them (see insert, Figure 2). They were loaded with a 9.8 N deadweight against the disk which was rotated at 1000 rpm. The pin slid on the disk at a radius of 2.5 cm giving it a linear sliding speed of 2.7 m/s. The test specimens were enclosed in a chamber so that the atmosphere could be controlled. Atmospheres of dry argon (<100 ppm H_2O), dry air (<100 ppm H_2O), or moist air (10 000 ppm H_2O) were evaluated.

Each test was stopped after predetermined intervals of sliding and the pin and disk were removed from the friction apparatus. The contact areas were examined by optical microscopy and photographed, and surface profiles of the disk wear track were taken. Locating pins insured that the specimens were returned to their original positions. Disk wear was determined by measuring the cross-sectional area on the disk

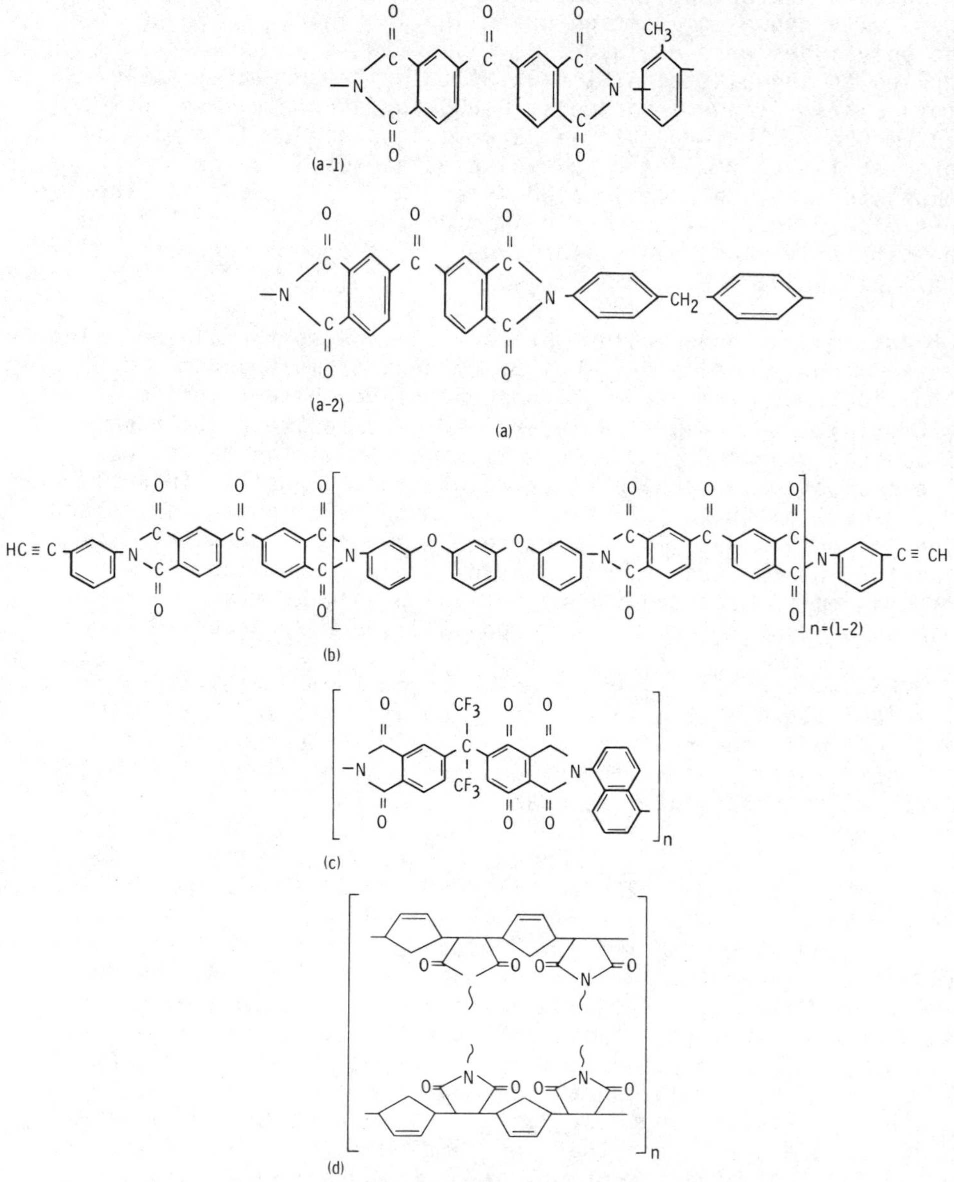

Figure 1. Structure and designation of non-proprietary polyimides.

(a-1) 20 percent of the recurring units
(a-2) 80 percent of the recurring units
(a) PIC-5 polyimide film
(b) PIC-6 polyimide film
(c) PIC-7 polyimide film and type "C" solid body
(d) Type "A" idealized structure polyimide solid body

wear track (from surface profiles), and rider wear was
determined by measuring the wear scar diameter on the
hemispherically tipped rider after each sliding interval and
then calculating the volume of material worn away.

RESULTS AND DISCUSSION

Polyimide Films

General wear mechanisms. The wear process of a polyimide
solid body or a composite body is one of gradual wear through
that body[20]. A film, however, can wear and lubricate by
either of two processes[21-22]. One, it can wear similar to a
solid body by gradually wearing away; or two, it can be quickly
worn away with the subsequent formation of a secondary film at
the substrate interface. In the second mechanism, shearing of
the secondary film provides the lubrication. Figure 3 gives
cross-sectional views of a film wear track, illustrating these
two mechanisms.

When gradual wear through the film occurs, generally no
measurable wear occurs to the metallic pin; but when the
lubrication process is of the secondary film type, wear of the
rider increases usually at a constant rate with sliding
distance. This is most probably due to the fact that some metal
contact occurs during the shearing of the film.

All seven polyimide films evaluated could provide
lubrication by either mechanism[22]. In general, the rider wore
gradually through the film and then a secondary film formed
which provided lubrication for an extended period of time. The
secondary film mechanism for the polyimide films was not studied
in great detail since lubrication by this mechanism is usually
improved by the addition of a solid lubricant such as graphite
or graphite fluoride. Also, if possible, the gradual wear
through film mechanism is preferred, since there is less chance
of adhesive metal to metal wear occurring which could lead to
catastrophic failure.

Friction and wear - 25° C. Polyimide film wear volume
tended to increase in a linear manner (from zero) as a function
of sliding distance. Figure 4 plots wear volume for a
representative polyimide film as a function of sliding
distance. Wear rates were determined by taking a linear
regression fit (least squares) of these data. Average wear
rates are given for each polyimide evaluated in Table I.

The table also gives the average "steady state" friction
coefficient obtained for each polyimide. The table indicates
that the three films that gave the lowest friction coefficients
also gave the highest wear rates. The other four films gave
higher friction coefficients but lower wear rates.

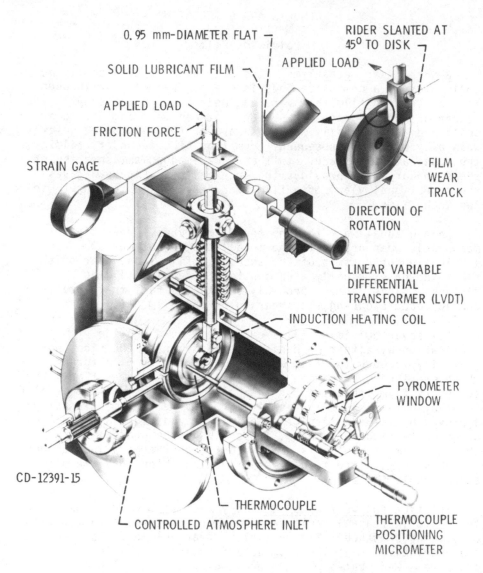

0.95 mm-DIAMETER FLAT

SOLID LUBRICANT FILM

APPLIED LOAD

FRICTION FORCE

STRAIN GAGE

RIDER SLANTED AT 45° TO DISK

APPLIED LOAD

FILM WEAR TRACK

DIRECTION OF ROTATION

LINEAR VARIABLE DIFFERENTIAL TRANSFORMER (LVDT)

INDUCTION HEATING COIL

PYROMETER WINDOW

CD-12391-15

THERMOCOUPLE

CONTROLLED ATMOSPHERE INLET

THERMOCOUPLE POSITIONING MICROMETER

Figure 2. Friction and wear apparatus.

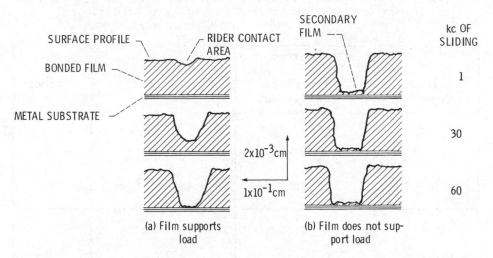

Figure 3. Cross sectional area schematics of the wear areas on a bonded film (polymer or other type of solid lubricant film) after 1,30, and 60 kc of sliding illustrating the two different types of macroscopic lubricating mechanisms. (Note that vertical magnifiation is 50 times horizontal magnification.)

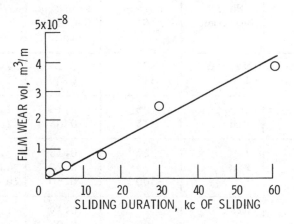

Figure 4. Representative example of polyimide wear as a function of sliding distance. (Data from PIC-1 polyimide film at 25°C in moist air.)(Taken from ref. 24.)

1059

Table I. - Classification of Polyimide Films into Two Friction and
Wear Groups, for Ambient Temperature Conditions
(22° to 27° C in 50% R.H. air).
(Taken from ref. 22)

Polyimide Type	Average "Steady State" Friction Coefficient	Average Film Wear Rate	Group
PIC - 1	0.13	40×10^{-14} m^3/m	
PIC - 4	0.13	80×10^{-14} m^3/m	I
PIC - 7	0.10	40×10^{-14} m^3/m	
PIC - 2	0.23	10×10^{-14} m^3/m	
PIC - 3	0.27	6×10^{-14} m^3/m	II
PIC - 5	0.30	6×10^{-14} m^3/m	
PIC - 6	0.28	12×10^{-14} m^3/m	

Table II. Comparison of Friction and Wear Data Obtained on GFRPI
Composites Using Different Experimental Apparatus and Geometries
(Taken from ref. 27)

Experimental Apparatus	Composite Wear Specimen	Temperature, °C	Average Friction Coefficient	Average Wear Rate, m^3/m
Self-aligning plain bearings	Molded liner	25 315	0.15 0.05	1.2×10^{-14} 1.2×10^{-14}
	Insert liner	25 315	0.15 0.05	2.0×10^{-14} 2.0×10^{-14}
Pin-on-disk	Pin	25[a] 25[b] 300[a]	0.15 0.27 0.50	$.25 \times 10^{-14}$ 4.1×10^{-14} 20×10^{-14}
	Disk	25 300	0.19 0.05	1.3×10^{-14} 1.5×10^{-14}

(a) Sliding distance of 1 km
(b) Sliding distance of 19 km

The polyimide films were thus classified into two groups[22]: Group I, low friction coefficients (<.13) - high wear rates (>40 x 10^{-14} m^3/m) and Group II, high friction coefficients (>.23) - low wear rates (<12 x 10^{-14} m^3/m).

Film wear surface morphology - 25° C. The polyimides were classified into the two groups not only because of their friction and wear properties, but because of similarities in film wear track surface morphology. Representative photomicrographs of each group are shown in Figures 5 and 6.

Group I wear surfaces were characterized by being covered with powdery, agglomerated, birefringent polyimide wear particles. The wear process was of an adhesive nature; but the surface layer appeared to be brittle and subsequent crumbling of it into relatively large particles readily occurred. Transfer to the pin was found to be thin and tended to plastically flow across the contact area.

Group II polyimides were characterized by a rough looking surface (Figure 6) that tended to plastically flow for a short distance, but then tended to break-up into very fine wear particles at the leading edge. The wear particles did not agglomerate on the film wear track as did Group I polyimides, although they had a tendency to build-up on the metallic pin.

One polyimide in Group II (PIC-2) produced wear surfaces that had only localized rough looking areas (Figure 7). For this polyimide, the wear surface, in general, was very smooth with striations in the sliding direction. But, even though these surfaces were smoother than other Group II polyimides, film wear rates and friction coefficients were not significantly different.

From a surface morphological point of view, however, PIC-2 films produced the most desirable wear tracks, a smooth, almost glass-like wear surface (Figure 7). Also the transfer films tended to be thinner and to coalesce and plastically flow across the pin contact area better than the other Group II polyimides. The other polyimides tended to form transfer films of compacted polyimide wear particles which slid across the metallic pin and tended to wear it.

Molecular relaxations. It is well known that most polymers have a glass-transition temperature (T_G) which marks their change from a rubbery state to a glassy state. It is not so well known that polymers also possess secondary transitions below their T_G. Some of these secondary transitions appear to influence the mechanical properties of polymers.

Both the glass transition and the secondary transitions can be related to one or more molecular relaxation processes (changes in motion within polymer chains). The glass transition

1061

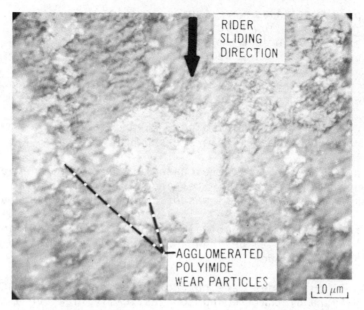

Figure 5. High magnification photomicrograph of the wear track on PIC-4 polyimide films illustrating typical surface morphology on Group I polyimides.

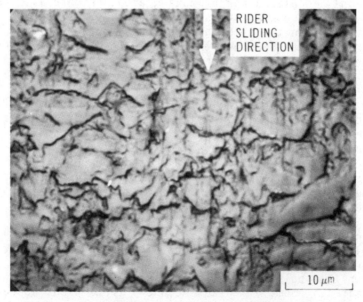

Figure 6. High magnification photomicrograph of the wear track on PIC-3 polyimide films illustrating typical surface morphology on Group II polyimides.

is related to the motion of longer segments of the main chain; while the secondary transitions are related to rotations or oscillations of side chains, subgroups, chain ends or short segments of the main chain.

The friction and wear properties of the polyimide film PIC-1 have been compared to the molecular relaxations that occur in this polyimide[23]. Figure 8 gives this comparison, where the logarithmic decrement (from torsional braid analysis (TBA)) and friction coefficient (obtained in dry argon) are plotted as a function of temperature for a film cured in air and for the same film heated to 500° C in dry argon.

The peaks in the logarithmic decrement curve (α, β, β_{H_2O}, γ) represent loss maxima or temperatures at which the molecular segment achieves its greatest degree of freedom. As the temperature is decreased below the loss maxima, the motion responsible for the loss mechanism is continuously hindered, until at some temperature it becomes completely frozen. This is the onset temperature of the peak.

A large transition in the friction coefficient of PIC-1 polyimide films occurred at approximately 30° to 40° (Figure 8). If the secondary molecular relaxation were responsible for this transition, it appears the onset of the peak rather than the peak maxima is the parameter which best correlates with the friction transition.

Figure 8 also indicates that, by heating the polyimide in dry argon to 500° C, the T_G of the polyimide was increased to that temperature, with the subsequent reduction in friction coefficient between the temperatures of 300° and 500° C.

Temperature and atmosphere effects. In addition to temperature, atmosphere can also markedly affect the friction and film wear properties of polyimide films[24]. Figure 9 plots average friction coefficient and film wear rate as a function of temperature for PIC-1 polyimide films evaluated in three different atmospheres; dry argon (<100 ppm H_2O), dry air (<100 ppm H_2O) and moist air (10 000 ppm H_2O).

In dry air and argon, a transition (from higher friction and wear to lower friction and wear) occurred between the temperatures of 25° C and 100° C. But in moist air, this transition has shifted to a higher temperature. It has been postulated that this shift in transition temperature in moist air is due to the β_{H_2O} molecular relaxation (Figure 8). The H_2O molecules are believed to hydrogen bond to the polyimide molecules, constraining their motion, and inhibiting their ability to plastically flow in thin surface layers.

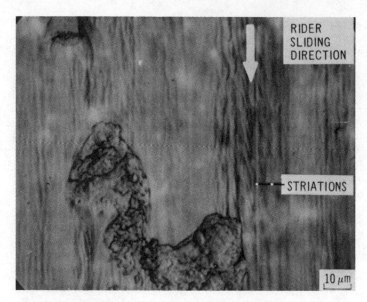

Figure 7. High magnification photomicrograph of the wear track on
PIC-2 polyimide films illustrating a variation of the wear surface
morphology for Group II polyimides.

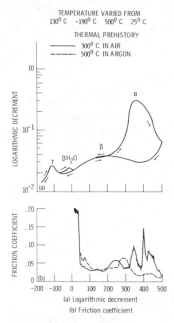

Figure 8. Comparison of the logarithmic decrement (from torsional
braid analysis) to the friction coefficient of polyimide (PIC-1)
films for a film cured at 300°C in dry air and for the same film
heated in argon to 500°C before testing.(Taken from ref. 23.)

Figure 10 presents a typical film wear surface for temperatures above the transition. The wear track becomes very smooth and the primary wear mechanism appears to be the spalling of a thin, textured layer of the polyimide (Figure 10). Similarly, the transfer films tend to flow across the pin in very thin layers and there is no tendency for them to buildup with sliding duration as they did at 25° C.

At temperatures below the transition, atmosphere did not markedly affect film wear rate (Figure 9), but H_2O in the atmosphere did tend to reduce the friction coefficient at this temperature. At temperatures above the transition, lower friction and wear were obtained in argon atmospheres than in air atmospheres, indicating there may be an oxidation effect at elevated temperatures.

Polyimide - Bonded Solid Lubricant Films

High contact stress effects. Solid lubricants are often added to polymer films to improve the friction and wear characteristics; but their presence can reduce the load carrying capacity of the polymer. In reference 25, molybdenum disulfide (MoS_2) and graphite fluoride ($(CF_{1.1})_n$) were added to PIC-1 polyimide films. The presence of the solid lubricants reduced the strength of the films to the point that they would not support the hemispherical-tipped pin sliding under a load of 9.8 N.

Figure 11 illustrates what happened to the polyimide-bonded $(CF_{1.1})_n$ film under the above sliding conditions[21]. A series of fine cracks developed on the film wear track which led to the eventual crumbling away of the film in less than 15 kilocycles of sliding. Once this took place, lubrication occurred by the secondary film mechanism. Polyimide and $(CF_{1.1})_n$ wear debris compacted into valleys between sandblasted asperties in the 440C stainless steel disk substrate, and flat plateaus were worn in the asperities (Figure 11(b)).

The mechanism of lubrication became the process of shearing very thin films of the polyimide - $(CF_{1.1})_n$ mixture between flat areas on the pin and on the metallic asperities. This process continued until the lubricant became depleted from the valleys.

Figure 12 plots wear life (defined as the number of sliding revolutions to reach a friction coefficient of 0.30) and average friction coefficient as a function of temperature for polyimide alone and for polyimide-bonded $(CF_{1.1})_n$ or MoS_2 films.

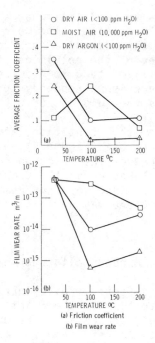

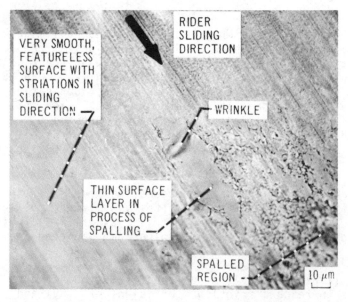

Figure 9. Effect of temperature and atmosphere on the friction coefficient and wear rate of polyimide films. (Specific type of polyimide film, PIC-1). (Taken from ref. 24.)

Figure 10. High magnification photomicrograph of a PIC-1 polyimide
film wear track at 150°C in dry air (<100ppm H_2O) illustrating the
sliding surface morphology at a temperature above the transition.

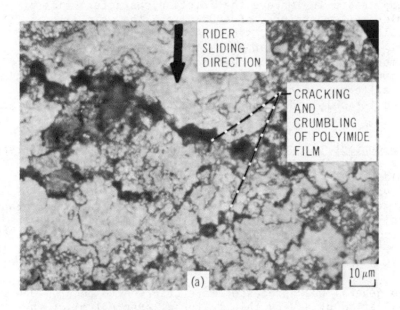

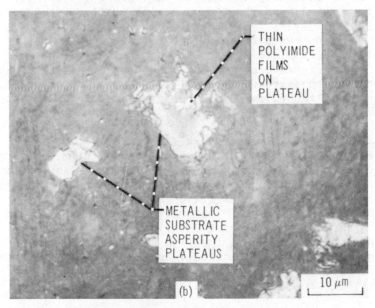

(a) Fracturing of film.

(b) Secondary film wear mechanism.

Figure 11. Photomicrographs of wear tracks on polyimide-bonded graphite fluoride films subjected to high projected contact stress (hemisphere under 9.8 N load (stress greater than 105 MPa)).

MoS$_2$ tended to improve the friction characteristics of polyimide to temperatures of 400° C, and the wear life at 25° C; but at 100° C and higher, shorter wear lives were obtained. (CF$_{1.1}$)$_n$ improved the wear life at all temperatures (but only improved the friction coefficient at 25° C and at temperatures of 400° C and above). Of course, one must realize that a different wear process (gradual wear through film) occurred for polyimide by itself at temperatures of 100° C and higher.

Low contact stress effects. It was discovered in reference 21, that if lower contact stresses were applied to the polyimide-bonded (CF$_{1.1}$)$_n$ films, the bonded film itself could support the load and the film would wear in a manner similar to polyimide alone. Figure 13 shows a photomicrograph on a polyimide-bonded graphite fluoride film subjected to a projected contact stress of 14 MPa. The figure illustrates the tips of the film asperities can support this load. Wear at this point is a process of gradual truncation of these asperities.

Two general wear processes were observed to occur. One was the spallation of a very thin textured layer (<1μm) of the film (Figure 14(a)) and the other was the brittle fracture of thicker surface layers (1 to 6 μm) of the film (Figure 14(b)). The occurrence of the brittle fracture wear was dependent on the contact stress, the contact area, and the build-up of thick transfer films[26]. Low contact stresses and large contact areas tended to promote the build-up of thick, nonflowing transfer films which increased adhesion and thus friction and wear.

Figure 15 gives a friction trace for a 0.95-mm diameter flat sliding against a 45-mm-thick polyimide-bonded film. The trace is divided into two regions: the first stage of lubrication where gradual wear through the film occurs and the second stage of lubrication where the lubrication occurs by the secondary film. In the first stage of lubrication, friction tends to increase gradually with sliding duration. This increase corresponds to truncation in the film asperities and the ensuing increasing contact area. Once full contact was obtained, the average friction coefficient during the first and second stages of lubrication were nearly equivalent (0.24).

Polyimide Solid Bodies

Friction and wear. In addition to use as films, polyimides have also been molded into solid bodies and their tribological properties evaluated[20]. Figure 16 compares average friction coefficients and wear rates at 25° C for 440C HT stainless steel hemispherically tipped pins sliding on two different polyimide solid bodies made from types "A" and "V" polyimides. In

1068

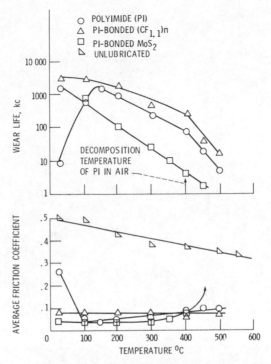

Figure 12. Friction coefficient and wear life as a function of temperature for three solid lubricant films run in dry air (moisture content, 20 ppm). (Taken from ref. 25.)

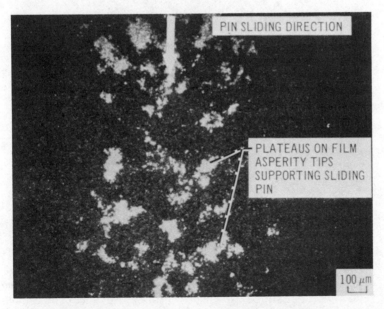

Figure 13. Photomicrograph of wear track on polyimide-bonded graphite fluoride film subjected to a low projected contact stress (14 MPa).

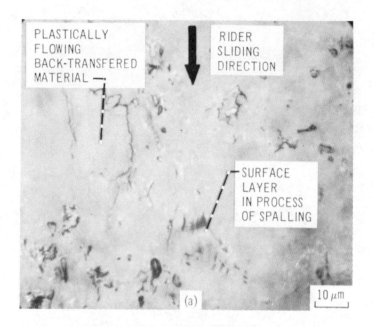

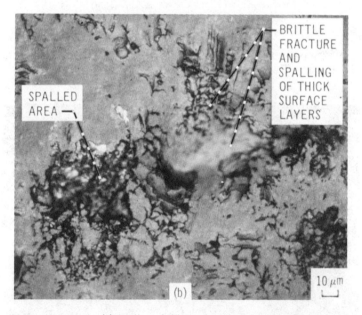

(a) Scaling of thin surface layers.

(b) Brittle fracture of thick surface layers.

Figure 14. Photomicrographs of surface morphology on the wear tracks of polyimide-bonded graphite fluoride films subjected to low contact stresses (14 MPa) illustrating wear mechanisms.

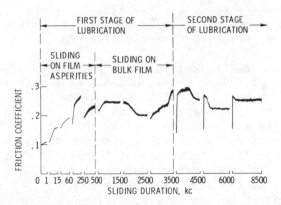

Figure 15. Friction trace for a 440C HT stainless-steel hemispherically tipped rider (with a 0.95 mm diameter flat on it) sliding on
polyimide-bonded graphite fluoride film. (Taken from ref. 21.)

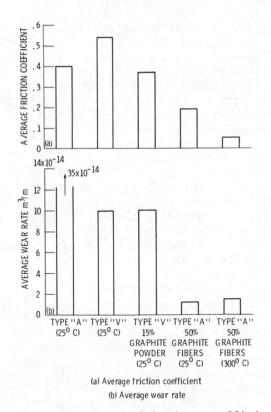

(a) Average friction coefficient
(b) Average wear rate

Figure 16. Comparison of average friction coefficients and average
wear rates of two different types of polyimides made into solid
bodies (disks) and of two composite solid bodies made from the
polyimides.

addition, the figure shows similar data for composites made by adding 15 percent graphite powder to type "V" polyimide and by adding 50 percent graphite fibers to type "A" polyimide.

The friction coefficients of both polyimides are rather high compared to the polyimide films, but the wear rate of type "A" polyimide fits in very well with Group II polyimide films (Table I). Additions of 15 percent graphite powder to type "V" polyimide reduced the friction coefficient but the wear rate (at least with this geometry) remained the same as the base polyimide (Figure 16). Additions of 50 percent graphite fibers to type "A" polyimide markedly improved the friction coefficient (0.40 to 0.19) and the wear rate (35×10^{-14} to 0.6×10^{-14} m^3/m) when compared to the base polyimide.

The graphite fiber reinforced polyimide (GFRPI) was also evaluated at 300° C. Friction coefficient was even lower at this temperature (0.05) but the wear rate was nearly the same (0.7×10^{-14} m^3/m) as that at 25° C.

Wear mechanisms. A photomicrograph of typical wear surface morphology at 25° C on type "A" polyimide solid bodies is shown in Figure 17. The wear surface was rough looking and consisted of thick plastically flowing surface layers (up to 3μm) that tended to spall and crumble. Even though wear rates were higher (Figure 16) films were similar to Group II polyimide films. That is, thick, nonshearing transfer films tended to build-up with increasing sliding duration.

Type "V" polyimide tended to wear similar to Group II polyimide films (Figures 6 and 7). The addition of 15 percent graphite powder did not markedly alter the wear process. Figure 18 gives a typical photomicrograph of the wear surface of 25° C on this composite. The graphite particles did not mix with the polyimide to form a surface layer; and in fact may have weakened the structure slightly, since pits and crumbling was observed on the wear surface (Figure 18).

Figure 19 gives a photomicrograph of a typical wear track surface in moist air for a low modulus graphite fiber reinforced polyimide composite. The fibers have mixed together with the polyimide to form a very thin surface layer. The wear process appears to be the spalling of this surface layer (Figure 19).

At 300° C, the fibers did not mix with the polyimide on the wear surface as they did at 25° C (Figure 20). Instead a very thin layer of the polyimide tended to flow across the fibers and dominate the friction process. As mentioned in a previous section, some polyimides possess transitions, above which very low friction occurs.

1072

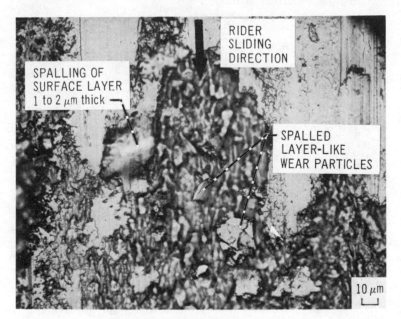

Figure 17. Photomicrograph of typical wear surface morphology at 25°C on the type "A" polyimide solid body.

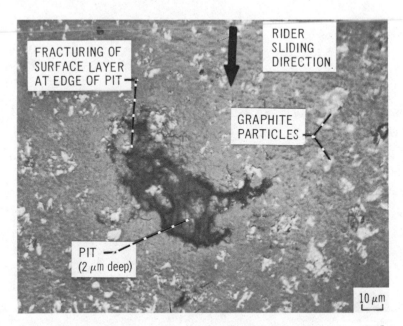

Figure 18. Photomicrograph of typical wear surface morphology at 25°C on type "V" polyimide composites with 15% graphite powder additions.

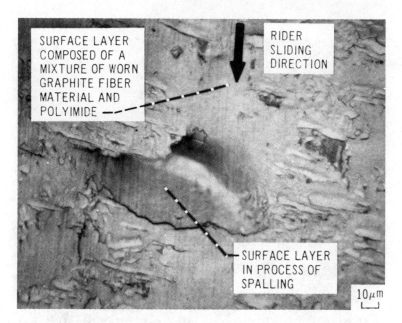

Figure 19. Photomicrograph of typical wear surface morphology at 25°C in moist air on the graphite fiber reinforced polyimide composite.

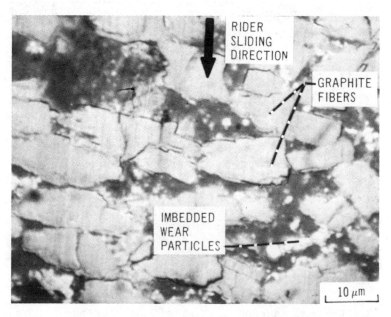

Figure 20. Photomicrographs of typical wear surface morphology at 300°C in moist air of GFRPI composites.

Geometry effects. In reference 27, the sliding specimen configuration was reversed and a hemispherically tipped GFRPI composite pin was slid against a 440C HT stainless steel disk counterface. The friction coefficients and composite wear rates were found to be highly dependent on sliding duration. Initial friction coefficients were slightly lower than those obtained on composite disks (0.15 vs 0.19) but as sliding duration increased, the friction coefficient rose to an average value of 0.27 or higher.

Average composite pin wear rates were initially (up to 20 km of sliding) lower than composite disk wear rates (2.5 x 10^{-15} m^3/m vs 13 x 13^{-15} m^3/m); but as sliding duration increased pin composite wear rate increased and after 19 km of sliding, composite pin wear rate was 41 x 10^{-15} m^3/m.

The reasons for these differences can be inferred from differences in wear surface morphology for a GFRPI pin sliding against a 440C HT disk after long sliding durations. Initially transfer films to the disk were very thin (similar to the opposite configuration), but as a sliding duration increased, thick nonshearing transfer occurred (Figure 21 (b)) which increased both friction and wear. There was considerable back transfer to the composite pin (Figure 21(a)), and the fibers and polyimide did not mix together to form a shear film as was the case for the opposite configuration.

The friction and wear data from pin-on-disk tests have been compared to data from an end use application, plain spherical bearing (Table II)[28-29]. It is interesting to note that results from the bearing tests compare very closely to the pin-on-disk tests when the metal pin slid against the GFRPI composite, but not when a composite pin slid against the metal counterface. These observations point out the importance of evaluating materials in a configuration that closely approximates the intended end use application.

Comparison to Other Polymers

The friction and wear properties of polyimide films, solid bodies, and composites are compared to some commercially available polymers in Table III. All were evaluated at Lewis Research Center under the same experimental conditions on a pin-on-disk tribometer.

Except for Ultra-High-Molecular-Weight Polyethylene (UHMWPE), the polyimide formulations are much superior to the others. UHMWPE is a material currently being used in artifical human joints[30]. It is an exceptionally good friction and wear material at ambient temperatures (25° C), but does not have high temperature stability.

Table III. Comparison of Average Friction Coefficients and Wear Rates of Some Commercially Available Materials to Polyimide Films, Solid Bodies, and Composite Materials Evaluated Under the Same Conditions.

[Pin-on-disk tribometer; hemispherically tipped pin, 440C stainless steel; films applied to 440C stainless steel disks or disks made of material evaluated; speed, 1000 rpm (2.7 m/s); load, 1 kg; controlled testing atmosphere of 50% RH air.]

Material Evaluated	Temperature (C°)	Film or Solid	Average Friction Coefficient	Average Wear Rate (m^3/m)
Polyphenylene sulfide ** 40% graphite fibers	25	solid	0.30	6200×10^{-16}
Poly (amide-imides) ** PTFE & Graphite powders	25	solid	0.37	1800×10^{-16}
UHMWPE **	25	solid	0.13	380×10^{-16}
Type "A" polyimide	25	solid	0.40	3500×10^{-16}
Type "V" polyimide **	25	solid	0.54	1000×10^{-16}
Type "V" polyimide ** 15% graphite powder	25	solid	0.37	1000×10^{-16}
Type "A" polyimide 50% graphite fibers	25 300	solid solid	.19 .05	130×10^{-16} 150×10^{-16}
Type PIC-1 polyimide	25 100 200 100*	film film film film	.13 .24 .07 .02*	4000×10^{-16} 3000×10^{-16} 400×10^{-16} 8×10^{-16}*
Type PIC-3 polyimide	25	film	.27	600×10^{-16}
PIC-1 bonded graphite fluoride powder ***	25	film	.22	30×10^{-16}

 * Test atmosphere, dry argon (<100 ppm H_2O)
 ** Commercially available materials
*** Lower contact stress (7 MPa)

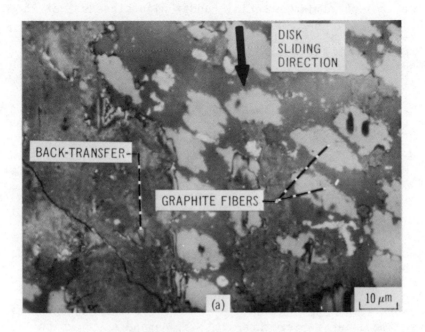

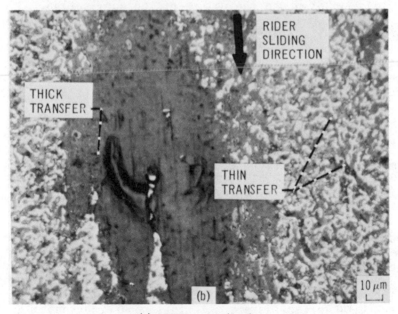

(a) GFRPI composite pin.

(b) Transfer to 440C HT steel counterface.

Figure 21. Photomicrographs of typical sliding surface morphology
for a GFRPI composite pin sliding against a 440C HT stainless steel
counterface in moist air at 25°C.

The best polyimide material (under high stresses) at 25° C was GFRPI. It had a higher friction coefficient than UHMWPE (0.19 vs 0.13) but wear rate was lower (130 x 10^{-16} vs 380 x 10^{-16} m^3/m). At 300° C, approximately the same wear rate for GFRPI was obtained, and the friction coefficient was much lower (0.05).

The best friction and wear properties (of any polymer) were obtained with the PIC-1 polyimide film at 100° C in dry argon. A friction coefficient of 0.02 and a wear rate of 8 x 10^{-16} m^3/m was obtained. Low wear rates were also obtained with PIC-1 polyimide-bonded graphite fluoride films at 25° C under low contact stresses (30 x 10^{-16} m^3/m).

CONCLUDING REMARKS

The tribological properties of polyimide films, polyimide solid bodies, polyimide-bonded solid lubricant films, and polyimide composites have been discussed and compared. The results indicate they have considerable promise for self-lubricating applications to temperatures of 350° C in air, however the upper temperature limit is dependent on which type of polyimide is used and the nature of the application.

In general, the polyimides tend to be brittle and wear by the brittle fracture of surface layers (up to 3 μm or more, depending upon the contact stresses applied). To obtain optimum lubrication with the polyimides it is believed that shear must be induced to occur in very thin surface layers (<1μm).

Some polyimides possess a transition temperature, above which the molecules obtain a degree of freedom necessary to plastically flow in these thin layers; but H$_2$O molecules from the atmosphere can hydrogen bond to the molecular chains and constrain their motion. That is, the transition is either masked in the presence of water vapor or translated to a higher temperature.

Thus, the tribological problem is to alter the polyimide in some manner to induce the formation of thin surface layers at temperatures below the transition. The addition of solid lubricants can help in this, but the solid lubricant must be compatible with the polyimide in order that they mix together to form very thin surface layers. In this regard, graphite fluoride (($CF_x)_n$) works well with polyimide in that the two mix together and shear in a very thin surface layer under light contact stresses. A disadvantage of adding powdered solid lubricants (such as (($CF_x)_n$) is that they tend to reduce the load carrying capacity of the film or composite, since they tend to agglomerate and have planes of easy shear.

Graphite fibers can be added to polyimide solids to increase the load carrying capacity by reinforcing the structure. In addition they can improve the friction and wear properties. Graphite fibers tend to dominate the tribological results, and different types of fibers can produce different results.

Polyimides, with or without solid lubricant additions, tend to produce thick, nonshearing transfer films which lead to increased friction and wear. It is believed this problem can be mitigated by proper additive formulation or by the proper sliding configuration design. Smaller contact areas and higher contact stresses tend to produce thinner transfer films.

REFERENCES

1. N. W. Todd and F. A. Wolff, Mater. Design Eng., 60 (2), 86 (1964).
2. C. E. Sroog, A. L. Endrey, S. V. Abramo, C. E. Berr, K. L. Olivier, and W. M. Edwards, J. Polym. Sci., Pt. A., 3 (4), 1373 (1965).
3. J. F. Hencock and C. E. Berr, SPE Trans., 5 (2), 105 (1965).
4. N. A. Adrova, M. I. Bessonov, L. A. Laius, and A. P. Rudakov, "Polyimides: A New Class of Thermally Stable Polymers," Progress in Materials Science Series, No. 7, Technomic Publishing Co., Inc., Stamford, Conn., 1970.
5. R. L. Fusaro, NASA TM-81381, 1980.
6. J. K. Lancaster, Tribology, 6 (6), 219 (1973).
7. F. J. Hermanek, Metal Prog. 97 (3), 104 (1970).
8. J. Theberge, in "ASLE Proceedings - International Conference on Solid Lubrication," SP-3, pp. 166-184, American Society of Lubrication Engineers, Park Ridge, IL., 1971.
9. R. D. Brown and W. R. Blackstone, in "Composite Materials; Testing and Design," ASTM STP-546, pp. 457, American Society for Testing and Materials, Philadelphia, PA., 1974.
10. H. E. Sliney and R. L. Johnson, NASA TN D-7078, 1972.
11. S. Bangs, Power Transm. Design, 15 (2), 27 (1973).
12. M. N. Gardos and B. D. McConnel, American Society of Lubrication Engineers Preprint No. 81-LC-3A-3, 1981.
13. J. P. Giltrow and J. K. Lancaster, in "International Conference on Carbon Fibres, Their Composites and Applications," Paper 31, Plastics Institute, London, 1971.
14. J. P. Giltrow and J. K. Lancaster, Nature, 214 (5093), 1106 (1967).
15. J. K. Lancaster, J. Phys. D., 1, 549 (1968).
16. R. A. Simon and S. P. Prosen, in "Twenty-Third Annual Technical Conference, SPI Reinforced Plastics Composite Division, Proceedings," Section 16-B, pp. 1-10, Society of the Plastics Industry, Inc., New York, 1968.
17. J. W. Herrick, in "Reinforced Plastics-Ever New; Proceedings of the Twenty-Eighth Annual Technical Conference," pp. 17-D1 to 17-D6, Society of the Plastics Industry, Inc., New York, 1973.

18. J. P. Giltrow and J. K. Lancaster, in "Tribology Convention, Pitlochry, Scotland, May 15-17, 1968, Proceedings," pp. 149, Institution of Mechanical Engineers, London, 1968.
19. M. N. Gardos and B. D. McConnell, American Society of Lubrication Engineers Preprint No. 81-LC-3A-4, 1981.
20. R. L. Fusaro and H. E. Sliney, ASLE Trans., 21 (4), 337 (1978).
21. R. L. Fusaro, ASLE Trans. 24 (2), 191 (1981).
22. R. L. Fusaro, NASA TP-1944, 1982.
23. R. L. Fusaro, ASLE Trans., 20 (1), 1 (1977).
24. R. L. Fusaro, ASLE Trans., 21 (2), 125 (1978).
25. R. L. Fusaro, NASA TN D-6714, 1972.
26. R. L. Fusaro, Wear, 75, 403 (1982).
27. R. L. Fusaro, American Society of Lubrication Engineers Preprint 82-AM-5A-2, 1982.
28. H. E. Sliney and T. P. Jacobson, NASA TP-1229, 1978.
29. H. E. Sliney, Lubr. Eng., 35 (9), 497 (1979).
30. D. A. Sonstegard, L. S. Matthews, and H. Kaufer, Sci. Am., 238 (1), 44 (1978).

ELECTROPHORETIC DEPOSITION OF POLYIMIDES: AN OVERVIEW

L. C. Scala, W. M. Alvino and T. J. Fuller

Westinghouse Electric Corporation
R&D Center
Pittsburgh, Pennsylvania 15235

Industrial aqueous electrodeposition (ED) of polymers has been known for at least 20 years. Although the process and concepts are well defined for aqueous systems, little has been published about non-aqueous electrodeposition. Since the utility of aqueous ED is limited to low performance polymers, new ways were sought to electrodeposit high performance resins so as to use them as coatings and films for a variety of electrical applications. In this paper, an overview is given covering the non-aqueous electrodeposition of polyimides and modifications thereof which include preformed imides, amic acids, ether-imides, amide-imides and ester-amide-imides.

Information is presented on emulsion preparation and properties, Faraday parameters (equivalent weight and coulombic yield), nature of the emulsion particle (size, charge, composition, mobility) and nature of the deposited coating. Practical applications of non-aqueous electrodeposition will also be reviewed.

The electrophoretic deposition of polyimides at constant voltage obeys Faraday's Laws of electrolysis. Deposition yield is a linear function of coulombs and is non-linear with respect to time. The yield and quality of the deposited polyimide are strongly dependent upon a number of parameters that are closely related to the emulsion composition.

These include polymer type and concentration, precipitant-solvent type, concentration and ratio, and type and concentration of surfactant.

The stability and nature of the emulsion are affected by the method of preparation.

The solids content of the deposited coating is 2,000 times greater than that of the starting emulsion. The solids content of the deposited coating increases with increasing voltage suggesting an electroendosmosis phenomenon.

Quantitative studies show that the effective equivalent weight of the deposited species ranges between 5,000-60,000 grams per Faraday and coulombic yields range from 50 to 300 milligrams per coulomb.

Successive depositions from the same emulsion result in a depletion of the electro-active species and show a linear decrease in yield. A decrease in coulombic yield and equivalent weight is also observed.

The emulsion particles can be visualized as micelles composed of many macromolecules. Both the micelles and the macromolecules are polydisperse in size and charge. Each micelle is distinguished by a certain precipitant, solvent and polymer concentration that is different from that of the bulk of the medium. These micelles range in size from $<0.1\mu$ to about 2μ. Electrophoretic mobility is of the order of 1×10^{-4} cm^2 v^{-1} sec^{-1} and micelle charge ranges from 5 to 120 charges per micelle.

The charged micelles migrate to the oppositely charged electrode to form a porous conductive coating whose porosity depends upon the polymer type and the emulsion components. The porosity in turn affects the resistance of the wet coating and hence the rate of film growth and final thickness.

Coatings of polyimide polymers have been electrodeposited onto a variety of electrical components, on metal circuit cores, uranium molybdenum compositions, and semiconductor devices.

INTRODUCTION

Although electrophoresis dates back to the early 1800's, industrial use of this concept has become a reality only within the past 20 years.[1-10] The electrophoretic deposition process consists of the migration of charged particles, molecules or micelles to the electrode of opposite charge under the influence of an electric field. Commercial processes (about 1,000 installations in the U.S. in 1980) are based exclusively on aqueous systems because of low cost, safety, ecological considerations and the ready availability of the required water dispersible polymers. The latter are, intrinsically, low performance materials (especially in terms of thermal stability), and, therefore, limit the scope of electrophoretic deposition. Our interest has been to develop an electrophoretic system capable of depositing high performance resins, specifically polyimide type polymers, so that their superior mechanical, electrical and thermal properties could be exploited for use as electrical insulation.

THEORY

In electrolysis the amount of chemical change produced by the application of electrical current to the electrolyte is proportional to the total amount of charge passed through the cell. Furthermore the amount of change produced is proportional to the equivalent weight of the substance undergoing that change. The chemical change produced by electrolysis may be the electrodeposition of a metal, the evolution of a gas, the oxidation or reduction of a species in solution or, in fact, any reaction that involves the transfer of a charge across an electrode interface. If the reaction is electrodeposition, then the amount of change is usually expressed as the weight of the deposit, and Faraday's Law may be expressed by the following Equation:

$$W = ItE/F \qquad (1)$$

where: W = weight of the deposit in grams
  I = current in amperes
  t = time in seconds
  E = equivalent weight of the substance undergoing deposition
  F = Faraday (96,500 coulombs).

The product "It" represents the total charge passed and is expressed in coulombs.

NON-AQUEOUS ELECTROPHORETIC SYSTEMS

Aqueous electrophoretic deposition as a coating technique has gained commercial acceptance largely because it provides a continuous uniform flaw free coating over the object to be coated. Some advantages are shown in Table I. The use of non-aqueous media offers a number of additional advantages (see Table II). One of the more important virtues is that water dispersibility is not required and a variety of polymer types can be electrodeposited.

Westinghouse has developed a non-aqueous electrophoretic system[11-14] that allows electrodeposition of a number of resins which could not be electrocoated otherwise. Table III is a partial list of the polymers that have been electrodeposited onto a variety of conductive substrates. All these polymers are polyimide-based and were thoroughly studied because of their importance in high performance electrical equipment as found in aircraft, for example. Typical heat-cured coatings on aluminum panels are shown in Figure 1. We are not limited to the polymers shown; in effect, if a suitable emulsion can be made, most polymers can be electrodeposited. Other polymer types investigated include polysulfones, epoxy, polyarylate, polyparabanic acid, polyhalostyrene, etc. The reason for the versatility of the non-aqueous system in terms of coatable resins rests on the fact that the process is nearly independent of the chemical structure of the resins but is strongly influenced by the characteristics of the emulsions made with them. Unlike aqueous systems, non-aqueous deposition can accommodate a variety of resin structures, is independent of pH, requires no special additives, does not involve gas evolution at the electrodes and produces films with high rupture voltage and thicknesses (0.1 to 5 mils).

PREPARATION AND PROPERTIES OF COATING SOLUTIONS

Non-aqueous electrophoretic deposition of polymers is generally carried out from emulsions. The emulsion is made by adding a solution of the resin to a precipitant, which is an organic liquid completely miscible with the solvent of the original resin solution, but which does not dissolve the resin. The result of this addition is the formation of micelles containing polymer, solvent and precipitant in various proportions depending on the nature of the ingredients. These dispersed droplets carry surface charges which are the result of adsorbed ions, dissociation of surface groups, impurities and frictional forces. Figure 2 is a schematic representation of an emulsion.

An electrical double layer is formed around the particles consisting of ions of one sign surrounded by ions of opposite sign. The existence of the double layer presupposes a charged particle

Table I. Advantages of Electrophoretic Deposition Over Dip or Spray Coating Procedures.

- Electric process easy to control, automate, program.
- Smooth films of constant thickness, no runs.
- Throwing power allows coating penetration into cracks, flaws, boxes, less accessible areas.
- Inherently a flaw-free process, gives 100% coverage.
- High resin utilization (up to 95%).
- Edge and point coverage.
- Negligible drag-out (short time between deposition and further handling).
- One shot coating.
- Low labor cost.

Table II. Advantages of Non-aqueous Electrophoretic Deposition of Polymers.

- Capability of depositing wide variety of high performance polymers.
- Simple bath composition, process and equipment.
- Capability for automated batch operation.
- Flaw-free coatings.
- Edge and point coverage.
- Deposited coatings free of conducting impurities.
- Recoverable and reusable bath components.
- No special metal substrate pre-treatment.
- High electrocoating efficiency.
- No gas evolution at electrode.
- Deposited films have high rupture voltage.

Table III. Resin Types.

Code Name	Polymer Type	Supplier
Pyre ML	Polyamic acid	DuPont
Ultem-1000	Polyether-imide	General Electric
PI-2080	Polyimide	Upjohn
Torlon	Polyamide-imide	Amoco
PDG-981	Polyamide-imide	P.D. George Co.
Omega	Polyester-amide-imide	Westinghouse
AI-10H	Polyamide-imide	Amoco

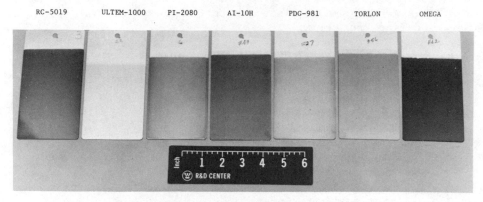

Figure 1.   Electrodeposited coatings on aluminum panels.

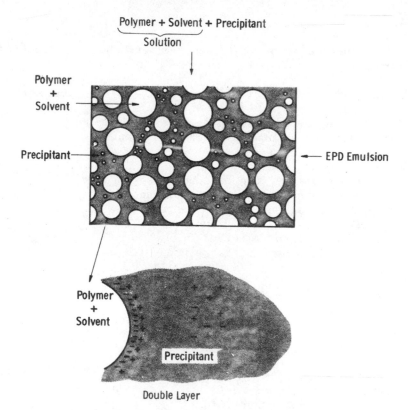

Figure 2.   Model of emulsion formation.

and is well documented in the literature and can be found in any physical or colloid chemistry textbook.[17] It essentially describes the electrical condition of a surface at the interface of two phases. As a result of this, a potential difference between the fixed and diffuse layers is created. When a potential is applied, both the particle and the inner portion of the double layer are affected causing the particle to move in one direction and the outer surface of the double layer in the other direction. The shear force created causes separation of the double layer and the charges in the double layer give rise to an electrical potential whose value at the shear surface is the zeta potential. This affects dispersion stability and electrophoretic mobility.

Simple equipment is required and the preparation of a typical emulsion is shown in Figure 3. Surface active agents are generally not needed so that the deposited coating contains only the original resin uncontaminated by conductive impurities. Some emulsion properties are shown in Table IV. In general, emulsion stability depends upon solids content, solvent/precipitant ratio and type, and preparation technique. Visually the emulsions are transparent or opaque and exhibit a variety of colors (see Figure 4).

MICELLE COMPOSITION

Analysis of the colloidal dispersions by light scattering, sedimentation and spectroscopic techniques indicates that the micelles are polydisperse in size ($0.03-2\mu$) and are solvent rich aggregates of high molecular weight polymer molecules. A dark field photomicrograph of the emulsion particles is shown in Figure 5. Magnification is 800X and the particle size range is between 0.4 and 1.6 microns. The colloidal particles are composed of from 12 to 60% resin solids with the remainder of the composition being divided between solvent and precipitant. Each micelle is estimated to have between 5 and 120 negative charges.

The size of a micelle is huge compared to that of an ion. Consider for example a chloride ion whose radius is 0.1 nm, and a macromolecule of hemoglobin (Mw = 68,000) whose radius is 3 nm. The radius of a micelle is about 100 nm (polyamide-imide emulsion). Given these dimensions, the total volume of a micelle would contain about 36,000 hemoglobin molecules (Figure 6). Any variation in the size of the micelles would have an enormous effect on the mass of material being deposited and hence on the apparent equivalent weight of the electroactive species. Table V presents some data on the micelle properties of various polyimide emulsions.

Figure 3. Emulsion preparation.

Table IV. Emulsion Properties.

Particle Size   — <0.1–2.0 μ
Particle Shape   — Spherical
Stability   — Hours to Months
Appearance   — Clear to Opaque White
Bath Conductance   — 10^{-3} to 10^{-6} mhos/cm
Apparent pH   — 4–12

Figure 4. Non-aqueous polymer emulsions.

Figure 5. Darkfield photomicrograph of emulsion particles.
Mag. 800X. Particle size range: 400-1700 nm.

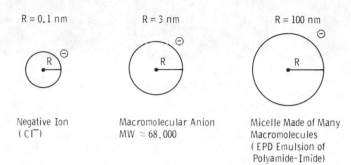

Negative Ion
(Cl⁻)

Macromolecular Anion
MW ≈ 68,000

Micelle Made of Many
Macromolecules
(EPD Emulsion of
Polyamide-Imide)

- Micelle Has High Mass to Charge Ratio

Figure 6.   Volume of 36,000 macromolecules = 1 micelle.

Table V.   Micelle Properties.

Resin	Radius, Å	No. of Apparent Negative Charges	Electrophoretic Mobility, cm^2/v-sec x 10^4	Zeta Potential, mV
AI-10H	220–1,800	15–65	0.5–1.1	11–29
Torlon	130–780	7–36	1.2–1.3	14–21
PI-2080	100–290	6–50	1.6–4.9	27–53
Pyre ML	120–660	5–120	0,7–5	12–55
Ultem	230–1,440	4–27	0.5	9
PDG-981	190–620	10–30	1.2	13

PROCESS CONDITIONS

Table VI lists ranges of operating conditions for the non-aqueous electrodeposition of polyimide type polymers. The experimental apparatus includes a DC power supply, emulsion tank, and assorted measuring equipment, such as voltmeter, ammeter, coulometer, recorder, etc. A schematic of the deposition cells we used is shown in Figure 7.

DEPOSITION FROM POLYMER SOLUTIONS AND EMULSIONS

An emulsion gives higher deposition yield. Electrocoated yields from polymer solutions are negligible (<3 mg) due to the highly passivating nature of the deposited polymer molecules and to the redissolution of the deposited coating. In addition, deposition from polymer solutions is limited only to those that contain ionizable groups since only electrically charged molecules are capable of electrophoresis in non-aqueous solutions. In emulsions the precipitant acts as a coagulant and retards the dissolution of the deposited coating. As with emulsion stability, the deposition yields are dependent on precipitant/solvent ratio and solids content. The amount of material deposited is negligible below the cloud point of the solution. As micelle concentration increases above the cloud point, the EPD yield increases and passes through a maximum. This maximum represents the optimum balance between precipitant, solvent and emulsion solids. Yields decrease at high P/S ratios and high solids due to instability of the emulsion caused by coagulation of the micelles which leads to precipitation of the particles from the emulsion.

DEPOSITION PARAMETERS

Effect of Coulombs. The electrodeposition of polymers, like that of metals, obeys Faraday's Law. Figure 8 shows the relationship between yield of polymer deposited and charge passed through the emulsion for a number of polyimide-based polymers.

Since the deposition process obeys Faraday's Law, we can calculate an effective equivalent weight (E) of the electroactive species being deposited. Table VII shows the range of E and Z (coulombic yield) values calculated for various polymer systems. These values are independent of the applied potential, provided the composition of emulsion does not change. They are, however, dependent on how the emulsion is prepared, because the interactions of the polymer with the dispersion medium affect both the micelle size and charge. In the emulsion we can have particles with few or many charges, depending on the properties of the emulsion medium.

Table VI. EPD Process Conditions.

Voltage	–	25–300 Vdc
Current Density, ma/in.^{2}	–	0.5–5
Coulombic Yield, mg/c	–	60–260
Deposition Time	–	0.4–30 Minutes
Electrode, Bias	–	Anodic
Electrode Material	–	Copper, Aluminum, Brass, Steel, Nickel

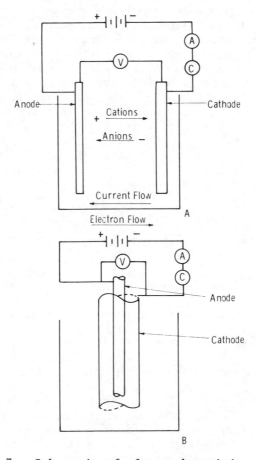

Figure 7. Schematic of electrodeposition cells.

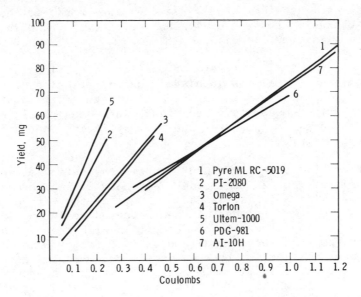

Figure 8. Effect of coulombs on yield of polymer deposited at anode.

Table VII. Faraday Parameters.

Polymer	E mg/Faraday	Z mg/Coulomb
Pyre ML	7,000	75
PI-2080	18,000	190
Ultem	22,000	230
AI-10H	7,000	75
Torlon	12,000	120
PDG-981	6,800	70
Omega	9,600	100

Assuming the mass of the particle is constant, the equivalent weight will be low if the particle has many charges (more coulombs passed per weight of deposit). On the other hand, if the particle has few charges, then the equivalent weight will be high. Thus the process is more efficient.

Effect of Time. The rate of polymer deposition is constant (see Figure 9), until the insulating characteristics of the deposited polymer start taking effect. As the film thickness of the deposit increases, film resistance increases resulting in a greater potential drop across the film affecting the rate of film growth (see Figure 10).

Effect of Voltage. Figure 11 shows the effect of voltage on the deposited yield at the anode. As the voltage increases the mass of material being deposited, as well as the rate of deposition, increase. However, because of the increased resistance of the deposited coating, the rate of deposition at higher voltages decreases from the theoretical amount by a factor whose magnitude depends upon the polymer type and emulsion composition. As an example, for the polyamic-acid polymer (Pyre ML) the observed slope values differ from the theoretical ones by a factor of about 1.2 (see Table VIII and Figure 11).

Effect of Solids. Figure 12 shows a linear relationship between the concentration of polymer in the emulsion and the yield of polymer deposited at the anode. On the other hand, successive depositions from the same emulsion result in decreased yields due to the fact that the micelle concentration is being depleted; this results in a change in the number of charge carrying species. Emulsion droplets might acquire different charge to mass ratios, and those with the highest ratio are expected to migrate and deposit faster than others. So what we might have is a fractionation based on charge to mass ratio. This depletion affects the Faraday constants E and Z. In Faraday's classic studies, metal ions lost by coating on one electrode are replenished by the other electrode. As a result the metal ion concentration in the plating solution remains constant. Resin is not replenished in emulsions after deposition, thus it is not surprising that electrodeposition of resins shows a concentration dependence.

Electrode Type. Aluminum, steel, brass, nickel, platinum, have all been used effectively as electrode materials. Each material affects the deposition process in terms of deposited yield and current. The actual shape and magnitude of the current decay is controlled by many factors (emulsion ingredients, polymer type, etc.). Typical curves are shown in Figure 13 for some polyimide type polymers electrodeposited on aluminum panels. Some measure

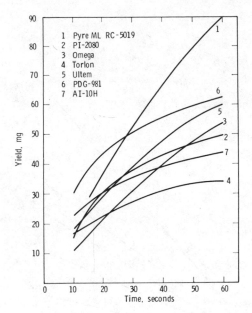

Figure 9. Effect of deposition time on yield of polymer deposited at anode.

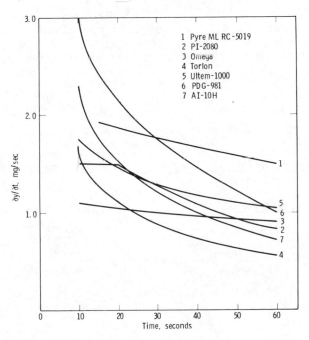

Figure 10. Effect of deposition time on the rate of polymer deposited at the anode.

1095

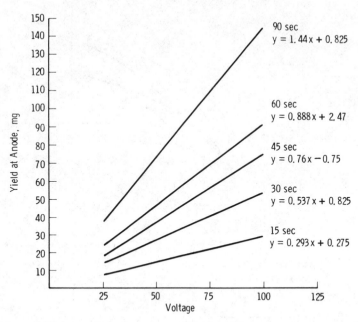

Figure 11. Effect of voltage on yield of polyamic acid deposited at anode for various times.

Table VIII. Deposition Rates From Figure 11.

Observed Slope	Theoretical Slope
0.293	0.293
0.537	0.586
0.76	0.879
0.888	1.172
1.44	1.758

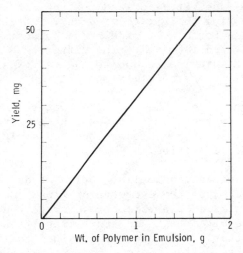

Figure 12. Yield deposited on anode versus weight of polymer in emulsion (Polyamic acid).

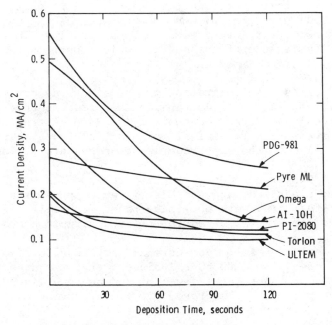

Figure 13. Current decay during deposition of polymers on aluminum panels.

of the porosity of the wet deposited coating can be obtained from these curves by noting their rate of decay. The effect of other electrode materials for a polyimide and a polyamide-imide is shown in Figure 14 and Table IX. The differences in the top and bottom curves in Figure 14 are due to the use of a different precipitant in preparing the emulsions.

Coating Composition. The deposited wet films of different resin emulsions vary greatly in solids content, from less than 15% to more than 50% (Table X). Electroendosmosis plays a part in non-aqueous electrodeposition.

The precipitant/solvent weight ratios of deposited coatings are greatly reduced from those of the original emulsions. When centrifugation was used to physically separate the suspended phase, a polymer phase resulted that had approximately the same solids concentration as that of wet electrodeposited coating. The wet deposit is a highly swollen, porous coating which can be fused into a continuous flaw free coating after elimination of precipitant and solvent. Nearly all systems studied show a marked increase in coated solids content as voltage is raised from 50 Vdc to 200 Vdc (Figure 15).

Coating Process. Figure 16 shows a conception of the deposition of charged micelles which migrate to the oppositely charged electrode to form a porous conductive coating that is high in solids compared to the original emulsion. The porosity of the coating depends upon the polymer type as well as the emulsion ingredients and their interactions. This porosity, in turn, affects the resistance of the coating which controls the rate of film growth.

The micelles after migration to the electrode coagulate and adhere to its surface by a destabilization process that overcomes the small repulsive forces on the micelles (concentration coagulation). A direct electrochemical reaction involving charge transfer from the micelle to the electrode is not necessary for coagulation.

Type of Substrates Coated. Using the non-aqueous EPD process a variety of electrical and non-electrical components have been coated satisfactorily with polyimide-based materials. Figure 17, for example, shows metal cores for printed circuit boards.

Other Possibilities. Some preliminary results of our experiments with non-aqueous EPD have indicated that the deposited coating is higher in molecular weight and narrower in MWD than the starting polymer. Therefore, EPD might provide a way of upgrading polymer properties.

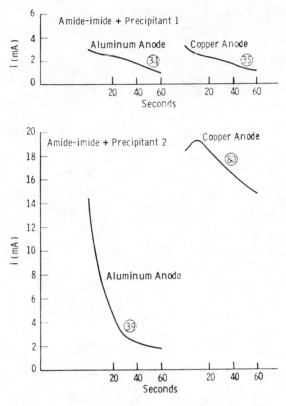

Figure 14. Typical current decay curves - nickel screen cathode, various anodes. (All curves: rod anodes, 100 Vdc); circled nos. are yields in mg. Two different precipitants used.

Table IX. Effect of Electrode Material.

Polymer	Electrode	Deposited Yield, mg	Z mg/c
AI-10H	Al	39	135
	Cu	80	81
	Ni	45	136
	Iron	40	119
	Brass	50	78
Pyre ML	Al	30	127
	Cu	34	140

Table X.   Coating Composition.

Polymer	Solids Content of Wet Deposited Coating, %
Pyre ML	12–33
PI–2080	21–32
Ultem	25–40
AI–10H	27–44
Torlon	38–60
PDG–981	48–57
Omega	45–53

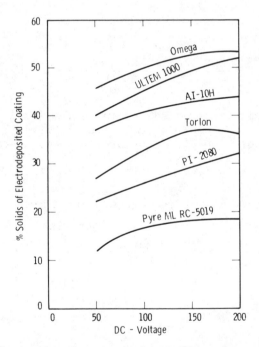

Figure 15.   Effect of voltage on the solids content of electro-deposited coating.

CHARGED EMULSION PARTICLES

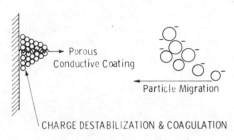

→ Porous
Conductive Coating

Particle Migration

CHARGE DESTABILIZATION & COAGULATION

Figure 16. Schematic of charged emulsion droplets.

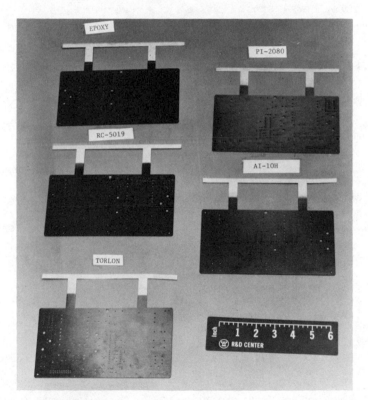

Figure 17. Electrodeposited coatings on metal core substrates.

CONCLUSIONS

   While the electrodeposition of polymers is not a new process,
deposition from non-aqueous media offers some very interesting
possibilities. It extends the range of polymers that can be used
beyond those used in aqueous systems. It offers a way of
increasing thickness of the deposited coating. It could provide
a way of upgrading the properties of film forming polymers. And
it offers a way of processing high temperature polymers not
processable by conventional means. Furthermore, it presents a
promising area for fundamental studies in the colloid and
electrochemical areas.

REFERENCES

1. H. U. Schenck, H. Spoor, M. Marx, Prog. Organic Coatings,
    7, 1 (1979).
2. J. Mazia, Metal Finishing, p. 109, February 1981.
3. R. H. Chandler, "Advances In Electrophoretic Painting,
    1975-1977", R. H. Chandler, Ltd., Braintree, Essex, England,
    1978.
4. G.E.F. Brewer, Encyclopedia of Polymer Science & Technology,
    Vol. 15, p. 178, Interscience, John Wiley & Sons, Inc., New
    York, N.Y (1971).
5. G. E. F. Brewer, in"Electrodeposition of Coatings", G.E.F.
    Brewer, Editor, Adv. Chem. Series No. 119, American Chemical
    Society, Washington, D.C., 1973.
6. B. A. Cooke, N. M. Ness, L. Palliel, "The Electrodeposition of
    Paint", Chapter 11 in Industrial Electrochemical Processes,
    Elsevier Pub. Co., 1971.
7. C. Fink, M. Feinleib, Transactions of Electrochemical
    Society, Vol. 94, p. 309, 1948.
8. A. R. H. Tawn, J. R. Berry, Oil & Colour Chemist's Associa-
    tion Journal, Vol. 48, p. 790 (1965).
9. P. Pierce, Z. Kovac, C. Higginbotham, Ind. Eng. Chem. Prod.
    Res. Dev., Vol. 17, No. 4, p. 317 (1978).
10. F. Beck, "Fundamental Aspects of Electrodeposition of Paint",
    in Progress in Organic Coatings, 4, p. 1, Elsevier Sequoia
    S.A., Lausanne, Netherlands (1976).
11. W. M. Alvino and L. C. Scala, J. Appl. Polym. Sci., 27, 341
    (1982).
12. W. M. Alvino, L. C. Scala, T. J. Fuller, J. Appl. Polym. Sci.,
    28, 267 (1983).
13. L. C. Scala, W. M. Alvino, T. J. Fuller, Paper Presented at the
    the IPC Meeting, Boston, MA, 1982. Paper No. IPC-TP-421.
    "A Non-Aqueous Electrophoretic Deposition Process". High
    Performance Polymers As Coatings For Metal Core PCB's.
14. D. C. Phillips, J. Electrochem. Soc., 119, No. 12, 1645 (1972).
15. S. Glasstone, "Elements of Phys. Chem.", D. Van Nostrand,
    Inc., 1946, NY, NY.

EVALUATION OF HIGH TEMPERATURE STRUCTURAL ADHESIVES FOR EXTENDED SERVICE

C. L. Hendricks and S. G. Hill

Boeing Aerospace Company
P.O. Box 3999
Seattle, Washington 98124

High temperature stable adhesive systems were evaluated for potential Supersonic Cruise Research (SCR) vehicle applications. The program was divided into two major phases; Phase I "Adhesive Screening" evaluated eleven selected polyimide (PI) and polyphenylquinoxaline (PPQ) adhesive resins using eight different titanium (6Al-4V) adherend surface preparations; Phase II "Adhesive Optimization and Characterization" extensively evaluated two adhesive systems, selected from Phase I studies, for chemical characterization and environmental durability. The adhesive systems which exhibited superior thermal and environmental bond properties were LARC-TPI polyimide and polyphenylquinoxaline both developed at NASA Langley. The latter adhesive system did develop bond failures at extended thermal aging due primarily to incompatibility between the surface preparation and the polymer. However, this study did demonstrate that suitable adhesive systems are available for extended supersonic cruise vehicle design applications.

INTRODUCTION

The increased demands upon aerospace vehicles for improvements in structural efficiency, performance, and durability has pushed to the limit the capabilities of conventional materials and designs. One area of materials and design that can provide significant improvements in airframe structures is the use of structural bonding. This method of joining is highly resistant to fatigue damage and offers the potential for lower production costs. For supersonic cruise vehicles that experience elevated temperatures to 450°F for extended times, this technology has lacked the materials and the data to adequately demonstrate its feasibility.

EXPERIMENTAL

The program described in this paper, funded by NASA Langley Research Center,[1] extensively evaluated all available high temperature stable resins as adhesives and all production practical titanium surface preparations to advance the state-of-knowledge and technology of high temperature metal bonding. Previously developed adhesive systems and studies failed to solve a combination of problems which included lack of adequate thermal stability, impractical process requirements and/or lack of ability to be processed into large area bond assemblies. This program had as its primary objective the evaluation, test, and demonstration of high temperature structural adhesives for supersonic vehicle applications. Efforts were concentrated into two phases; Phase I, "Adhesive Screening," and Phase II, "Adhesive Optimization and Characterization." Phase I evaluated eleven adhesive resins and eight Titanium (6Al-4V) surface preparations. The candidate adhesives studied were LARC-13 (NASA Langley), NR150A2 (DuPont), NR150B2 (DuPont), NR056X (DuPont), HR-602 (Hughes), LARC-13 with amide-imide modification (Boeing), LARC-13 with methylnadic capping and meta phenylene diamine modification (Boeing), Polyphenylquinoxaline (NASA Langley), PPQ with boron powder modification (Boeing), LARC-TPI (NASA Langley) and FM-34 (American Cyanamid) used as a baseline to verify the test values of the experimental adhesives.

The titanium surface preparations evaluated were 5 volt chromic acid anodize (Boeing BAC 5890), 10 volt chromic acid anodize (Boeing BAC 5890), phosphoric acid anodize (Boeing BAC 5555), Pasa-Jell 107, Phosphate fluoride (Boeing BAC 5514), Phosphate fluoride (Picatinny Modified), phosphate fluoride with grit blast (Boeing BAC 5514 modified), Turco 5578 alkaline etch, and the British RAE H_2O_2/NaOH process. Lap Shear Strength and Crack extension were the primary tests used to judge adhesive bond properties. In addition, titanium surface preparation characterization and test coupon adhesive failure analyses were conducted using SEM, STEM, and Auger techniques. These analyses were performed to more fully understand the mechanisms for bond or disbond.

RESULTS AND DISCUSSION

Test results from Phase I (based upon a combination of bond strength, processibility, and availability) were used to select LARC-13, PPQ, NRO56X, and LARC-TPI for additional evalua- tion in Phase II.

Phase II initial activities were to chemically charac- terize each of the adhesive resins. Three different batches of LARC-13, PPQ, and NRO56X resins were analysed using High Pressure Liquid Chromatography (HPLC) and Infrared Techniques. In general, all the adhesive resins exhibited relatively low variability from batch-to-batch.

To further characterize each adhesive resin, lap shear bonds were prepared using three different cure cycles. These coupons were tested at initial ambient and at 505K after extended (to 3000 hours) exposure to 505K. The cure cycles selected concentrated on varying upper dwell temperature. In this study the dwell temperatures were 588K (600°F), 602K (625°F), and 616K (650°F).

At this point in the program both LARC-13 and NRO56X were eliminated from any future study. LARC-13 was dropped because Phase I long term 505K exposure data showed a steady reduction in lap shear strength attributed to thermal degradation. NRO56X was withdrawn from the market by DuPont and this made further evaluation questionable since it could not support future potential production demand. The 602K cure cycle was selected as most optimum for PPQ (highest strength values) and the 616K cure cycle was selected for LARC-TPI (also highest strength values).

The next major effort in Phase II was to determine the environmental durability of the remaining PPQ and LARC-TPI systems.

Lap shear and crack extension coupons were prepared for these systems using their respective selected cure cycles and subjected to environmental exposure and test.

Unstressed thermal exposure was conducted on the two remaining systems. For this study lap shear coupons were exposed to 505K for up to 10,000 hours. The PPQ adhesive system was tested after 100, 500, 1000, and 5000 hours. A separate set of coupons (5 each set) was tested at 219K (-65°F), ambient, 422K (300°F) and 505K after exposure. The 10,000 hour 505K coupons were too weak to test.

All test values for PPQ showed a relative rapid decline during the first 5,000 hours exposure. The horizontal dashed line in plots (Figures 1-6) represents the minimum acceptable structural bond strength for design ($\sim$14 MPa).

The PPQ failures were characterized as gradually changing from cohesive to adhesive. This type of thermal failure also appeared in Phase I exposure and test. Surface failure analysis had shown that an incompatibility between PPQ resin and the titanium oxide layer had occurred. The PPQ resin was pulling clear from the oxide. Conclusions at this point were that the polymer was thermally stable but an unknown factor was creating an incompatibility with the anodized prepared oxide.

The results for unstressed thermal (505K) aging for LARC-TPI were much different. This adhesive exhibits stability to 9500 hours exposure. The data for expected 20,000 hours exposure have not been completed at the writing of this paper. All LARC-TPI lap bond failures have been cohesive.

This adhesive system is considered to be stable at extended 505K exposure conditions based on these test results.

Stressed bond exposure testing was performed on each system. PPQ and LARC-TPI lap shear coupons were subjected to 219K, ambient, and 505K while stressed at 25% of ultimate. Additional coupons were stressed at 30% of ultimate at 219K and ambient temperatures, and at 50% while exposed to 505K. Stressed coupons from each group were tested to failure at their exposure temperature after 100, 1000, 3000, and 5000 hours. For PPQ the 219K and ambient exposed shear strengths were relatively unchanged from initial test values. However, the 505K bond strengths exhibited a steady decline in strength and at the 5000 hour test had dropped to about 11 MPa (below structural 14.5 MPa requirements).

The test results for LARC-TPI were comparable to PPQ at the 219K and ambient exposure conditions. At 505K the lap shear strengths at all stress levels remained significantly above 14.5 MPa.

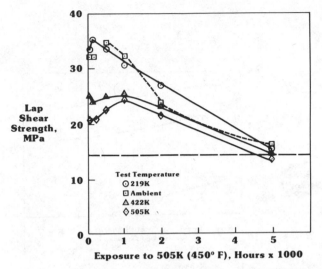

Figure 1. Unstressed Lap Shear Strength of PPQ Exposed to 505K as a Function of Time.

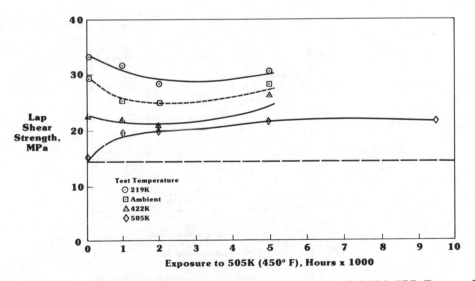

Figure 2. Unstressed Lap Shear Strength of LARC–TPI Exposed to 505K as a Function of Time.

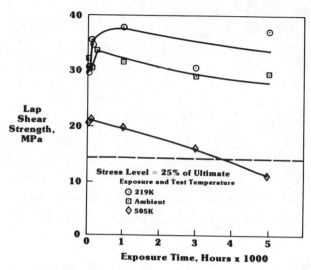

Figure 3. Stressed Lap Shear Strength of PPQ Exposed to 505K as a Function of Time. Stressed at 25% of Ultimate.

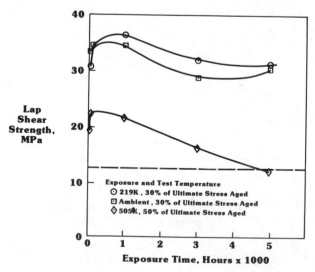

Figure 4. Stressed Lap Shear Strength of PPQ Exposed to 505K as a Function of Time. Stressed at 30 & 50% of Ultimate.

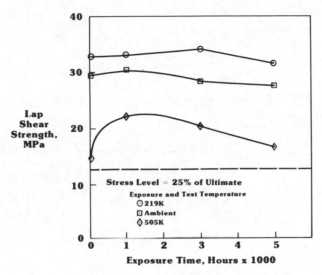

Figure 5. Stressed Lap Shear Strength of LARC-TPI Exposed to 505K as a Function of Time. Stressed at 25% of Ultimate.

Conolusions from these tests were that 219K and ambient exposure has no significant effect upon PPQ and LARC-TPI bond strengths. PPQ is affected by the 505K exposure after 5000 hours exposure whether stressed or unstressed. LARC-TPI exhibited bond stability at all temperatures and stress levels.

Humidity exposure was conducted also in this study (Figures 7-9). The adhesive systems, PPQ and LARC-TPI, were subjected to 332K/95% relative humidity for up to 2000 hours using lap shear and crack extension coupons. The coupons were tested (lap shear) or measured (crack extension) after 1000 and 2000 hours exposure. Both PPQ and LARC-TPI do not show any shear strength degradation when tested at 219K. LARC-TPI strengths appear to drop more rapidly than PPQ when tested at ambient after exposure. Both systems drop in strength to the same degree when tested at 505K after exposure. Crack extension results appear to show about the same crack growth rate for PPQ and LARC-TPI at the 1000 to 2000 exposure times. General conclusion from these test results is that both adhesive systems show moderate bond strength loss when tested at 505K after high humidity exposure.

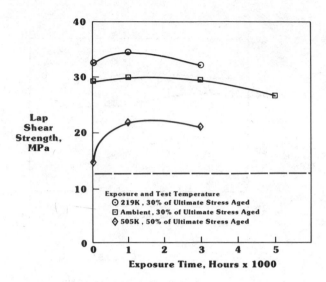

Figure 6. Stressed Lap Shear Strength of LARC-TPI Exposed to 505K as a Function of Time. Stressed at 30 and 50% of Ultimate.

To further establish the durability of PPQ and LARC-TPI adhesive bonds; both systems were exposed to various fuels and fluids (Figures 10-13) using lap shear and crack extension coupons. The fuels and fluids used were JP-4 (ambient exposure), MIL-H-5606 (ambient and 334K exposure), skydrol (ambient and 334 exposure) MIL-H-7807 (ambient exposure), and de-icing fluid (ambient exposure). Test coupons were tested (lap shear) or measured (crack extension) after 30 days, 60 days, and 5000 hours exposure.

As shown in Figures 12 and 13 the only significant crack growth occurred on exposure to skydrol. This is not unexpected since this fluid is known to attack a wide variety of polymers. PPQ was especially sensitive to hot skydrol. LARC-TPI did show some crack growth in hot skydrol but at much lower levels than for PPQ.

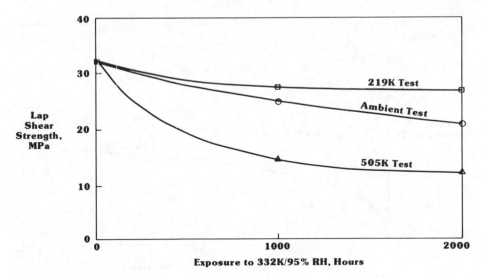

Figure 7. Lap Shear Strength of PPQ on Exposure to Humidity as a Function of Time.

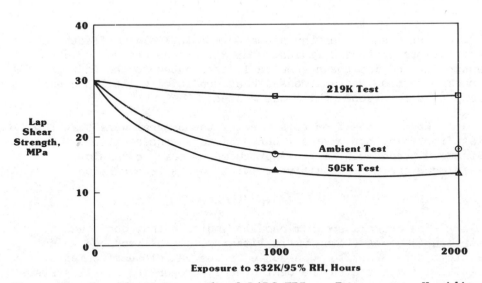

Figure 8. Lap Shear Strength of LARC-TPI on Exposure to Humidity as a Function of Time.

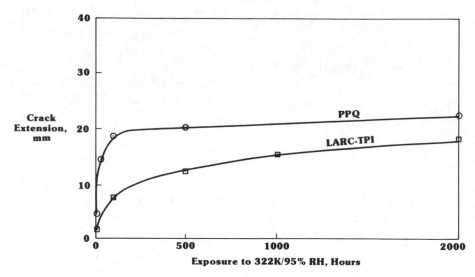

Figure 9. Crack Extension of PPQ and LARC-TPI on Exposure to Humidity as a Function of Time.

Lap shear strength values were relatively unaffected by the various fuels and fluids. The difference between crack extension and lap shear results is attributed to the fact that crack extension was a stressed bond condition while lap shear coupons were exposed while unstressed.

Conclusions from this part of the program were that PPQ is affected by hot skydrol; LARC-TPI to a lesser extent. Stressed bonds such as those produced by crack extension specimens are more sensitive to fluid exposure conditions.

CONCLUSION

The overall program results indicate that good quality structural bonds (titanium-to-titanium) can be produced which exhibit durability to supersonic cruise vehicle environments. LARC-TPI demonstrated the best overall properties of all systems evaluated. PPQ with proper metal surface preparation could possibly also perform as desired in these environments.

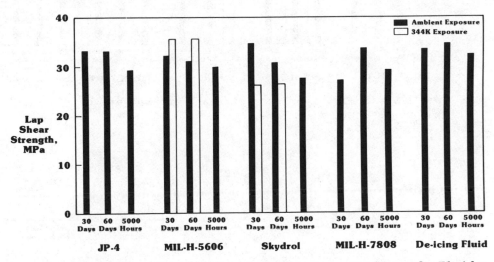

Figure 10. PPQ Lap Shear Strength on Exposure to Aircraft Fluids as a Function of Time.

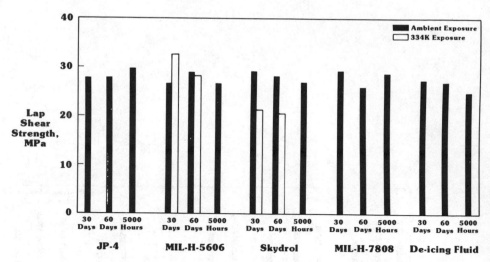

Figure 11. LARC–TPI Lap Shear Strength on Exposure to Aircraft Fluids as a Function of Time.

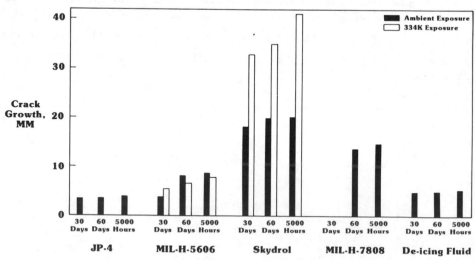

Figure 12. Crack Extension of PPQ Exposed to Aircraft Fluids as a Function of Time.

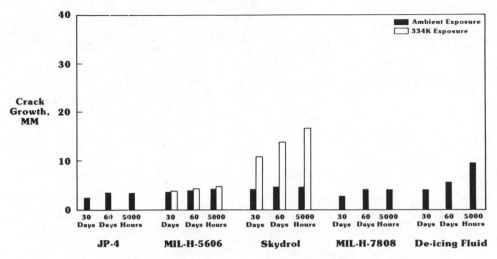

Figure 13. Crack Extension of LARC-TPI Exposed to Aircraft
Fluids as a Function of Time.

ACKNOWLEDGEMENTS

Support of this work by the NASA Langley Research Center
is gratefully acknowledged.

REFERENCES

1.   C. Hendricks, S. G. Hill and P. D. Peters, "Evaluation of
       High Temperature Structural Adhesives for Extended Services",
       NASA Contractor Report 165944, July 1982.
2.   C. L. Hendricks and S. G. Hill, "Evaluation of High Tempera-
       ture Structural Adhesives for Extended Service", SAMPE
       National Symposium, Vol. 25, pp. 39-55, May 1980.

POLYIMIDE PROTECTIVE COATINGS FOR 700° SERVICE

R. J. Jones, G. E. Chang, S. H. Powell and H. E. Green

TRW Energy Development Group
One Space Park
Redondo Beach, CA 90278

TRW has discovered that partially fluorinated polyimide resins prepared from 2,2-bis[4-(4-aminophenoxy)-
phenyl]hexafluoropropane (4-BDAF) possess great potential
for long term service in highly oxidative environments at
temperatures up to and including 700°F. The 4-BDAF
monomer and polyimides therefrom were discovered in 1975
on NASA/ Lewis Research Center Contract NAS3-17824.

In 1981, TRW initiated a joint study with the Alameda, California Naval Air Rework Facility to assess
whether the polyimides may be suitable as a salt spray
corrosion resistant coating for jet engine compressor
components. A polyimide prepared from 4-BDAF was evaluated as a protective coating for Inconel and titanium specimens possessing representative plasma sprayed metallic
coatings. The polyimide coated metallic test specimens
were subjected to a two hour muffle furnace bake at 900°F
in air, followed by a five hundred hour salt spray exposure at ambient room temperature. At the conclusion of
the Navy Test, inspection showed that essentially zero
corrosion occurred on specimens coated with the partially
fluorinated polyimide.

Promising results were also obtained on assessment
of the polyimides as a thermo-oxidated protective coating
for commercially available polyimide composites intended
for ≥ 5000 hour service at temperatures up to 550°F in
air. During the composite coating assessment, a solution
spray process for the polyimides was developed.

1. INTRODUCTION

Linear aromatic/heterocyclic condensation polyimides have been an item of commerce for almost two decades. These polymers have found widespread acceptance for many aerospace applications including matrices for high performance specialty seals, radomes, wire coatings and other electrical product applications. The commercially available linear polyimides are very tough in final form versus their thermosetting counterparts. They are normally suggested for use in air to a maximum of 450°F to 600°F for several hundred hour durations. Until recently, the upper temperature stability of these polymers was suitable to meet most product application requirements. However, advanced aircraft, engine and missile designs require that a second generation of linear polyimides be developed that are capable of a minimum of a hundred hour service in air at 700°F and thousands of hours at 450°F to 600°F. A new class of linear polyimides under development at TRW have demonstrated initial promise as a viable second generation candidate. The results of initial coating product application studies on the new TRW partially fluorinated polyimides constitute the subject of this paper.

A series of linear condensation polyimides have appeared on the market, beginning in the 1960's, for high performance resin applications. The polymer systems that have endured as viable commercial systems include such resins as Skybond polyimides from Monsanto,[1] and the Vespel and Kapton polyimide products available from Du Pont,[2,3]. It is these polyimides which best exemplify the generic type of resin which include the TRW partially fluorinated polyimides.

For reference purposes, other linear condensation polyimides such as high performance NR-150B materials from Du Pont,[4] and the flexible polyimides developed at TRW,[5] have shown novel combination properties. However, because neither is generally available as an item of commerce, no further consideration will be given to these polyimides. Because high performance thermosetting polyimides are generally not considered for coating applications, these materials will also not be elaborated on.

Linear aromatic/heterocyclic condensation polyimides can best be described as the reaction products of an aromatic tetracid or derivative (the dianhydride or diesters are commonly used) and an aromatic diamine. The solvents commonly employed to promote the reaction are aprotic in nature which normally include dimethyl formamide, dimethyl acetamide, hexamethyl phosphoramide and N-methyl pyrrolidinone. The reaction is normally run at or near ambient room temperature employing an ingredient to solvent loading of up to 40% by weight of the aromatic ingredients. The linear condensation polymer formed in solution from the equal molar quantities of a dianhydride and a diamine is known as a polyamide-acid. The polyimide is then formed by solvent removal and imidization which typically occurs at or below 400°F, depending upon whether or not reduced pressure is employed. A representative general reaction employed to form linear condensation polyimides is presented in Equation 1.

The most frequently used aromatic dianhydrides are pyromellitic dianhydride and benzophenone tetracarboxylic acid dianhydride. The dianhydrides are commonly abbreviated PMDA and BTDA, respectively. The TRW partially fluorinated polyimides employ these two anhydrides either alone or in combination to tailor specific resin properties.

Commercially available linear condensation polyimides normally employ one and two ring aromatic diamines such as *meta*-phenylene diamine, methylene dianiline and oxydianiline. The TRW polymers are based upon a novel four-ring aromatic diamine which is described below.

TRW has discovered that use of partially fluorinated aromatic diamine 2,2-bis[4-(4-aminophenoxy)phenyl]hexafluoropropane 4-BDAF see structure below) yields linear condensation polyimides possessing greater promise for long term service in highly oxidative environments at temperatures up to 700°F versus lower thermo-oxidative stability possessed by their commercially available counterparts. The 4-BDAF monomer and polyimides therefrom, discovered in 1975 on NASA Lewis Research Center Contract NAS3-17824,[6] are described and claimed in U.S. Patent Numbers 4,203,922 and 4,111,906,[7,8].

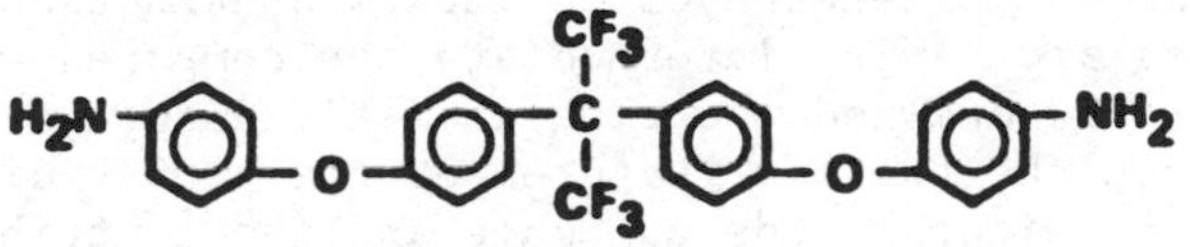

4-BDAF

Beginning in 1981, Morton Thiokol has conducted process and purity assessments on the diamine. This effort has prompted Morton Thiokol to obtain an exclusive license from TRW to commercially produce the diamine under the art taught in U.S. Patent Number 4,203,922. The product is now available from Morton Thiokol.

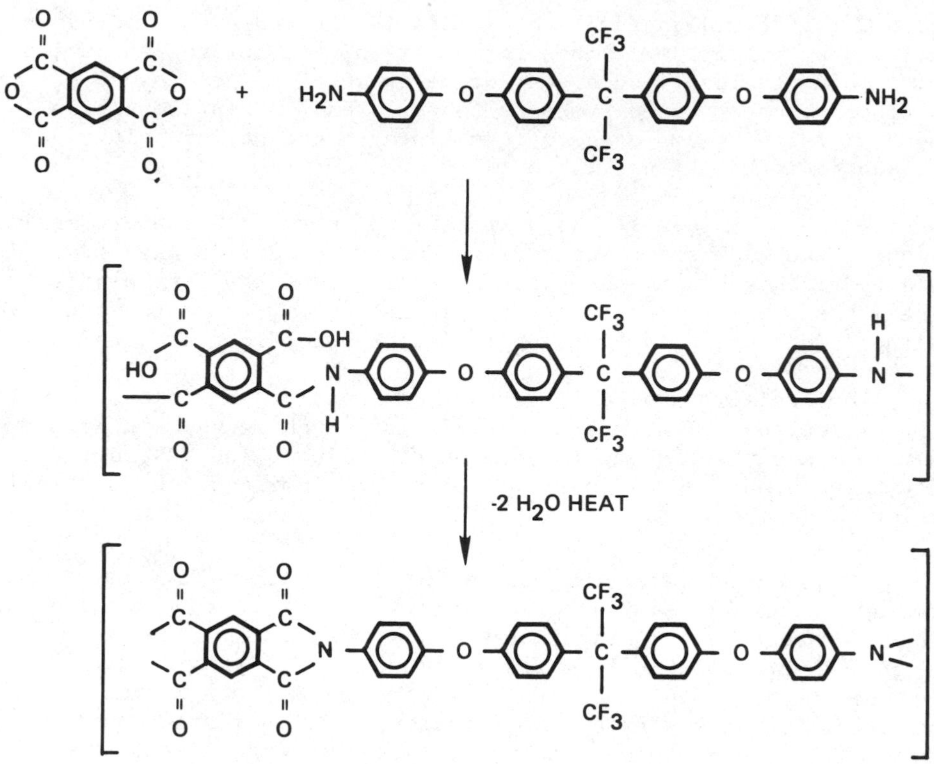

Equation 1. Representative Reaction Employed to Form a Linear Condensation polyimide.

The 4-BDAF diamine possesses a novel combination of properties versus high volume aromatic diamines normally used to produce high performance aromatic/heterocyclic linear condensation polyimides. Key properties of the TRW intermediate are compared with those of commonly employed one and two ring aromatic diamines in Table I. It is felt that the two key features of the 4-BDAF diamine are the high molecular weight of the molecule and the contribution of this factor to reduce the toxicity contributed by the two amino groups as exemplified by the LD_{50} versus lower molecular weight aromatic diamines. Another potential advantage of the 4-BDAF diamine is that fewer imide linkages per repeat unit are produced from this compound versus other commonly used diamines. It is felt that this factor will contribute to potential higher toughness, process-ability advantages and long-term hydrolytic stability versus competitive products. A comparison of volatiles produced in imidization and imide linkages per unit repeat segment is presented in Table II. The very high thermo-oxidative stability of polymers produced from the 4-BDAF diamine is ascribed to the presence of the perfluoroisopropylidene linkage.

1120

Table I. Comparison of 4-BDAF Diamine with other Commonly Employed Diamines.

DIAMINE	MOLECULAR WEIGHT (g/mole)	% NH$_2$/Mole	MELTING POINT (°C)	TOXICITY LD$_{50}$ (g/Kg)
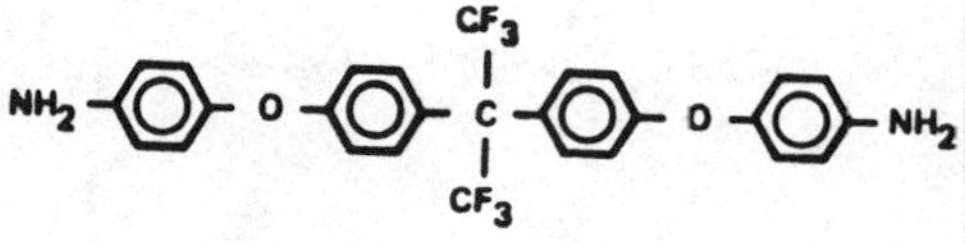 4 – BDAF	518.46	6.2	164	1.37
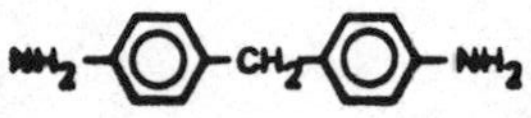 MDA	198.29	16.2	93	0.19
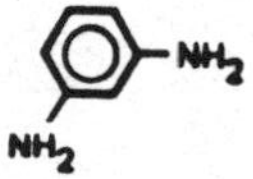 MPDA	108.15	29.6	63	0.3

Table II. Relative Volatile and Imide Content of Representative Polyimides.

POLYAMIDE-ACID REPEAT UNIT	MOLECULAR WEIGHT OF POLYAMIDE-ACID REPEAT UNIT (g/mole)	THEORETICAL AQUEOUS VOLATILE MATTER EVOLVED IN CONVERTING AMIDE-ACID TO IMIDE (%)	MOLECULAR WEIGHT OF IMIDE LINKAGES PER REPEAT UNIT (%)
4 BDAF/PMDA 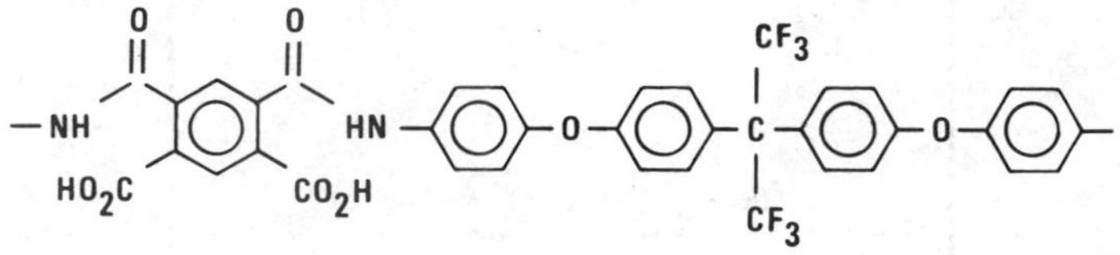	737	4.9	20.0
MDA/PMDA 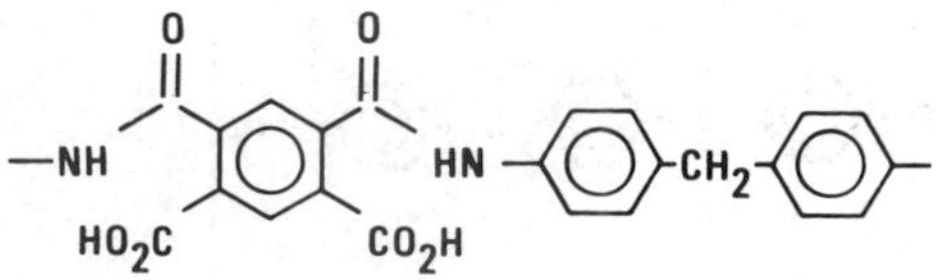	416	8.7	36.8
MPDA/PMDA 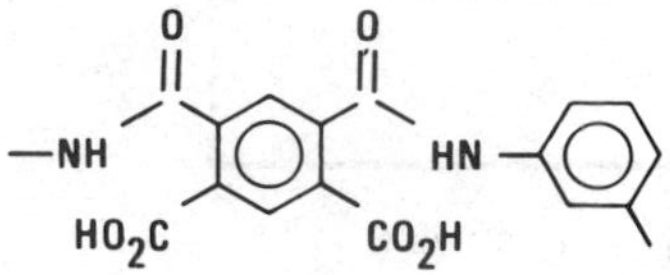	326	11.0	48.3

2. STUDIES ON PARTIALLY FLUORINATED POLYIMIDES

Although, as previously stated, the linear condensation polyimides from 4-BDAF diamine were discovered in 1975, directed
research and development activities on the diamine and polyimides
therefrom was not undertaken at that time due to lack of clearly
identifiable markets for products. In 1980, in response to urging
by the jet engine primes and government agencies, TRW initiated
studies to determine whether the partially fluorinated polyimides
produced from 4-BDAF diamine could serve to replace currently
available condensation polyimides as well as serve as a potential
substitute for the very high performance linear resins offered by
Du Pont as the NR-150B family of polyimides.

2.1 Initial Polyimide Matrix Resin Studies

The 1980 studies on the TRW partially fluorinated polyimides
were structured to assess basic thermo-oxidative stability, glass
transition temperature and processability characteristics related
to potential use as a jet engine compressor stage stator vane bushing at a service life temperature of 675°F. This investigation was
jointly sponsored by the GE Aircraft Engine Group, Evendale, Ohio
and the Bearings and Applied Technology Divisions of TRW. In this
study, a number of significant discoveries were made as summarized
below.

Two candidate polyimide polymers were prepared from pyromellitic
dianhydride (PMDA) and benzophenonetetracarboxylic acid dianhydride
(BTDA) employing TRW's 4-BDAF as the diamine ingredient at the
Redondo Beach, California facility. Solutions of the amide-acid
precursor were supplied to the TRW Bearings Division, Jamestown,
New York for processing studies. The amide-acid solutions were converted into polyimide molding compounds employing Celanese's GY-70
high modulus chopped graphite at a 50% by weight level as the reinforcement. The two candidate materials were compression molded into
resin discs of dimensions approximately one-inch in diameter by one-
inch in length. Prototype jet engine compressor stator vane bushings were machined from the discs to 0.001-inch tolerances. A
photograph of the prototype bushings is provided as Figure 1.

Resin disc samples of each of the two candidate polyimides were
provided to GE for oxidative and thermo-mechanical testing. The two
polyimide candidates were subjected to the GE oxidative screening
test which consists of a one hundred hour exposure at 675°F under a
four atmosphere flow of compressed air. The results from this test
and related thermal measurements are presented in Table III.
These data show that the 4-BDAF/BTDA formulation was
definitely unsuitable for further consideration due to its weight
loss at 675°F and the surprisingly low glass transition temperature
of ∿ 590°F. Conversely, although the weight loss was relatively

Figure 1.   Prototype Jet Engine Stator Vane Bushings.

Table III. Partially Fluorinated Polyimide Thermal Analysis
Results.

Property Measurement	Experimental Results	
	4-BDAF/PMDA	4-BDAF/BTDA
Glass transition temperature (Tg)[a]	390°C (734°F)	310°C (590°F)
Onset temperature of oxidative degradation[b]	495°C (923°F)	490°C (914°F)
Isothermal aging in air (% weight loss)[c]	20	34

[a]Determined on a Du Pont Model 990 thermalmechanical analysis attachment operating in a penetration mode.

[b]Determined on a Du Pont Model 990 thermalgravimetric analysis attachment employing a 3°C/minute heat-up rate and 100cc/minute air flow.

[c]Determined on composite discs containing 50% by weight GY-70 chopped fiber reinforcement; aging conditions consisted of a 675°F temperature for 100 hours employing a four atmosphere compressed air flow.

high for the 4-BDAF/PMDA formulation, the highly desirable Tg of this candidate of $\sim 740°F$ was held to be very promising and the material was deemed worthy of further study.

Subsequent joint GE/TRW studies on polymer development have yielded neat molded 4-BDAF/PMDA specimens which give a weight loss of only 8% at 675°F. A key factor in this significant product improvement is thought to be the current high quality of the 4-BDAF diamine product by Morton Chemical on up to 50-1b/run batches. The initial work employed diamine synthesized at TRW on a much smaller scale. The stator vane bushing development effort is continuing employing diamine produced by Morton Thiokol.

The initial promise of the GE/TRW study on the partially fluorinated polyimides prompted TRW to investigate other product applications for the resins. Beginning in 1981, TRW instituted a R&D program in which resin application assessments in areas other than compression molded bushings was a key objective. Joint evaluation participation by several government agencies and industrial organizations was structured and implemented. A summary of progress in product application assessments is provided below.

In 1981 and 1982, TRW initiated evaluation of the partially fluorinated polyimide resins produced from 4-BDAF diamine in a number of product application areas which emphasize very high thermo-oxidative stability, plus, in some instances, a combination of other features such as salt spray resistance, high lubricity and wear resistance and high adhesion to metallic and polymeric surfaces. Because the property requirements varied in each product application, either the 4-BDAF/PMDA formulation or the 4-BDAF/BTDA formulation was selected for evaluation. A list of several key product assessment studies conducted are as follows:
- Salt water resistant coating for plasma sprayed metallic jet engine components
- Oxidative barrier coating for polyimide composites
- Matrix resin for high performance rotary engine and hydraulic seal systems

The results of the coating evaluation studies completed to date are discussed in the remainder of this paper. Initial lubricity and wear test results are the subject of a NASA Lewis Research Center technical publication,[9]. Additional tribological property evaluations are being conducted by TRW on NASA Lewis Research Center Contract NAS3-23054,[10]. Assessment of the partially fluorinated polyimides as a matrix for high performance hydraulic seal back-up rings is planned for study at TRW on USAF Contracts F33615-81-C-5092 and F33615-82-C-5021,[11,12]. A summary of properties other than thermal measurements determined in these and related property assessment studies is provided in Table IV. Further data on coating applications are provided below.

Table IV. Representative Partially Fluorinated Polyimide Properties.

Property Measurement	Results
Specific Gravity	~ 1.4[a]
Hydrolytic Stability	No degradation observed in two hours water boil[b]
Solvent Resistance	No degradation noted in methylethyl ketone and jet reference fuel soak[b]
Filler Compatibility	Amenable to fibrous and powder filling to at least 50% by weight[a]
Filler Wetting	Excellent (photomicroscopy)[a]
Surface Smoothness	10 to 30 microinches per inch (root mean square) on steel and titanium substrates[c]
Machinability	Excellent (0.001 to 0.002-inch tolerance[a]
Spraying Behavior	Excellent ($\leq$ 0.0005-inch tolerance coatings are reproducibly obtained employing a solvent mixture and conventional spray hardware)
Salt Spray Resistance	No degradation noted after 500 hour salt spray exposure at RT
Adhesion (Qualitative)	Adheres tenaciously to polyimide/graphite composite, titanium, steel and plasma sprayed metal surfaces

a) Determined on compression molded discs.
b) Determined on films of nominal 0.005-inch thickness.
c) The metallic surface roughness was approximately 150 microinches/inch prior to coating.

2.2 Polyimide Coating Studies

In mid-1981, TRW was contacted by the United States Navy inquiring whether the polyimides may be suitable as a salt spray corrosion resistant coating for jet engine components containing plasma sprayed metallic surfaces. This type of jet engine hardware is particularly susceptible to corrosion. Because initial thermal and moisture characterization of the partially fluorinated polyimides in thin cast film and compression molded specimens looked promising, it was thought that the resins may possess attractive features as a protective coating.

A 4-BDAF/PMDA resin sample was prepared and submitted for evaluation as a high temperature protective coating at the Naval Air Rework Facility, Alameda, California. The resin varnish, prepared at 25% solids loading solution, was diluted to $\sim$ 10% solids concentration, then brush coated onto steel and titanium specimens possessing representative plasma sprayed metallic surfaces. The resin coating was processed as follows:
- Evaporate solvent and imidize resin for three hours at 400°F
- Postcure samples for sixteen hours at 800°F

The coated test specimens were then subjected to a Navy accelerated screening test that assesses the probability for coating service life of 100 hours at 600°F to 700°F in a compressed air environment. The 4-BDAF/PMDA coated metallic test specimens were subjected to a two hour muffle furnace bake at 900°F in air, followed by a five hundred hour salt spray exposure at ambient room temperature. At the conclusion of the Navy test, inspection showed that essentially zero corrosion occurred on specimens coated with the 4-BDAF/PMDA resin. Conversely, specimens coated with a high performance silicone resin and uncoated controls demonstrated severe salt induced corrosion. The test specimens resulting from the salt spray corrosion testing are shown in Figure 2.

The initial salt spray corrosion resistance results discussed above were assessed to be promising and prompted TRW to undertake further work in this product application area. Because brush coating is a relatively uncontrolled application process in terms of coating thickness uniformity, it was decided to investigate whether the partially fluorinated polyimides were amenable to a conventional spray deposition process. This was judged to be particularly important for coating relatively smooth surfaces normally encountered in many metallic and non-metallic components.

For investigation of a spray coating process, TRW selected polyimide/graphite composites for testing. This decision was prompted by an expressed interest from Boeing Aerospace Company to evaluate the partially fluorinated polyimides as an oxidation

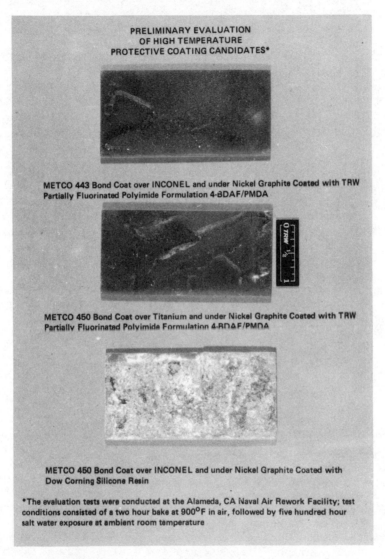

Figure 2. Test Specimens Resulting from United States Navy Salt Spray Corrosion Resistance Screening.

barrier coating to extend the useful life of PMR-15,[13] as a graphite composite matrix for $\geq$ 2000 hour service at 550° in air. This application requires a coating that could be applied to a reproducible $\pm$ 0.0005 thickness tolerance which is very difficult via a brush technique. Because the Boeing thermo-oxidative requirement was < 600°F, the 4-BDAF/BTDA formulation was selected as the partially fluorinated polyimide candidate.

Combinations of solvents were identified for the 4-BDAF/BTDA that would spray deposit a varnish containing $\geq$ 10% by weight solids loading of 4-BDAF/BTDA amide-acid to a reproducible 0.0005-inch or greater thickness that presents minimal problems during drying and imidization. The solvent mixture identified is void of components susceptible to peroxide formation and of those thought to present severe human hazard risks.

The partially fluorinated polyimide spray process discussed above was employed to coat a total of fourteen PMR-15 graphite reinforced composite panels. Both unidirectional Celion-6000 graphite fabric reinforced panels were included in the coating investigation. A representative unidirectional fiber reinforced panel is shown in Figure 3. Each composite panel was coated with the 4-BDAF/BTDA polyimide to a thickness of 0.004 $\pm$ 0.0005-inch employing the spray process.

The 4-BDAF/BTDA polyimide coated composites were delivered to Boeing Aerospace Company and subjected to isothermal aging in air at 550°F. The air flow employed in the aging study was two cubic feet/minute. Selected composite panels were removed from the isothermal aging test after two thousand hours of exposure and property measurements were conducted on 4-BDAF/BTDA coated specimens as well as uncoated control panels. The property test results completed to date are summarized in Table V. Representative fabric reinforced composites coated with 4-BDAF/BTDA after the two thousand hour aging period at 550°F are shown in Figure 4.

As can be seen from the initial data presented in Table V, the 4-BDAF/BTDA coated composites show a definite improvement in weight retention and selected mechanical properties over the uncoated controls after the two thousand hour aging period in air at 550°F. The coating remained very flexible after isothermal aging for five thousand hours at 550°F. Because the 4-BDAF/BTDA polyimide coating continued to show promise at the end of the five thousand hour aging period, it is felt that the partially fluorinated polyimides will find significant future trade acceptance as a superior oxidative barrier coating for protecting PMR-15 polyimide matrix structural composites employed in high temperature airframe and engine applications in advanced military and commercial aircraft.

Figure 3.   Representative 4-BDAF/BTDA Resin Coated on a PMR-15
Unidirectionally Reinforced Composite.

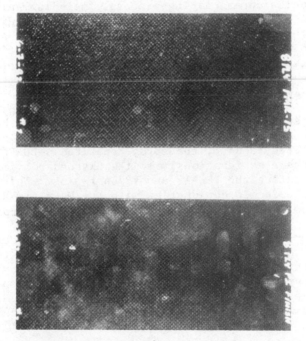

Figure 4.   Representative PMR-15/Celion-6000 Graphite Fabric
Composites Coated with 4-BDAF/BTDA Polyimide after Two
Thousand Hour Aging in Air at 288°C.

Table V. Initial Property Test Results on PMR-15 Composite Coated with 4-BDAF/BTDA[a]).

Property Measurement	Property Value	
	Uncoated Control	Coated
Flexural Strength (ksi)	45.7	54.8
Flexural Modulus (Msi)	4.9	4.5
Weight Loss (%)	5.2	2.8

[a]The property measurements were conducted at ambient room temperature on a representative PMR-15/Celion-6000 (polyimide sized) graphite fabric composite coated to a nominal 0.004-inch thickness with 4-BDAF/BTDA polyimide.

A similar composite effort has been initiated with Hitco Materials to investigate the partially fluorinated polyimides as an oxidative protective barrier coating for V378A bismaleimide composites,[14]. Preliminary test results indicate that the TRW composite spray coating process, including initial cure of the 4-BDAF/BTDA resin at 550°F, does not significantly deteriorate the initial V378A composite properties. A summary of coated and uncoated V378A/T-300 graphite unidirectional panel properties is presented in Table VI. It is planned to continue this investigation including long-term aging in air at 450°F. This aging study is currently in progress.

Further studies are in progress to investigate the 4-BDAF/BTDA resin as a coating for Skybond 703/Quartz glass composites. In work completed to date, the spray deposited 4-BDAF/BTDA resin appears to adhere as tenaciously to the Skybond polyimide as was observed in coating the PMR-15 and V378A resins. This evaluation study is continuing in conjunction with the Navy Air Rework Facility, Alameda, California and Boeing Military Airplane Company.

Upon completion of development of the spray process method for coating composites, TRW applied this methodology to representative plasma sprayed metal substrates. This study was undertaken to compare spray coating behavior with the brush coated specimens prepared at the Naval Air Rework Facility, Alameda which were summarized at the beginning of the technical discussion.

Test specimens were obtained that consisted of nickel aluminide/nickel graphite plasma sprayed coatings on titanium. Prior to initiating spray coating with the 4-BDAF/PMDA polyimide,

Table VI. Comparison of Initial Properties of Uncoated and 4-BDAF/BTDA Coated V378A/T-300 Composites[a].

| Property Measurement | Property Results | | | |
	RT Uncoated	450°F	RT Coated	450°F
Flexural Strength (ksi)	290	175	259	165
Flexural Modulus (Msi)	18.5	17.4	15.6	15.1
Short Beam Shear (ksi)	16.8	7.8	15.2	7.7
Density	1.57	--	1.56	--
Resin Content (%)	28.4	--	33.1	--
Fiber Volume (%)	63.9	--	59.3	--
Void Content (%)	1.2	--	0.2	--

[a]The properties were determined by Hitco Materials

the metallic specimens were degreased with 1,1,1-trichloroethane, then dried at 900°F for two hours to off-gas any air trapped in voidy areas of the plasma sprayed coating.

The test specimens were spray coated with the 4-BDAF/PMDA polyimide employing essentially the identical process developed for composite coating. In this study, the substrates were coated to a nominal 0.001-inch and 0.004-inch in order to assess the effect of thickness on coating performance. After solvent drying, imidization and postcure of the resin coating in the range of 250°F up to 700°F, the polyimide coated specimens were isothermally aged in air at 700°F as discussed below.

Duplicate polyimide spray coated test specimens at 0.001-inch and 0.004-inch thickness on two representative plasma sprayed metallic coatings were isothermally aged at 700°F for one hundred hours employing an air flow of 10 cubic centimeters/minute. In this aging test it was determined that the 0.001-inch thick coating gave the most satisfactory results. A resin weight loss of only fifteen percent was obtained on the 0.001-inch coating after the aging period. Photomicroscopy analysis of sectioned specimens determined that a uniform coating was spray deposited on the very rough, irregular plasma sprayed metallic surface. Representative 4-BDAF/PMDA coated specimens after sixty-five hours aging at 700°F are shown in Figure 5.

Representative photomicrographs of the 0.001-inch and 0.004-inch coatings of 4-BDAF/PMDA are presented in Figures 6 and 7, respectively. These photomicrographs were taken on resin coated specimens that were isothermally aged at 700°F for sixty-five hours in air. As can be seen from the photomicrographs, a highly

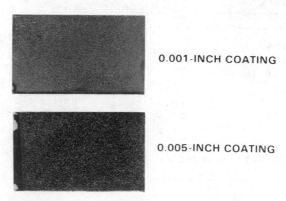

0.001-INCH COATING

0.005-INCH COATING

Figure 5. Representative Plasma Sprayed Metallic Substrates Coated with 4-BDAF/PMDA after Sixty-five Hours Aging in Air at 371°C (700°F).

desirable complete coating with polyimide of the plasma deposited metallic surface was obtained, along with penetration of resin into the highly irregular surface of the nickel/aluminide substrate.

The 0.004-inch coated specimens each showed a tendency for the multi-coated layers to delaminate interfacially at or near the metal surface and higher resin weight losses were determined. This infers that further development work is required to define drying and imidization cycles which optimize the cohesive strength of successive multi-coated layers when coatings of thickness > 0.001-inch are desired.

Because the 0.001-inch thick 4-BDAF/PMDA coating displayed excellent thermo-oxidative stability at 700°F as well as adhesion to the plasma sprayed metallic surfaces, work is planned to continue spray coating evaluations on this type of metallic substrates.

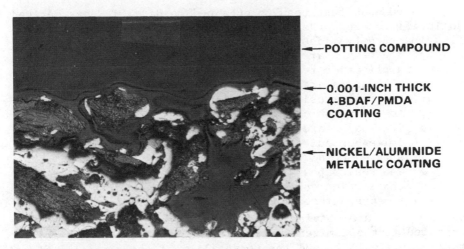

POTTING COMPOUND

0.001-INCH THICK
4-BDAF/PMDA
COATING

NICKEL/ALUMINIDE
METALLIC COATING

Figure 6.   Photomicrograph of 0.001-Inch 4-BDAF/PMDA Polyimide
Coated on Nickel/Aluminide after Sixty-five Hours Aging at 371°C
(700°F) in Air (400x).

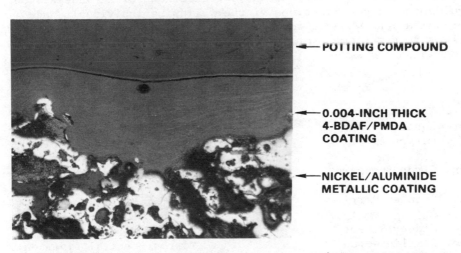

POTTING COMPOUND

0.004-INCH THICK
4-BDAF/PMDA
COATING

NICKEL/ALUMINIDE
METALLIC COATING

Figure 7.   Photomicrograph of 0.004-Inch 4-BDAF/PMDA Polyimide Coated
on Nickel/Aluminide after Sixty-five Hours Aging at 371°C (700°F) in
Air (600x).

3. CONCLUSIONS

Based upon test results obtained to date, TRW has derived the following conclusions regarding the application of partially fluorinated polyimides as a protective coating:

- The resins are amenable to a practical spray coating application technique
- The resins are suitable for use as an oxidative barrier and/or corrosion inhibiting coating at temperatures up to a minimum of 700°F
- The resins are promising as protective coatings for polyimide composites, high performance metal alloys and plasma sprayed metal surfaces
- The resins possess lubricity properties and wear resistance which enhance the oxidative and corrosion resistance features stated above.

Because the data obtained to date employed neat resin, it is expected that optimized coatings containing selected formulary ingredients will extend the versatility of the partially fluorinated polyimides as high performance, protective materials.

4. ACKNOWLEDGEMENTS

TRW wishes to express its appreciation for the assistance, guidance and contributions of the following individuals and organizations whose efforts to date have made significant contributions to development of the partially fluorinated polyimide technology:

Individual	Organization
Dr. Tito Serafini	NASA Lewis Research Center
Dr. Robert Fusaro	NASA Lewis Research Center
Mr. Norman Amdur	Naval Air Rework Facility, Alameda, California
Mr. Clyde Sheppard	Boeing Aerospace Company, Seattle, Washington
Mr. Stanley Harrier	General Electric Aircraft Engine Group
Dr. Sidney Street	Hitco Materials
Dr. Phillip Weyna	Morton Chemical Division of Morton Thiokol
Mr. Raymond Hoffman	Morton Chemical Division of Morton Thiokol
Dr. David Booth	Morton Chemical Division of Morton Thiokol
Mr. Skip Stuart	Morton Chemical Division of Morton Thiokol
Dr. Keith Baucom	PCR Division of SCM

5. REFERENCES

1. Technical Bulletin No. 6236A, Skybond 703, High Heat Resistant Resins, Monsanto, St. Louis, MO.
2. Exploring Vespel Territory, The Properties of Du Pont Vespel Parts, Du Pont Company, Wilmington, DE.
3. Kapton® Polyimide Film, Summary of Properties, Du Pont Company.
4. H. H. Gibbs, 17th National SAMPE Symp., 17, III-B-6 (1972).
5. R. J. Jones and H. N. Cassey, Final Contractor's Report, Contract F33615-69-C-5134, AFML Report No. TR-71-24, Part 1, 1971.
6. G. Zakrezewski, M. K. O'Rell, R. W. Vaughan and R. J. Jones, Final Report Contract NAS3-17824, NASA CR-134900, 1975.
7. R. J. Jones, M. K. O'Rell and J. M. Hom, U.S. Patent 4,203,922 (1980).
8. R. J. Jones, M. K. O'Rell and J. M. Hom, U.S. Patent 4,111,906 (1978).
9. R. L. Fusaro, "Tribological Evaluation of Composite Materials Made from a Partially Fluorinated Polyimide," NASA Technical Memorandum 82832, April, 1982.
10. Contract NAS3-23054, Unpublished Results.
11. Contract F33615-81-C-5092, Unpublished Results.
12. Contract F33615-82-C-5021, Unpublished Results.
13. T. T. Serafini, P. Delvigs and G. R. Lightsey, J. Appl. Polym. Sci., 16, 905 (1972).
14. S. W. Street, 25th National SAMPE Symp., 25, 366 (1980).

POLYIMIDES AS INSULATING LAYERS FOR IMPLANTABLE ELECTRODES*

H. K. Charles, Jr., J. T. Massey and V.B. Mountcastle†

The Johns Hopkins University
Applied Physics Laboratory
Laurel, Maryland 20707

Planar neuron microprobe structures with single and multiple (4) independent electrodes or sensor sites distributed over a linear distance of several millimeters are described. The neuron microprobes make extensive use of polyimide for base dielectric, top protection, and window or sensor area defining layers. A typical probe fabrication sequence is presented with special emphasis on the details of the polyimide processing (e.g., substrate preparation, deposition, photolithography, and curing). To date in acute experiments, single sensor site probes have been implanted trans-pia and recordings made.

---

* This work was supported under Public Health Service Grant NS 07226
† The Johns Hopkins University, School of Medicine, 725 North Wolf Street, Baltimore, Maryland 21205

INTRODUCTION

The human brain is capped by the cerebral cortex whose size and complexity are dominant features of the brain. In man it occupies about 60-65% of the total volume of the brain, if spread flat would cover more than 2000 square cm and contains between 20 and 40 billion neurons (information processing and storage nerve cells) although it is only 2 to 3 mm thick on the average. These cortical nerve cells are organized by specific connections into a large number of brain systems which are in turn connected with one another. Neurons communicate with one another and process information by passing signals along interneuronal connections - axons and synapses. The generation of such a signal (an action potential) in a nerve cell body and its conduction along a connecting process, an axon, is accompanied by a small extracellular electrical field.

For the last four decades fine-tipped microelectrodes inserted into brain tissue have been used for recording extracellular electrical fields (action potentials) generated by individual nerve cells (neurons).[1,2] This method of successive single neuron observations has contributed to many significant advances in understanding the function of the brain. Since 1965, this analysis of the electrical activity of single nerve cells during the performance of a precisely defined behavioral act by a trained, alert monkey, has lead to the conclusion that information processing in the nervous system (brain cortex) is done by populations of neurons distributed spatially.[3] However, the contribution of successively observed individual neurons to the formation of a neural processing function can be taken into account only by the construction of a dynamic model for combining the observed details of the spatio-temporal patterns of electrical activity within neuron populations. Conclusions drawn from these models based on data taken from the same animal preparation at different times or from different animal preparations may give misleading impressions of casual relationships. Furthermore, the possible involvement of specific neurons in contributing to multiple neural processing functions is difficult to assess under these conditions. Thus, these dynamic relations can only be determined by recording (or stimulating) simultaneously from selected populations of neural elements in the alert, behaving animal.[4]

Two major approaches to the simultaneous recording of neural signals have been taken. The first, historically, has been the fabrication of bundles of separate, single electrode microprobes[5] which had either a fixed three dimensional geometrical configuration[6,7] or a fixed two dimensional (lateral) geometrical configuration with adjustable depth[8]. These arrays have been developed to a point where their capability to record simultaneously has been demonstrated. The second approach has utilized the application of microelectronic technology for the fabrication of planar multielectrode microprobes. Although this approach has been under development for over a decade[9] so called "printed circuit" microprobes remain custom fabricated prototypes and have been tested only in very limited experimental situations.[10]

The planar microprobes described in this study use molybdenum as a substrate material, polyimide for dielectric layers and gold as the electrode material. These structures are currently being fabricated with up to four independent electrodes or sensor sites distributed over a linear (in depth) distance of several millimeters. Within the same geometrical outline of the probe body it is envisioned that sensor site density can be increased by a factor of from four to five as the manufacturability improves and the biological implications of probe site placement, impedance and geometry become fully verified. Although, similar probes have been reported using molybdenum substrates[11,12] with various combinations of dielectrics and electrode structures. The molybdenum-polyimide combination reported here is new and we believe represents a significant improvement in probe structure flexibility and hence ultimate reliability during implantation and subsequent recording.

PROBE STRUCTURE

The neuron microprobe consists of a planar structure with single or multiple independent electrodes or sensor sites distributed over a distance of several millimeters. The basic microprobe structure consists of: (1) a metal supporting layer or substrate: (2) a dielectric isolation layer; (3) a patterned metallization or electrode layer; and (4) a final dielectric layer with windows exposing the individual sensor sites. Since one of the goals in the initial development effort was the direct penetration of the membranes surrounding the brain (dura mater and pia mater), molybdenum was selected as the optimum substrate material for its strength and its compatibility with both biological systems and microelectronic processing techniques. Currently, several molybdenum probe systems are under development using 99.99 percent pure molybdenum foils with thicknesses of $50.8\,\mu m$ (for direct membrane penetration in acute experiments) and both $25.4\,\mu m$ and $18\,\mu m$ (for implantation after surgical membrane removal in chronic experiments).

Polyimide is used as the first dielectric layer because of its superior adhesion, dielectric integrity, processing compatibility[13] and lack of biological system impact.[14] Polyimide has been used in many biological applications with particular success in cochlear implants.[15,16] Current studies are underway to further document the survivability and biocompatibility of polyimide. Sputtered gold layers appear to be the most acceptable electrode material to date as far as maintaining dielectric integrity and total process compatibility, especially with the polyimide. Other electrode structures using both evaporated chromium-gold films and sputtered titanium layers have been tried but with inferior results to the gold. Eventually, the gold can also be platinized (by the sputter deposition of a platiunum layer only[18] on the exposed sensor sites using reverse lift-off techniques and its subsequent chemical conversion[19] to platinum black) to increase sensor site effective area. Both photoresist and polyimide have been used as a top or window defining dielectric. Probes, currently in process, and future designs will use polyimide for all dielectric layers.

POLYIMIDE PROCESSING

The polyimide resins used in the neuron microprobe development included both Hitachi PIQ[20] and DuPont PI-2555.[21] They have been used relatively interchangeably with approximately equal success, although DuPont PI-2555 appeared to produce a more pinhole free dielectric on the molybdenum surface. Before neuron microprobe processing could begin, many experiments were performed to determine the adhesion of the polyimide film during coating, pre-curing photolithography, post curing and subsequent film deposition and/or top level processing. An example of poor polyimide adhesion to the molybdenum surface, after photolithography, is shown in Figure 1. The light striations in the photomicrograph are from the fibrous structure of the rolled

Figure 1. Photomicrograph of Lifted Polyimide Film Pattern (Square Film Pattern is Approximately 1.25mm x 1.25mm).

molybdenum foil. The polyimide adhesion was improved by polishing the molybdenum surface using a phosphoric-nitric acid solution (1 part H_3PO_4, (concentrated), 3 parts HNO_3, (concentrated), and 2 parts H_2O). A polyimide coupler was used in all cases (Hitachi PIQ coupler with the PIQ resin and DuPont VM-651 with the PI-2555). The PIQ coupler was dispensed from a hypodermic on the polished molybdenum wafers, spun at 3,000 rpm for 30 seconds, and baked at 130°C for 15 minutes. A solution was made from the VM-651 coupler, methanol and deionized water (200 ml of absolute methanol, 10 ml of deionized water and 0.1 ml (2 drops) of VM-651. The VM-651 solution was dispensed onto the polished molybdenum wafers, spun at 5,000 rpm for 30 seconds, and air dried. Following coupler drying, the respective polyimide resins were applied and spun at 3,000 rpm for 30 seconds producing approximately 2 μm thick layers on our molybdenum surface. After polyimide application and prior to photolithography, the polyimide was pre-cured using a five step air-nitrogen cycle as shown in Table I. The maximum temperature reached during the pre-cure was 120°C.

Table I. Polyimide Cure Cycle

o PRE-PATTERNING CURE

      30°C for 30 minutes
      50°C for 30 minutes
      80°C for 30 minutes
     100°C for 30 minutes
     120°C for 60 minutes   (in a vented oven)

o POST ETCHING CURE

     220°C for 60 minutes
     350°C for 60 minutes

Photolithography of the polyimide layer was performed using a combined photoresist patterning and polyimide etch step which eliminates the normal photoresist developing, baking, and separate polyimide etching (usually hydrazine) process. The basic technique involves the application of Shipley AZ1350J photoresist[22] on top of the pre-cured polyimide layer. The photoresist was applied at a spin speed of 3,000 rpm for a duration of 30 seconds. After infrared baking for 30 minutes, the photoresist was exposed with an ultraviolet light source through the appropriate neuron microprobe mask. The exposed photoresist was then developed using a solution of Shipley MF-312 developer (1 part MF-312 and 1 part deionized H_2O). The developing of the photoresist, using the MF-312 solution, also etches the underlying pre-cured polyimide film, thus eliminating a separate and potentially hazardous etching step. Similar behavior of positive photoresist developer and polyimide layers has been observed by other experimenters.[15,23] The resolution of the technique appears comparable to that of AZ1350J itself with excellent self alignment and minimal undercutting. Control of undercutting can be achieved using various dilutions of the MF-312 developer. Dilutions in the range of 1 to 10 parts deionized H_2O to 1 part developer were tried. In general, the more dilute the solution the less undercutting, but greater the tendency to leave undissolved polyimide in tight pattern areas. In order to understand the polyimide etching behavior of the MF-312 developer solution, one must consider its basic chemical composition. The MF-312 developer is a metal free solution containing no organic solvents with tetramethyl ammonium hydroxide $((CH_3)_4NOH \xrightarrow{\hspace{2cm}} (CH_3)_4N^+ + OH^-)$[24] as the active ingredient. In a patent by Lindsey[25] polyimide was shown to be effectively etched using a variety of basic compounds such as sodium hydroxide, potassium hydroxide and quarternary ammonium hydroxides. Similarly in a patent by Fish[26] the tetralkylammonium hydroxides, dissolved in dipolar aprotic solvents, were found to rapidly dissolve Kapton® at rates between 4 μm/minute and 25 μm/minute depending upon temperature (25°C to 80°C). Thus it appears that the tetramethyl ammonium hydroxide in the MF-312 developer is responsible for the etching of the polyimide layer during the developing of the AZ1350J photoresist. No attempt has been made to determine the polyimide etch rate of the developer solution, although it appears that the polyimide layer disappears in approximately 20-30 seconds (consistent with a 4 μm/minutes rate and the 2 μm/film thickness). Following the polyimide patterning the resist is stripped and the polyimide is cured at high temperature as shown in Table I. Scanning electron photomicrographs of cured patterned polyimide layers on the molybdenum illustrating the sharpness of the pattern edge and the tendency for the polyimide film to smooth the striations in the molybdenum foil are shown in Figures 2 and 3.

MICROPROBE FABRICATION

The neuron microprobes are fabricated either eight (in the case of the single site acute probe) or four (for the four sensor site chronic probe) to a molybdenum substrate (2.5 cm x 2.5 cm) using the processing sequence outlined in Table II.

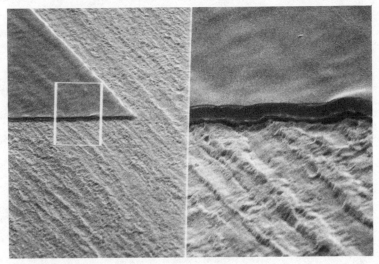

Figure 2. Scanning Electron Photomicrograph of Patterned
Polyimide Layer on Molybdenum Surface (Polyimide
Layer Approximately 2 μm thick).

Figure 3. Scanning Electron Photomicrograph of Patterned
Polyimide Layer on Molybdenum Surface. Enlarge View
of Film Edge Shown in Figure 2. (Polyimide Layer is
Approximately 2 μm thick).

Table II. Neuron Probe Fabrication Sequence .

- o   Molybdenum Preparation
- o   Gold Film Masking Layer (Reverse Side - Cr/Au Evaporation)
- o   Polyimide Application (Front Side - Spin Coating)
- o   Polyimide and Reverse Side Photolithography
- o   Polyimide Cure
- o   Gold Film Electrode Layer (Front Side - Sputter Deposition)
- o   Gold Electrode Layer Photolithography
- o   Polyimide Application (Front Side - Spin Coating)
- o   Polyimide Window (Sensor Site) Photolithography
- o   Probe Separation (Molybdenum Etching)
- o   Cleaning
- o   Test and Measurement

Prior to fabrication the molybdenum substrates are cut to size and polished in the phosphoric-nitric acid solution described above until the surface has a smooth matte-like finish with the surface striations reduced to a minimum (typically 45 seconds). Next a chromium-gold backside masking layer (approximately $2 \mu m$ in thickness) is deposited by physical vapor deposition from a resistance heated source. The polyimide coupler, followed by the resin, is applied to the remaining, exposed molybdenum surface. The polyimide is then pre-cured according to the recipe shown in Table I. Next, both sides of the substrate are coated with AZ1350J photoresist (spun at 3,000 rpm for 30 seconds) and infrared baked for 30 minutes. The photoresist layers on the front and back of the substrate are then exposed through the neuron microprobe separation mask pattern (back and front sides are mirror images of the four or eight individual probe outline geometries). The exposed photoresist layers are developed using the MF-312 solution producing the separation pattern in the back photoresist layer and in both the photoresist layer and the polyimide film on the front surface. Following photoresist baking at 90°C for 30 minutes, the back chromium-gold layer is etched (potassium iodide-iodine etch for the gold and HCl with a zinc strike for the chromium). After stripping the photoresist, the polyimide is given its final cure as described in Table I. A one percent potassium hydroxide solution is used to remove any coupler residue after polyimide curing. Gold (approximately 0.6 μm in thickness) is then sputter deposited onto the patterned polyimide and etched into the conductor lines and sensor sites using a standard photolithography cycle with Kodak 809 photoresist.[27] Etching was accomplished again using the potassium iodide-iodine solution. Following electrode patterning, the top polyimide dielectric was applied, patterned and cured according to the methods described above. The probes are finally separated by immersing the substrates in a molybdenum etch (1 part H_2SO (concentrated), 1 part HNO_3 (concentrated), 5 parts H_2O). The fibrous nature of the molybdenum has led to incomplete etching and residual metal projections on the etched probes as shown in Figures 4 and 5. It is also obvious that the probe pattern has been applied both parallel and perpendicular to the molybdenum striations or fibers as shown in Figures 4 and 5, respectively. No difference in performance has been observed with either direction, but less projections (at the tips) seem to occur for properly etched probes in the perpendicular direction.

A portion (blades of three acute probes) of a processed substrate is shown in Figure 6. Figures 7 and 8 present views of individual probe blade and the 10 μm x 10 μm sensing site. This neuron probe design is intended for use in acute experiments to demonstrate the viability of direct, through membrane implantation, and the recording ability of a blade type, single-sided, sensor structure (in contrast to the currently used drawn wire, glass insulated, round electrodes). To facilitate this implantation a quick change, pluggable contact housing was developed for use with the current drawn electrode microdrive implantation tube. View of a mounted acute probe in this housing are shown in Figures 9 and 10.

A portion of a processed substate containing four chronic probes with four sensor sites is shown in Figure 11. This microprobe is designed for use in long term chronic experiments. Methods for lead attachment and ultimate placement and holding of the probe structure on the moving brain mass within the skull cavity are under development.

IMPLANTATIONS

The first probes processed for acute animal trials had the same relative size and shape as the acute probes described above with a blade thickness of approximately 50 μm, a 2 mm implantable length, a 10 μm x 10 μm exposed sensor site located

Figure 4.   Photomicrograph of a Neuron Probe Tip After Molybdenum
Etching Illustrating Incomplete Tip Etch.   Probe
Pattern Applied Perpendicular to Molybdenum Fibers
(Electrode Line Width Approximately 10 μm).

Figure 5.   Photomicrograph of Neuron Probe Tip After Molybdenum
Etching Illustrating Incomplete Tip Etch.   Probe
Pattern Applied Parallel to Molybdenum Fibers
(Electrode Line Width Approximately 10 μm).

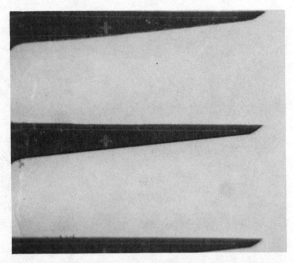

Figure 6.   Photomicrograph of Completed Single Sensor Site Acute
Neuron Probe Tips After Molybdenum Etching.   Distance
Between Probe Tips is Approximately 2 mm.

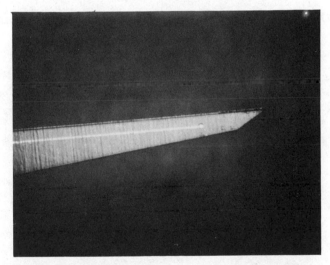

Figure 7.   Photomicrograph of Individual Acute Probe Tip.   Probe
Pattern Perpendicular to Molybdenum Fibers (electrode
Line Width is Approximately 10 μ m).

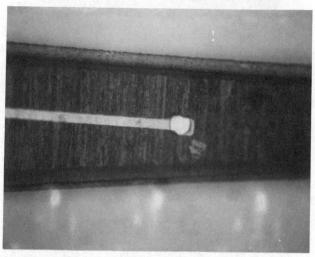

Figure 8. Photomicrograph of Individual Sensor Site on Acute
Probe Tip. Probe Pattern Perpendicular to Molybdenum
Fibers (Electrode Line Width is Approximately 10 μm).

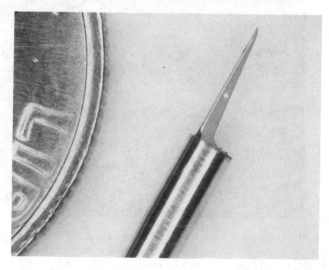

Figure 9. Photomicrograph of Mounted Acute Probe (Blade View)
in the Quick Disconnect Housing. The Housing Diameter
is Approximately 1.75 mm. The coin is a U.S. Dime.

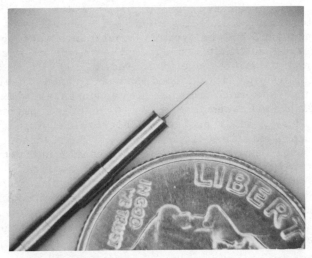

Figure 10. Photomicrograph of Mounted Acute Probe (Edge View) in
the Quick Disconnect Housing. The Housing Diameter is
Approximately 1.75 mm. The coin is an U.S. Dime.

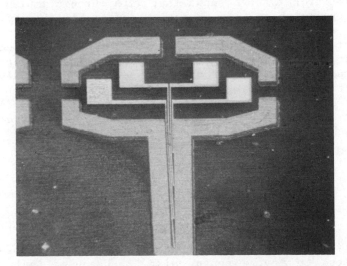

Figure 11. Photomicrograph of Completed Four Sensor Site Chronic
Probe After Molybdenum Etching. Distance Between
Probe Body and Surrounding Substrate Metallization is
Approximately 0.5 mm.

400 μm from the probe tip, and a blade width of 125 μm at the electrode location. These probes were mounted in the quick disconnect housing, and in typical experiments were inserted into the cortex of an anesthetized cat. The impedance was determined to be in the 5 to 7 MΩ range (measured at 1000 Hz in a reference saline solution) and it dropped somewhat upon insertion. The microprobe tip was much rougher in appearance (when viewed at 400x magnification) when compared to a drawn platinum-iridium probe. Platinum-iridium probes exhibit typical impedances in the 1 to 5 MΩ under similar circumstances. Penetration of the pia posed no problem, but the recording of neural activity typically proved disappointing with only a few small amplitude (factors of 2 to 5 less than round wire signals) neural spikes were observed. At the conclusion of the experiments a lesion producing current of two microamperes was passed through the probe (for a period of one second). Some change in probe impedance was then observed after this pulse.

Several reasons have been advanced for the low signal activity including increased sensor impedance, one sided, limited signal collection volume (when compared to spherical volume of round wire probe), trauma of the implant, and the actual brain damage due to the physical size and shape of the planar implant. Several researchers have shown that with chronic implantation of microelectrodes there is a gradual recovery from injury, so that hours or days after the initial electrode implantation cells become active that were previously silenced by the trauma of the initial implant. Also, the activity of the same single cerebral neurons can be observed for days and weeks, continuously. Thus, future activity will be directed not only at experiments designed to answer questions about sensor site placement, size, impedance, etc., but also towards the development of minimum sized chronic probes such as shown in Figure 11.

DISCUSSION

Current neuron microprobes are single, round-wire electrodes that are fabricated by drawing wires (encapsulated in glass tubes) to a very sharp point. Subsequent individual tip conditioning and exposure techniques are then employed to produce the final recording electrode. The extension of these methods to multielectrode structures with spatially distributed multiple sensor sites is extremely difficult.[28] An alternate method for producing spatially distributed multiple sensor site probes is to produce planar structures using techniques developed by the electronics industry for the processing of integrated circuits.[10] Such structures can make use of the uniformity, high resolution, and the simultaneous multiple probe fabrication (several devices on a single substrate) capability of modern photolithographic processing. Since planar photolithography techniques are independent of pattern shape and size (over relatively wide ranges), those methods developed initially for a single electrode site are easily extendable to multiple sensor sites; thus eliminating the problem encountered by the current round electrodes.

Polyimide has played an integral role in our ability to develop a planar neuron probe structure on a metal substrate. The dielectric integrity, adhesion, and metal film processing compatibility of the polyimide has allowed successful probe implementation and membrane penetration. The self-etching feature of the polyimide during the development of the overlying photoresist layer, while not unexpected when the detailed chemical composition of the developing solution is explored, provides a rapid processing sequence with excellent resolution and a minimum of undercutting. This technique should be applicable to any polyimide layer requiring photolithographic patterning regardless of substrate and could be of particular value to the semiconductor industry in minimizing processing time and wafer handling.

Currently chronic implantation experiments are underway with the four sensor site probe. Future plans call for the extension of these chronic experiments with probes containing eight and perhaps even sixteen electrode sites. Preliminary design sequences are underway for these structures.

ACKNOWLEDGEMENTS

The authors gratefully acknowledge the efforts of all the individuals who have made the realization of a polyimide based microprobe possible. A special thank you is given to P. R. Falk and S. P. Kelly who withstood the problems encountered in polyimide processing and molybdenum etching. Both D. P. Glock and K. J. Mach deserve special praise for their efforts in microassembly of the completed probes. In the areas of probe testing and concept development the efforts of Drs. Georgopoulasm Johnson, and Poggio are especially appreciated. A special thank you is given to V. S. Stephens, K. M. Eichstedt and J. I. Bittner for typing the manuscript.

REFERENCES

1. J. E. Rose and V. B. Mountcastle, J. Comp. Neurol., $\underline{97}$, 441 (1952).

2. K. Frank and M. C. Becker, in "Physical Techniques in Biological Research", W. L. Nastuk, Editor, Vol. 5, Chapter 2, Academic Press, New York, 1964.

3. E. V. Evarts, in "Methods in Medical Research", R. F. Rushmer, Editor, pp. 241-250, Year Book Medical Publishers, Chicago, 1966.

4. e.g., Neurosciences Research Program Workshop – Multi-microelectronic Recording in the Central Nervous System, Rockefeller University, New York, February 17-19, 1982.

5. These current single, round-wire microelectrodes are fabricated by drawing platinum-iridium alloy wires (encapsulated in glass tubes) to a very sharp point. Subsequent individual tip conditioning and exposure techniques are then employed to produce the final recording electrode.

6. J. Kruger and M. Bach, Exp. Brain Res., $\underline{41}$, 191 (1981).

7. J. Kruger, J. Neurosci. Methods, $\underline{6}$, 347 (1982).

8. H. J. Reitback and G. Werner, Experimentia, $\underline{39}$, 339 (1983).

9. K. D. Wise, J. B. Angell and A. Starr, IEEE Trans. on Biomedical Engg., $\underline{BME-17}$, 238 (1970).

10. R. S. Pickar, J. Neurosci. Methods, $\underline{1}$, 301 (1979).

11. H. D. Mercer and R. L. White, IEEE Trans. on Biomedical Engg., $\underline{BME-25}$, 494 (1978).

12. M. Kuperstein and D. A. Whittington, IEEE Trans. on Biomedical Engg., $\underline{BME-28}$, 288 (1981).

13. Polyimide has been extensively used in the semiconductor industry for the fabrication of multilevel electrode structures. For example see: K. Sato, et al., IEEE Trans. on Parts, Hybrids and Packaging, $\underline{PHP-9}$, 176 (1973).

14. Once cured, polyimides form a group of stable resins which have been marketed by DuPont under the trademark Kapton®. The biocompatibility of Kapton® has been shown previously. See for example: J. Polymer Science, $\underline{3A}$, 1373 (1965), and Reference 16 below.

15. S. A. Shamma-Donoghue, G. A. May, N. E. Cotter, R. L. White and F. B. Simmons, IEEE Trans. on Electron Devices, $\underline{ED-29}$, 136 (1982).

16. R. L. White, IEEE Trans. on Biomedical Engg., $\underline{BME-29}$, 233 (1982).

17. D. Basiulis, "Adhesion Studies" Presentation at the Thirteenth Neural Prosthesis Workshop, National Institutes of Health, Bethesda, Maryland, October 14-15, 1982. Information may be obtained from Workshop Coordinator, T. Hambrecht at NIH.

18. The reverse lift-off technique involves the deposition of film on a patterned photoresist layer and the subsequent removal of the unwanted film material by stripping the photoresist. For example see: W. G. Hanse, L. H. Donohue and A. J. Speth, IEEE Trans. on Electron Devices, $\underline{ED-29}$, 360 (1979).

19. Platinum metal layers can be platinized by immersion in a 3% platinum chloride solution containing 0.025% lead acetate. For example see: E. Gileadi, E. Kirowa-Eisner and J. Penciner, "Interfacial Electrochemistry: An Experimental Approach", p. 216, Addison-Wesley, Reading, Massachusetts, 1965.
20. Hitachi Chemical Company, Ltd., Tokyo, Japan.
21. DuPont Company, Wilmington, Delaware.
22. Shipley Company, Inc., 2300 Washington Street, Newton, Massachusetts 02612.
23. S. Miller, Circuits Manufacturing, 54, April 1977.
24. U. S. Department of Labor, Occupational Safety and Health Administration Material Safety Data Sheet "Developer MF-312".
25. W. B. Lindsey, U.S. Patent 3,361,589, January 2, 1968.
26. J. G. Fish, U.S. Patent 3,871,930, March 18, 1975.
27. Eastman Kodak Co., Rochester, New York.
28. P. Pochay, K. D. Wise, L. F. Allard and L. T. Rutledge, IEEE Trans. on Biomedical Engg., BME-26, 199 (1979).

NOVEL METAL-FILLED POLYIMIDE ELECTRODES

T. A. Furtsch[a], H. O. Finklea and L. T. Taylor*

Department of Chemistry
Virginia Polytechnic Institute and State
   University
Blacksburg, Virginia 24061

Palladium-coated polyimide films are evaluated as electrochemical electrodes. The film electrodes exhibit essentially identical behavior compared to bulk palladium electrodes. In aqueous 0.5M H_2SO_4, current peaks due to oxide formation, oxide stripping, hydrogen adsorption, and H_2 oxidation are observed. The ferri/ferrocyanide redox couple is grossly irreversible in the same electrolyte. Reversible electrochemical behavior is obtained for $Fe(EDTA)^{1-/2-}$ in 1M KCl/H_2O, and for ferrocene/ferricenium in 0.1M TEAP/dimethylacetamide.

----

[a]Visiting Professor, Tennessee Technological University, Cookeville, TN.
*To whom correspondence should be addressed.

INTRODUCTION

Recent studies involving the doping of selected polyimides with metal complex species have resulted in the formation of a number of thin films which have interesting electrical properties. Some of these films have been produced with a thin metallic surface on the polyimide substrate. Electron spectroscopy of film surfaces corroborates the fact that the oxidation state of the metal in these cases is zero valent.[1]

Metal-coated polyimides have been made from several different monomer combinations, doped with $Pd((CH_3)_2S)_2Cl_2$ (PDS), and exposed to air during the thermal curing process. Specifically, our films are produced from 3,3´,4,4´-benzophenone tetracarboxylic acid dianhydride (BTDA) or pyromellitic dianhydride (PMDA) and the diamines 4,4´-oxydianiline (ODA), 3,3´-diaminobenzophenone (DABP) or 3,3´-diaminodiphenylcarbinol (DABPC). These polyimides are typically doped with metal complex on a 4:1 polymer to metal mole ratio.[1] When cured at 300°C in a forced-air oven, the polymers containing PDS produce a film with a definite metallic surface on the side which is exposed to the atmosphere. Upon removal from the glass plate (used for casting the films) the dark non-metallic side of the film (which was against the glass during the curing step) can be converted to a metallic surface by simply reheating at 300°C. In the work reported here, however, only one side of the film was metallized.

We wish to report here a feasibility study regarding the application of these films as electrodes in electrochemical systems. The thrust of the study has been aimed at comparing the electrochemistry of polyimide films with thin metallic layers to that of more conventional palladium wire or foil electrodes and of platinum wire electrodes. This effort serves as an unusual extension of the many other uses that have been found for polyimides.

EXPERIMENTAL

Materials

Polymer films used as electrode material in this study were prepared and characterized as previously described.[1] Electrolyte solutions were prepared from reagent-grade materials and conductivity grade water (obtained from a Barnstead Nanopure System). Tetraethylammonium perchlorate (Kodak) (TEAP) was recrystallized three times from water and dried under vacuum overnight.

Other electrode materials were either platinum wire,

palladium wire or palladium foil all obtained from Alfa products.

When polymer films were used as electrodes, they were mounted in a Teflon electrode holder which exposed both sides of a _ca_. 2 mm length of a 3 mm wide film strip to the electrolyte. Electrical contact to the films was accomplished by pressing the palladium-coated side against a stainless steel rod.

Apparatus

Cyclic voltammetry experiments were carried out with a Bioanalytical Systems Model CV-3 potentiostat and a Moseley X-Y Recorder. The electrochemical cell used a conventional three electrode system with a saturated calomel electrode as reference and a platinum wire as counter electrode. All experiments were carried out at ambient temperature. All potentials are relative to the saturated calomel electrode (SCE). Platinum and palladium wire or foil electrodes were cleaned either by soaking in hot (80°C+) concentrated sulfuric acid or by soaking in cold chromic acid.[2] The palladium doped-polyimide films dissolve in hot H_2SO_4 and soaking the films in cold chromic acid removes most of the palladium surface after more than just a few minutes soaking. Films were, therefore, usually dipped quickly in the chromic acid and then electrochemically cleaned (see below) to remove any oxidizable organics from the surfaces.

RESULTS AND DISCUSSION

Electrode Characteristics in 0.5 M H_2SO_4

Cyclic voltammetry using a PDS-BTDA/ODA film strip as a working electrode was performed in 0.5 M H_2SO_4 as supporting electrolyte. Figure 1 shows typical cyclic voltammograms of the polymer strip compared with those for a Pd foil and a Pt wire. The polymer film electrode surfaces are cleaned electrochemically by successively setting the potential in the oxide formation region (+1.5V vs SCE) for 1 minute and then near the hydrogen evolution for 1 minute followed by repetitive cycling between these two regions. The cathodic potential was usually about 0.0V vs SCE which was sufficiently negative to reduce the oxide formed during oxidation but not to the point of generating molecular hydrogen. Hydrogen evolution occurs at about -0.25V vs SCE near the thermodynamic reversible potential. Voltages more negative than this, sustained for any period of time, produce enough H_2 to saturate the palladium metal on the film surface and subsequent positive sweeps are accompanied by very large anodic currents. It is necessary to maintain more positive potentials for several minutes in order to oxidize all the adsorbed hydrogen. Excessive cleaning of the electrodes results in visible loss of palladium[2] and greatly increased electrode resistance.

Several successive electrochemical cleaning cycles as described above result in a cyclic voltammogram which resembles that obtained using a platinum wire under the same electrolyte conditions (Figure 1). Oxide-formation peaks occur at +0.70V and 0.80V, and a concomitant oxide reduction peak appears at +0.44V. The similarity is especially noticeable in the range 0.00V to 0.20V where adsorption waves due to atomic hydrogen are visible (Figure 1c and d). It has not been possible to produce a duplicate cyclic voltammogram in this region using a palladium wire or foil. The appearance of the small hydrogen adsorption and desorption waves appears to be unique to the polyimide films where palladium electrodes are concerned[2,3]. Oxidation of hydrogen dissolved in the palladium becomes a prominent feature of the CV when the potential is swept slightly beyond the hydrogen adsorption waves.

We were not able to find a suitable electrochemical couple which, in 0.5M H_2SO_4, would suitably demonstrate the utility of the polyimide films in carrying out simple electrochemical experiments. The ferri/ferrocyanide system, which shows at least quasi-reversibility in 0.5 M H_2SO_4 with platinum as working electrode (Figure 2a), shows irreversible waves with the palladium electrodes (Figure 2b). The use of a fresh and uncleaned BTDA-ODA/PDS film electrode produces on the initial scan both cathodic and anodic waves with large peak separations (Figure 2c). These disappear on successive cycles and revert to the wave shown in Figure 2d. Pre-cleaned polymer films (either chemical or electrochemical cleaning) always produce the irreversible CV's shown in Figure 2d.

Electrode Characteristics in 1 M KCl

The apparent working range for a 1 M KCl solution (pH=3.5) with a PDS-BTDA/ODA film as working electrode is approximately +0.6V to -0.85V vs SCE. The cyclic voltammogram for the PDS polyimide film in this region is featureless and quite usable for the study of any electrochemical couple which falls within this voltage range.

In contrast to the solution containing 0.5 M H_2SO_4 as supporting electrolyte, 1.0 mM Fe^{+2} chelated with an unbuffered solution 10 mM in EDTA provides a very good example of a reversible electrochemical couple with this film. Figure 3 shows the cyclic voltammogram for the reversible one electron wave. $E^{0'}$ for this couple is -0.11V vs SCE and shows an average separation of

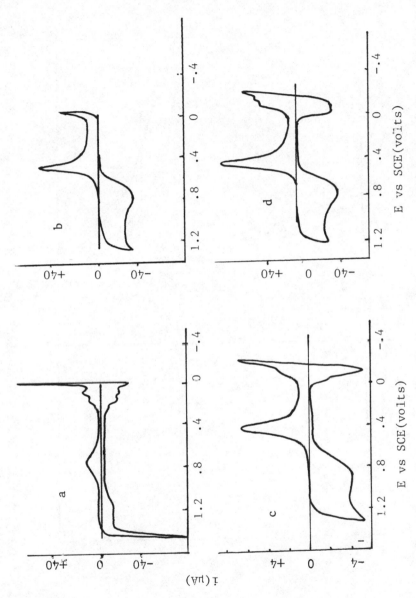

Figure 1. Cyclic voltammograms in 0.5 M H_2SO_4: Sweep rate 100 mV/s, a) Pt wire electrode (sealed in glass); b) Pd wire electrode; c) BTDA-ODA/PDS film (before cleaning); d) BTDA-ODA/ PDS film after cleaning.

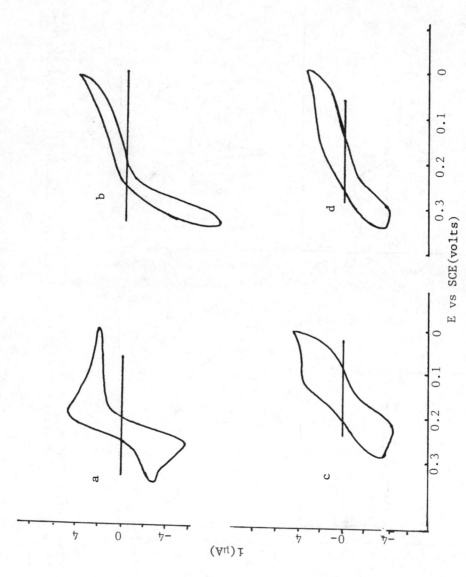

Figure 2. Cyclic voltammograms of ferrocyanide in 0.5 M H_2SO_4: Sweep rate 100 mV/s, a) Pt wire; b) Pd wire; c) BTDA-ODA/PDS film, fresh surface; d) BTDA-ODA/PDS film, after severe cycles.

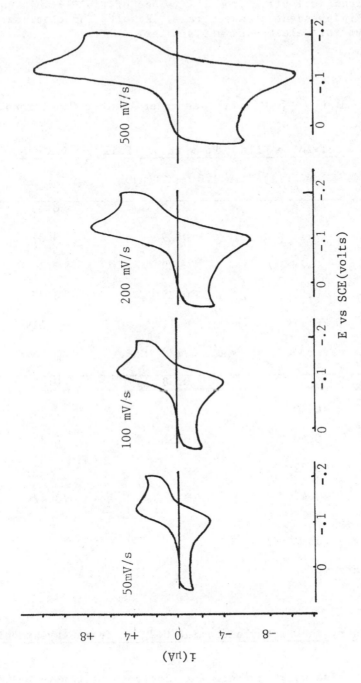

Figure 3. Cyclic voltammograms of iron EDTA complex in 1M KCl at various sweep rates using BTDA-ODA/PDS film electrode: a) 50 mV/s; b) 100 mV/s; c) 200 mV/s; d) 500 mV/s.

anodic and cathodic peaks of 70 mV. Further evidence of reversibility is found in Table I from the values of $i_p \nu^{-1/2}$ which are essentially independent of sweep rate.[4] Results for platinum wire and palladium foil electrodes are similar.

Table I. $i_p \nu^{-1/2}$ for Electrodes under Various Conditions.[a]

ν (mV/s)[b]	Polyimide Film	Pd wire or (foil)	Pt Wire
	$Fe(EDTA)^{2-}$ in 1.0 M KCl/H_2O		
50	0.26	0.29	0.38
100	0.27	0.29	0.37
200	0.28	0.29	0.36
500	0.29	0.28	0.35
E^0	−0.11V	−0.11V	−0.11V
	Ferrocene in DMAC (0.1 M TEAP)		
50	0.44	0.58	0.38
100	0.40	0.55	0.34
200	0.40	0.51	0.33
500	0.42	0.53	0.32
E^0	+0.47V	+0.47V	+0.47V

a)Units are $(\mu A)/(mV/s)^{1/2}$.

b)Units are mV/s.

Electrochemistry in Dimethylacetamide 0.2M in Tetraethylammonium Perchlorate

The effective working range for electrochemistry in DMAC/0.1M TEAP is −1.7V to +0.7V vs SCE. The ferrocene/ferrocenium system is a suitable one electron electrochemical couple to demonstrate the utility of polyimide Pd-coated films as electrode material.

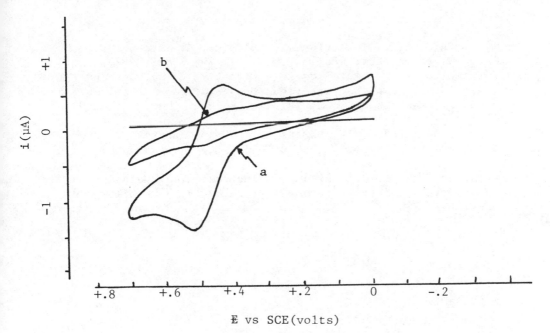

Figure 4. Cyclic voltammograms of ferrocene in 0.1 M TEAP/di-
methylacetamide using BTDA-ODA/PDS film electrode:
a) fresh Pd film; b) aged Pd film.

The constant values for $i_p v^{-1/2}$ in Table I indicate that the
reversibility of the ferrocene system is independent of electrode
material. Fresh films which have not been used before generally
give reversible results (Figure 4a). A film which has undergone
much use, however, gradually produces quasi-reversible cyclic
voltammograms (i.e. peak separation increases; Figure 4b). Used
films darken in color from the original bright metallic luster.

SUMMARY

The palladium-coated polyimide films display all the features
of bulk palladium metal electrodes: irreversible behavior of
$Fe(CN)_6^{3-/4-}$in H_2SO_4, reversible electrochemistry for $Fe(EDTA)^{1-/2-}$
in KCl and ferrocene/ferricenium in dimethylacetamide. Hydrogen
adsorption waves are observed for the first time with a palladium
electrode. The ease of preparation of these films suggests possible applications in the area of reversible hydrogen electrodes.

ACKNOWLEDGEMENT

We thank the National Aeronautics and Space Administration
(Grant NSG 1428) for sponsoring this research and Anne K. St. Clair
for supplying the thermal data.

1165

REFERENCES

1. T. A. Furtsch, L. T. Taylor, T. W. Fritz, G. Fortner and E. Khor, J. Polym. Sci: Polym. Chem. Ed. 20, 1287 (1982).
2. Pd electrodes lose metal during repetitive oxidations and reductions at a higher rate than most electrode materials. See O. A. J. Rand and R. Woods, J. Electroanal. Chem., 35, 209 (1972).
3. D. T. Sawyer and J. L. Roberts, Jr., "Experimental Electrochemistry for Chemists", pp. 66-67, Wiley Interscience, NY, 1974.
4. A. J. Bard and L. R. Faulkner, "Electrochemical Methods: Fundamentals and Applications", pp. 218-219, John Wiley and Sons, NY, 1980.

ABOUT THE CONTRIBUTORS

Here are included biodata of only those authors who have contributed to this volume. Biodata of contributors to Volume 1 are included in that volume.

Jerry M. Adduci is Associate Professor in the Department of Chemistry and the Materials Science and Engineering Program, Rochester Institute of Technology, Rochester, N.Y. where he has been since 1966. He received his Ph.D. degree in Organic Chemistry in 1963 from the University of Pennsylvania and was a Research Chemist with Fabrics and Finishes Department, DuPont, Wilmington (1963-1965) followed by postdoctoral research at the University of Maryland. During 1979-1980 he spent a year with Prof. Jorgen Kops, The Technical University of Denmark, Lyngby, Denmark. He has been active in the Rochester Section of the American Chemical Society. His interest and experience is in the synthesis of thermally stable materials, their modification and characterization in order to achieve new materials suitable for a variety of applications.

Hellmut Ahne has since 1978 been head of a laboratory and the radiations-reactive insulating materials and coating project at the Research Center of Siemens AG in Erlangen, W. Germany which he joined in 1969. He has been involved in a number of R&D areas and since 1975 he has focussed on radiation-reactive polymers for electron and photolighography. He studied Chemistry at the University of Erlangen-Nuremberg and obtained his degree in 1964. Subsequently, he went to the Technical University of Darmstadt becoming a Dr. rer. nat. in 1969.

William M. Alvino is a Fellow Scientist at the Westinghouse R&D Center in Pittsburgh which he joined in 1966. He was previously with Bell Telephone Laboratories, Murray Hill. He obtained his M.S. degree in Organic-Polymer Chemistry from Seton Hall University, South Orange, NJ in 1966. His research interest cover a broad area of polymer chemistry which includes synthesis of high temperature polymers, stability of polymers, ion exchange resins, and electrodeposition of polymers. He has 11 issued patents and 30 publications.

Eugene S. Anolick is presently an advisory physicist with IBM in the General Technology Assurance Laboratory in Poughkeepsie, N.Y. He joined IBM in 1968 and has worked extensively in reliability physics. He has done investigative work on models for humidity, voltage and temperature testing, and has also been involved in alpha particle effects, liquid cooling, dielectric breakdown, and electromigration. Before coming to IBM, he was with General Electric Company and carried out research on magnetic materials, x-ray electrophotography and thermoplastic recording materials. He received an M.S. in Physics from Carnegie Institute of Technology in 1950.

Francis V. Bedetti is currently a Staff Engineer at IBM, E. Fishkill, N.Y. and is working in packaging development. He joined IBM in 1964 and has worked in reliability and engineering groups. He studied as a Chemistry major at Lowell Technological Institute and has an A.A.S. from Orange County Community College in Middletown, N.Y.

Justin C. Bolger cofounded the Amicon Corporation in 1962 and currently manages the division producing epoxy and other formulated products. He received Sc.D. degree in Chemical Engineering from MIT. His present activities and approximately 80 publications and patents center on the development of new adhesives and insulation compounds. He is the author of eight book chapters on structural adhesives and on interfacial bonding mechanisms. He was Chairman of the 1970 Gordon Research Conference on the Science of Adhesion and a member of the National Materials Advisory Board's Committee on Aerospace Structural Adhesives.

R. L. Burke is with the M.I.T. Lincoln Laboratory, Lexington, MA.

Glenn Eu Chung Chang is a member of the professional staff in the Organic and Polymer Chemistry Section at TRW, Redondo Beach, CA. He is currently involved in NASA sponsored R&D on advanced polymeric materials such as the partially fluorinated polyimides. His past experience as a research chemist has included extensive work on gas and solution phase reaction kinetics, organic photochemistry, computer modeling of complex reaction kinetics, and synthesis. He received his Ph.D. degree in Physical Synthetic Organic Chemistry from the University of Oregon, and has published several articles.

Clinton C. Chao is currently a project manager at the Computer Research Center, Hewlett Packard Laboratories, Palo Alto. Prior to joining Hewlett Packard in 1977, he was an engineering supervisor at American Microsystems, Inc. He received his Ph.D. degree in Materials Science from Caltech. He is the author of a number of papers on materials processing and characterization, and high density circuit packaging.

Harry K. Charles, Jr. is a member of the Principal Professional Staff at the Applied Physics Laboratory of Johns Hopkins University, Laurel, MD, which he joined in 1972. He is currently Assistant Supervisor of the Microelectronics Group in the Technical Services Department. He received his Ph.D. degree in Electrical Engineering from Johns Hopkins University. He is a specialist in materials processing and analysis with particular emphasis on thin films, photolithography, and scanning electron microscopy. He has published over 35 technical papers, holds one patent, and has several patent applications pending. He is listed in Who's Who in the East, American Men and Women of Science, Dictionary of International Biography, and Men of Achievement.

George Chiu is an Advisory Engineer at IBM, East Fishkill, N.Y. He received his Ph.D. degree in Chemical Engineering from MIT. He has four patents and has published many technical and scientific papers.

Gary C. Davis is an organic chemist at the General Electric R&D Center in Schenectady, N.Y. He did his graduate research at the State University of New York at Albany. While at GE his work has centered on organic reaction mechanisms and process development. He has published in the areas of organic nitrations as well as hydrogen generation, and recently he has been interested in high temperature organic thin film dielectrics for electronic applications.

David R. Day is presently a Research Associate in the Department of Electrical Engineering at MIT. He received his Ph.D. degree in 1980 which was carried out as a joint project both at Case Western Reserve University and the University of Mainz, W. Germany. His research interest have included single crystal diacetylene polymerization, diacetylene monolayer polymerization, and monolayer structure analysis. His current research involves polymers for use in microelectronics and integrated-circuit dielectric property-sensors for chemical reaction monitoring.

Harry O. Finklea is Assistant Professor, Department of Chemistry, VPI&SU, Blacksburg, VA. He received his Ph.D. degree in 1976 from Caltech and was postdoctoral Fellow at Royal Institute of Great Britain (1975-1976), University of North Carolina (1976-1978), and Florida Atlantic University (1978-1979).

George M. Fohlen is a Research Scientist in the Chemical Research Projects Office at NASA's Ames Research Center, Moffett Field, CA. He received his Ph.D. degree in 1944 from the University of California, San Francisco. He has had wide experience in organic and polymer chemistry in industry before joining NASA in 1967. The principal areas of his polymer experience have included development of continuous processes for the production of phenolic resins, polyaromatic amines bisphenols, epoxy resins, polyisocyanates. For NASA he has developed novel intumescent coatings, fire-resistant transparent polymers, and high temperature useful polymers of the polycarbonate, epoxy, polyimide, phthalocyanine and phsophazene classes. He has numerous papers and patents to his credit.

T. J. Fuller is presently a Research Chemist at Allied Chemical Corporation. He received his Ph.D. degree in Organic Chemistry from Pennsylvania State University. He has experience in the organic synthesis of polyphosphazenes, phenol-formaldehyde and steroids. He has also worked in organo-metallic chemistry as applied to the modification of polymers, in the application of chemical kinetics to elucidate reaction mechanisms, and in the area of electrodeposition. He holds one U.S. patent and is the author or coauthor of 13 scientific publications.

Thomas A. Furtsch is Professor in the Department of Chemistry, Tennessee Technological University, Cookeville, TN. He received his Ph.D. degree in 1970 from the University of Texas.

Robert L. Fusaro is in the Lubricants Research Section of the Tribology Branch at the NASA Lewis Research Center in Cleveland, OH which he joined in 1967. He received an M.S. degree in Physics from Kent State University. His primary research area is in the field of solid lubrication and his research is directed at developing high temperature solid lubricant films and composite materials for aerospace applications. He is the author of over 35 technical reports and in 1979 received the American Society of Lubrication Engineer's (ASLE) Walter D. Hodson Award for his paper "Effects of Atmosphere and Temperature on the Wear, Friction and Transfer of Polyimide Films". Currently, he is the Mideastern Regional Vice President, Chairman of the ASLE Lubrication Fundamentals Committee and Chairman of the Aerospace Industries Council of ASLE.

Pieter Geldermans is a Senior Engineer at the IBM East Fishkill Development Laboratory and is working on advanced packaging technology. He received the B.S. degree in Analytical Chemistry in The Netherlands, and joined IBM in 1961 after working for several years with N.V. Philips, Eindhoven. He has worked primarily on magnetic ceramics, ceramics packaging, and Josephson packaging and has held a number of managerial positions in IBM. He has a number of patents in these fields and has coauthored a number of articles.

Gennady Sh. Gildenblat is an electrical engineer at the General Electric R&D Center in Schenectady. He did his undergraduate research at the Leningrad Electrical Engineering Institute. Before joining G.E., he was associated with the State Institute of Resistors and Capacitors in Leningrad, USSR. He is presently involved in device physics and VLSI process development, and has published in these areas.

C. C. Goldsmith is an Advisory Chemist at IBM's General Technology Division in East Fishkill, N.Y. He received his graduate degree in Chemistry from Wichita State University. His research Interests include the application of x-ray diffraction techniques to the study of materials properties.

Ben-Min Gong has been Associate Professor, Department of Applied Chemistry, Shanghai Jiao Tong University, Shanghai, PRC since 1982. He has been with this University since 1959, serving as Teaching Assistant (1959) and Lecturer (1978). He has published 14 papers and because of his research work dealing with photoresists, he received the first grade prize of Shanghai Jiao Tong University in 1979, and third grade prize of the National Defence Science and Technology of China in 1980. His interests are UV photoresists, E.B. and x-ray resists, x-ray mask technique, and synthesis and application of polyimides in LSI.

Howard M. Green is currently Manager of the Chemical Research and Applications Department at TRW, Redondo Beach, CA. He received his Ph.D. degree in Physical Organic Chemistry from UCLA. Since joining TRW in 1970, he has served as Project Manager or Principal Investigator on a variety of contractual studies in the preparation, characterization, application and testing of advanced materials. He is currently Project Manager on a NASA project to develop polyimide resin apex seals for advanced rotary airplane engines. His primary areas of specialization are the synthesis and application of polymers and organic intermediates.

Jeremy Greenblatt is currently a postdoctoral Fellow with IBM Corporation, Hopewell Junction, N.Y. which he joined in July 1981. He received his Ph.D. degree in Organic Chemistry in 1980 from the Hebrew University of Jerusalem, Israel and subsequently carried out postdoctoral research at Wayne State University. He was the recipient of various honors and awards based on his scholastic excellence. He is interested in physical and synthetic organic chemistry, plasma deposition and surface characterization of polymers, adhesion of coatings, and use of polymers in microelectronics, and has published about 10 papers.

Hartmut Haertner is currently head of a development department at E. Merck Darmstadt, W. Germany (which he joined in 1972) and is concerned with the further development and improvement of processes for the production of polyamides and light sensitive substances. He received Dr. rer. nat. (Doctor of Science) degree in

1972 from the Technical University of Munich.

Robert Haushalter is presently a Staff Chemist with Exxon Research and Engineering Company, Clinton, N.J. Prior to joining Exxon, he was at the Argonne National Laboratory. He received his Ph.D. degree in Inorganic Chemistry in 1979 from the University of Michigan. His research interests are synthesis and properties of new solid state materials.

Barbara A. Heath is presently with Inmos Corporation, Colorado Springs, CO. Before her current position, she was at the General Electric R&D Center in Schenectady. She is a Physical Chemist and did her graduate work at Brandeis University. She has published in the areas of electron nuclear resonance, multiphoton ionization spectroscopy and dry processing for integrated circuit fabrication.

Carl L. Hendricks is currently lead engineer in the Adhesives and Nonmetals Technology group in the Materials and Processes organization within Boeing Aerospace Company. In this position, he is responsible for materials and processes IR&D and contracts relative to thermoplastics and thermosets as applied to composites, composite matrices, moldings, and adhesives. He has 19 years of experience directly related to development, evaluation, and test of aircraft structural composites, molded plastics and adhesives. He received B.S. in Chemistry from the University of Washington, Seattle in 1961 and an MBA from the City University, Seattle in 1981, and has written several papers, documents and specifications.

Terry O. Herndon has been with MIT Lincoln Laboratory since 1958 where he has been concerned with circuit design, metallization, line generation and structural designs for magnetic thin film memories. He is presently responsible for photolithography, etching, metallization and packaging of LSI digital and memory devices. He did his undergraduate work at Antioch College and his graduate work at MIT.

Silvester G. Hill is currently a senior specialist engineer in the Nonmetallic Group of the Boeing Aerospace Company Materials and Processes organization which he joined in 1972. He is responsible for materials and processes on IR&D and R&D contracts relative to thermoplastic and thermoset composites, composite matrices, sealants, finishes, moldings and adhesives. He joined Boeing in the Structures Technology – Materials Organization in 1965 and worked in the area of new materials. In this position, he evaluated high temperature structural resins such as polyimides and polyquinoxalines. He has a B.S. in Chemistry from Lincoln University, Jefferson City, MO.

Satoru Isoda has been since 1978 in the Research Staff of the

Central Research Laboratory of Mitsubishi Electric Corporation, Hyogo, Japan. Received Ph.D. degree in 1978 for research thesis on Molecular Aggregation of Solid Aromatic Polymers from the University of Tokyo, and is the coauthor of some publications.

Tokio Isogai has been with the Production Engineering Research Laboratory of Hitachi, Yokohama since 1975. He is presently a manager of the 2nd Department of the Laboratory and is engaged in R&D of materials and process for microelectronic devices. In 1957 he joined Hitachi Research Laboratory where he was engaged in research and development of insulating materials. He received his Ph.D. degree in Electronic Engineering from Nagoya University in 1974.

Robert J. Jones is currently Business Area Manager, Advanced Chemical Technology, TRW, Redondo Beach, CA. In this capacity he is responsible for planning, coordination and solicitation of government contracts related to new and improved polymer/plastics R&D. On prior assignments at TRW, he was responsible for directing contract and TRW work in development of new monomers, polymers and polymerization processes. He has served as Program Manager or Principal Investigator of many DOD and NASA contracts. He has more than 25 patents encompassing new monomers and polymers, including the partially fluorinated polyimides to his credit. He received his Ph.D. degree in Organic Chemistry from the University of Missouri-Columbia.

Hirotaro Kambe has been Professor Emeritus, University of Tokyo since 1981. From 1937-1981 was Professor, Polymer Research Division, Institute of Space and Aeronautical Science, University of Tokyo. Received Dr.Sci. in 1967 from the University of Tokyo. Recipient of the Mettler Award in Thermal Analysis in 1977. Has authored three books: Rheology (1959); Thermal Stability of Polymers (1970); and Thermal Analysis (1975).

Fumio Kataoka is with the Production Engineering Research Laboratory, Hitachi Ltd., Yokohama, Japan which he joined in 1979. He received his Ph.D degree in Organic Chemistry in 1980 from Hokkaido University. He is presently engaged in research and development of organic insulating materials and resist materials for microelectronic devices.

Rudolf Klug has been a Chemist at E. Merck Darmstadt, W. Germany since 1959. He received his Dr. rer. nat. (Doctor of Science) in Organic Chemistry from the University of Darmstadt in 1958. He is presently involved in the research and development of polymers especially in the field of light sensitive polyamides.

Masakatsu Kochi is presently at the Institute of Interdisciplinary Research, Komba Campus, University of Tokyo, Tokyo,

Japan. He was a Research Associate at the Institute of Space and Aeronautical Science, University of Tokyo (1969-1981), and was lecturer at Toho University (1972) and at the Tokyo University of Agriculture and Technology in 1982. He received his Ph.D. in 1969 from the University of Tokyo for a thesis entitled "Photoemission from Organic Crystals in Vacuum Ultraviolet Region."

Mitsumasa Kojima is with the Hitachi Chemical Company, Yamazaki Works, Hitachi, Japan which he joined in 1979. He recieved his M.S. degree in 1979 from Shizuoka University. He is presently engaged in the development of organic materials for microelectronic devices.

Vitali A. Korobov is an Associate Engineer at IBM Corporation, Poughkeepsie, N.Y. He received M.S. in Theoretical Physics in 1978 from Moscow State University and another M.S. in Engineering and Applied Science in 1980 from Yale University.

L. J. Krause is currently with 3M Company, St. Paul, MN.

Hans Krüger has since 1980 been Head of the development Group for Liquid Crystals in the Siemens Tube Factory, Erlangen, W. Germany. He studied Physics at the Technical University of Munich and obtained his diploma in 1968. Subsequently, he joined the Central R&D Department, Siemens Company in Munich. His activities began in the field of crystal growth and in 1969 he switched to the liquid crystal field.

R. H. Lacombe is currently with IBM Corporation, Hopewell Junction, N.Y. He received his Ph.D. degree in the Department of Macromolecular Science at Case Western Reserve University with a theoretical thesis on polymer chain dynamics. This was followed by postdoctoral research in the Department of Polymer Science and Engineering, University of Massachusetts where he was involved in the problem of polymer solution thermodynamics and polymer-polymer compatibility. He is currently involved in the experimental characterization of the thermal-mechanical properties of high temperature polymers, laboratory automation, and theoretical modeling of the stress distribution and adhesion in multilevel microelectronic structures.

Walter J. Landoch has been with MIT Lincoln Laboratory since 1960 where he has been concerned with metallization, magnetic thin film memories, and printed circuit fabrication. He presently supervises photolithography and etching in digital integrated circuit processing. He graduated from Lowell Technological Institute and attended Merrimack College.

J. R. Lloyd has since 1978 been employed at IBM Data Systems Division at the East Fishkill Facility, N.Y. as a Reliability

Engineer where he has specialized in electromigration related problems. He received his Ph.D. in Materials and Metallurgical Engineering in 1978 from Stevens Institute of Technology. Much of his research was performed in cooperation with scientists at Bell Labs, Murray Hill. He also worked for a time for AMAX Base Metals R&D in Carteret, NJ.

Steve Lytle is presently employed as a linear I.C. development engineer with Motorola Semiconductor's Bipolar Technology Center in Mesa, AZ. He received his M.S. in Ceramic Science in 1981 from the Pennsylvania State University and joined Motorola directly after graduate school. For his graduate study, he investigated surface diffusion in MgO for which he was presented a Xerox Research Award.

Daisuke Makino has been with the Hitachi Chemical Co. Ltd. since 1970. From 1970-1974 he was engaged in the development of insulating materials for electric parts, and since 1974 he has been working on the development of new materials for semiconductor devices, liquid crystal displays, etc. He received his Dr. of Science for his research on the spectroscopic studies of polymers for Osaka University in 1973.

Joseph Mario is presently employed at Digital Equipment Corporation. He received his B.S. in Electrical Engineering from MIT in 1981.

Joe T. Massey is a member of the Principal Professional Staff and a Senior Fellow of the Applied Physics Laboratory of the Johns Hopkins University which he joined in 1946. He also holds a joint appointment as Associate Professor, Department of Biomedical Engineering, Johns Hopkins Medical School. At APL, he has worked in microwave physics, spectroscopy, electrical discharge plasmas, and laser research. He helped establish a collaborative biomedical engineering program with the Hopkins Medical Institutions and was it manager from 1967-1981. He received his Ph.D. degree in Physics from the Johns Hopkins University.

Hans-Joachim Merrem is with the E. Merck Darmstadt, W. Germany where he is in charge of the development and application of photoresists. He received Dr. rer. nat. (Doctor of Science) degree in Chemistry from the Freie Universität Berlin in 1978 followed by postdoctoral research in Toulouse, France.

Itaru Mita has been since 1973 Professor at the Institute of Space and Aeronautical Science, University of Tokyo, Tokyo. Received Doctorat (Universite de Strasbourg, 1959) and Ph.D. in Polymer Chemistry in 1972 from the University of Tokyo. Research interests are in the area of polymer science, particularly the thermal and photochemical reactions of polymers, chain dynamics and thermally stable polymers, and has published about 70 original papers and two review articles.

Kashmiri Lal Mittal* is presently employed at the IBM Corporation at Hopewell Junction, N.Y. He received his M.Sc. (First Class First) in 1966 from Indian Institute of Technology, New Delhi, and Ph.D. in Colloid Chemistry in 1970 from the University of Southern California. In the last ten years, he has organized and chaired a number of very successful international symposia and in addition to these volumes, he has edited eighteen more books as follows: Adsorption at Interfaces, and Colloidal Dispersions and Micellar Behavior (1975); Micellization, Solubilization, and Microemulsions, Volumes 1 & 2 (1977); Adhesion Measurement of Thin Films, Thick Films and Bulk Coatings (1978); Surface Contamination: Genesis, Detection, and Control, Volumes 1 & 2 (1979); Solution Chemistry of Surfactants, Volumes 1 & 2 (1979); Solution Behavior of Surfactants – Theoretical and Applied Aspects, Volumes 1 & 2 (1982); and Physicochemical Aspects of Polymer Surfaces, Volumes 1 & 2 (1983); Adhesion Aspects of Polymeric Coatings, (1983); Surfactants in Solution, Volumes 1, 2 & 3 (1984); and Adhesive Joints: Formation, Characteristics, and Testing (1984). Also he is Editor of the series, Treatise on Clean Surface Technology, a multi-volume work in progress. In addition to these books he has published more than 50 papers in the areas of surface and colloid chemistry, adhesion, polymers, etc. He has given many invited talks on the multifarious facets of surface science, particularly adhesion, on the invitation of various societies and organizations in many countries all over the world, and is always a sought-after speaker. He is a Fellow of the American Institute of Chemists and Indian Chemical Society, is listed in American Men and Women of Science, Who's Who in the East, Men of Achievement and many other reference works. He is or has been a member of the Editorial Boards of a number of scientific and technical journals. Currently, he is Vice-President of the India Chemists and Chemical Engineers Club.

Vernon B. Mountcastle, Jr. is currently University Professor of Neuroscience and Direction of the Philip Bard Laboratory of Neuroscience, the Johns Hopkins School of Medicine, Baltimore, MD. He received Doctor of Medicine degree from the Johns Hopkins University School of Medicine in 1942. Following military service he joined the faculty of the Johns Hopkins School of Medicine which he served as Direction of the Department of Physiology from 1964 to 1979. His major research interest is research in the physiology of the control nervous mechanisms in sensation and perception.

Kiichiro Mukai has been since 1971 with the Central Research Laboratory, Hitachi Ltd., Tokyo where he has worked on the development of multilevel interconnection technology for monolithic LSI,

---

*As the editor of this two-volume set.

and he is also interested in the electrical properties of polymers. He received his M.S. degree in Solid State Physics from Tokyo Institute Technology, Tokyo in 1971.

Takashi Nishida has been since 1974 with the Hitachi Central Research Laboratory, Hitachi Ltd. Tokyo where he initially worked on field emission cathode, and during 1975-1978 he was engaged in R&D of dielectric film deposition assisted by laser irradiation and InSb thin film Trs. He is currently working on the multilevel interconnection technology for VLSI's. He received his M.S. degree in Metallurgical Engineering from the University of Tokyo in 1974.

Isao Obara has since 1975 been with the Production Engineering Research Laboratory of Hitachi, Yokohama, Japan. He joined Hitachi Ltd. in 1962. He received his B.S. degree in Chemistry from Shizuoka University in 1962. He is presently engaged in research and development of organic insulating materials for microelectronic devices.

Erich Pammer has been employed at the Siemens Semiconductor Works, Erlangen, W. Germany since 1957. He studied chemistry at Munich University and graduated in 1954. In 1956, he was awarded his doctorate (Dr. rer. nat.) with the first ever synthesis of bromyl fluoride BrO_2F from Munich University. As Head of the Chemical Development Laboratory, he has been involved in Ge and Si epitaxy, passivation processes, CVD methods, plasma etching, polyimide coating and bump technology. He has applied for more than 50 patents in these areas.

John A. Parker is Chief of the Chemical Research Projects Office, NASA, Ames Research Center, Moffett Field, CA. He received his Ph.D. degree in Chemistry in 1951 from the University of Pennsylvania. Prior to joining NASA as a Research Scientist in 1962, he was Manager of the Chemistry Department of Armstrong Cork Company, Lancaster, PA. He was awarded the NASA Exception Scientific Achievement Medal in 1968 for his pioneering research on the ablation of heat shield materials. The results of this work have been used to provide protection from fire with a wide range of commercial applications. He has many published papers and patents and serves on several national advisory committees. He is a retired Captain in the U.S. Air Force Reserve.

Leonard M. Poveromo is a Technical Specialist for Advanced Composites at Grumman Aerospace Corporation, Bethpage, N.Y. (employed since 1969) and has been responsible for the development of high temperature polyimide materials and processes. He is presently Project Manager on Grumman's NADC funded Gr/BMI development program. He also was responsible for the PMR-15/Fiberglass dielectric insulation effort in support of Princeton's Tokamak Fusion Reactor Program. In addition, he has also coordinated the materials and processing efforts on a number of other programs. He graduated from Lehigh University with a B.S. in Chemical Engineering in 1969 and holds an MBA from Hofstra University.

Stephen H. Powell is a member of the professional staff in the TRW Chemical Research and Applications Department, Redondo Beach, CA. His major responsibilities are the area of material and process development of polyimides and special elastomers, and is currently the lead engineer in developing spray process methods for the TRW partially fluorinated polyimide coatings. He received his B.S. degree from Whittier College.

Stephen James Rhodes has been with Plessey Research (Caswell) Ltd. at the Allen Clark Research Center, Towcester, England since 1978 where he has worked on the characterization and introduction of magnetron sputtering. He has also established the use of poly-imide in a multi-level metallization process and is currently lead-ing a project developing new techniques for finer geometry process-ing of multilevel metallization. He has spent the last 17 years covering a wide range of engineering activities in semiconductor wafer processing with a number of companies including 5 years with Texas Instruments Ltd., Bedford as a Senior Product Engineer. He received an H.N.D. in Physics in 1965.

Deidra Ridley is presently employed at the Digital Equipment Corporation. She received her B.S. in Electrical Engineering from MIT in 1982.

Ronald Rubner has since 1981 been assistant manager of the head of Central Research and Development, Siemens AG, Erlangen, W. Germany which he joined in 1968. He studied chemistry at the Technical University of Munich and obtained his degree in 1964, and became a Dr. rer. nat. in 1967. At Siemens, he invented the salt free photoreactive polyimide precursors in 1973 and up to 1981 was responsible for polymer chemistry including radiation sensitive systems, cable and wire insulation and reactive resins.

Atsushi Saiki is with the Central Research Laboratory, Hitachi Ltd., Tokyo which he joined in 1967 and has been engaged in the research and development of the process technology of IC and LSI. Since 1970, he has worked on the development of multilevel interconnection technology. He received his M.E. degree in Applied Chemistry from Waseda University, Tokyo in 1967.

Gay Samuelson is currently a Senior Scientist with a joint Motorola/Shell venture called Solavolt, where she is developing materials and processes for interconnection and packaging of photovoltaic solar cells. Her tenure at Motorola has included synthesis and physicochemical characterization of liquid crystals followed by research and development of dielectric thin films for

MOS and bipolar applications. She received her Ph.D. in Bio-
chemistry in 1969 from the University of Wisconsin, Madison.

L. C. Scala is Manager, Polymer Chemistry, Westinghouse R&D
Center in Pittsburgh. He received his Dr.Sc. in Organic Chemistry
from the University of Bologna, and was postdoctoral Fulbright
Fellow at MIT. He has 34 years of experience in chemical research,
specializing in the formulation of high-temperature resistant polymers for electrical insulation as films and as laminated structures. He has done extensive work with ultrathin Langmuir films of
fatty acid, fatty acid salts, and polar polymers. Most recently he
has worked on the electron beam sensitive resist as applied to
integrated circuits, reverse osmosis membranes for water desalina-
tion, special battery separators, UV polymerization techniques, and
electrophoretic nonaqueous deposition techniques for flaw-free
insulation. He holds 15 patents and is the author or coauthor of 26
scientific publications.

Stephen D. Senturia is Professor of Electrical Engineering at
MIT which he joined after receiving his Ph.D. in Physics from the
same institution in 1966. His research interests have included NMR
and other physical studies in semiconductors. A parallel thread in
his research has been the application of the methods of physical
measurement to practical problems. These have included high-pre-
cision thermometry, oil-well logging, NMR instrumentation methods,
and the automated reclamation of resources from urban solid waste.
His most recent research development is the charge-flow transistor,
a new solid state device that permits the fabrication of MOS
integrated-circuit sensors for moisture measurement and chemical
reaction monitoring. He is the coauthor of Electronic Circuits and
Applications, a widely used introductory electronics text.

Tito T. Serafini is presently Head of the Polymer Matrix
Composites Section of the NASA Lewis Research Center where he has
been since 1963. He received his Ph.D degree in Physical Chemistry
from Case Institute of Technology; was a chemistry instructor at
Case in 1961, and a postdoctoral research Fellow at the University
of Turin (Italy) in 1962. In 1976, he received NASA's highest
science award, the Exceptional Scientific Achievement Medal, for
his studies which led to the development of the class of high
temperature polymers known as PMR polyimides. He is currently
responsible for the conception and development of advanced polymer
matrix composites for aerospace propulsion systems.

Wei-Hai Shen is Introductor at the Shanghai Jiao Tong University, Shanghai, China. She graduated from Fu Dan University in 1970
and in 1980 graduated (M.S.) from Shanghai Jiao Tong University.
During 1970-1978, she was Vice-Head of the Central Laboratory
An-Chin Petrochemical Works, Anhui, China and during 1978-1979 was
a teaching assistant at the Chinese University of Science and

Technology.

Fusaji Shoji has been with the Production Engineering Research Laboratory of Hitachi, Yokohama, Japan since 1977. He joined Hitachi Ltd., Hitachi Research Laboratory in 1969. He received his B.S. degree in Chemistry in 1964 from Yamagata University. He is presently engaged in research and development of organic insulating materials for microelectronic devices.

Anne St. Clair is a Senior Research Scientist in the Materials Division at NASA Langley Research Center, Hampton, VA. In 1969, she received a B.A. in Chemistry from Queens College and was a graduate of the Student Honors Program in Chemistry from Argonne National Laboratory. She received an M.S. in Inorganic Chemistry from VPI&SU in 1972. Her research interests at NASA have included the development of high-temperature polymers for applications on advanced aircraft and spacecraft.

Terry L. St. Clair is a Senior Scientist at NASA Langley Research Center, Hampton, VA where he has been since 1972. He was previously employed by E. I. duPont, Waynesboro, VA. He obtained a Ph.D. in Organic Chemistry from VPI&SU in 1972. His reasearch at NASA has involved the development of high-temperature resins for adhesive and matrix resin applications.

Hiroshi Suzuki has been since 1977 with the Hitachi Chemical Company Ltd. Where he has been working on the development of new materials for semiconductor devices, liquid crystal displays, etc. He received M.E. degree in 1977 from the Electric Communication University.

Issei Takemoto has since 1975 been with the Production Engineering Research Laboratory of Hitachi, Yokohama, Japan. He joined Hitachi Ltd., Yokohama Research Laboratory in 1974. He received his M.S. degree in Chemistry in 1974 from Osaka University. His principal interests have been in integrated circuit processing, especially photolithography.

Larry T. Taylor is Professor of Chemistry at VPI&SU, Blacksburg, VA. He recieved his Ph.D. degree in Inorganic Chemistry in 1965 from Clemson University and spent two years (1965-1967) as Research Associate at Ohio State University. He was NASA-ASEE Faculty Fellow at Langley Research Center in 1976. He has co-authored about 100 papers in the areas of synthesis and characterization of metal-containing polymers, modification of polymer properties, small molecule reactivity with transition metal complexes, and analytical techniques for the speciation of complex synfuel mixtures.

John Tkach has been with Commonwealth Scientific Corporation,

Alexandria, VA since 1981 and has specialized in technical marketing of ion beam etching and deposition equipment. He has been
working in the area of high vacuum and thin film processing since
1979, and his current interests lie mainly in new semiconductor and
thin film processing technologies. He received his B.S. degree in
Physics from the University of Toronto in 1979 and is currently
working on his master's degree in systems analysis and management
at George Washington University.

Dharmendra S. Varma is Professor of Fiber Science, Head of
the Dept. of Textile Technology, and Dean of Students at the Indian
Institute of Technology, New Delhi where he has been since 1966. He
received his Ph.D degree in 1965 from Glasgow University. He was
Visiting Professor at the University of California, Davis, 1979-
1981. He is a Fellow of the Institute of Engineers, India and a
Fellow of the Textile Institute. He is the author of more than a
hundred papers in textiles and related journals. His areas of
research include structure-property relationships in polymers,
modifications of polymers, fibers and composite materials.

Indra K. Varma is Professor of Polymer Science and Head of
the Centre for Material Science and Technology at the Indian
Institute of Technology, New Delhi, India. She received her D.Phil.
degree in 1961 from Allahabad University, India and Ph.D. in 1965
from Glasgow University. She was a NRC-NASA Research Associate at
Ames Research Center, Moffett Field, CA, 1979-1981. She has published more than ninety papers and has three U.S. Patents issued or
pending. Her areas of research activity include the synthesis and
characterization of thermally stable polymers, modification of
vinyl polymers, structure-property relationships in polymers and
mechanistic studies of the thermal degradation of vinyl polymers.

Wen-Chou V. Wang is a member of the technical staff at
Hewlett-Packard Laboratories, Palo Alto, CA which he joined in 1981
after receiving his Ph.D. degree in Materials Science and Engineering from Cornell University. He has authored several papers on
glass transition temperature, mechanical properties, stress
analysis, and environmental crazes of polymer materials.

Arthur M. Wilson is presently New Product Development Manager
for the Electronic Chemicals Division of Allied Chemical Corporation. Previously he was the manager of a photolithography and thin
film engineering section of the Linear Circuits Division and a
Senior Member of Technical Staff of Texas Instruments in Dallas. He
was a staff member of the Texas Instruments Semiconductor Research
and Development Laboratory and his main interest was the development of multilevel interconnect systems for very large scale
integrated circuits. Before moving to SRDL he was with the Quality
and Reliability Assurance Department. Before coming to Dallas, he
was a member of the Metals and Controls R&D Laboratory in Attle-

boro, MA where he developed fast charging NiCad batteries and circuits. He received his Ph.D. in Analytical Chemistry in 1958 from Northwestern University and from 1958-1966 was an assistant professor at Emory and Wayne State Universities and his research interest, <u>inter alia</u>, was solvent extraction with surface active agents. He holds five patents and is the author of several publications.

An-Ji Yo is Associate Professor at the Shangai Jiao Tong University, Shanghai, PRC. Graduated from Beijing University in 1956, and during 1957-1981 was Research Associate, Institute of Chemistry, the Academia Sinica. Research interest is in polymer chemistry.

Hitoshi Yokono has since 1977 been with the Hitachi Production Engineering Research Laboratory, Yokohama, Japan. He joined Hitachi Research Laboratory in 1954. He graduated from Ibaraki Technical College in 1960. His principal interests have been in organic insulating materials for electronic devices.

Rikio Yokota is a staff member in the Engineering Department at the Institute of Space and Astronautical Science, Administration of Education. He received his B.S. in Chemical Engineering in 1964 from Nihon University and was a postdoctoral Fellow at Clarkson College of Technology in 1974. He is a polymer chemist with interest in the properties of aromatic heat resistant polymers and their composites.

INDEX